Tandem Techniques

SEPARATION SCIENCE SERIES
Editors: Raymond P. W. Scott, Colin Simpson and Elena D. Katz

Quantitative Analysis using
Chromatographic Techniques
Edited by **Elena D. Katz**

The Analysis of Drugs of Abuse
Edited by **Terry A. Gough**

Liquid Chromatography Column Theory
by **Raymond P. W. Scott**

Silica Gel and Bonded Phases
Their Production, Properies and Use in LC
by **Raymond P. W. Scott**

Capillary Gas Chromatography
by **David W. Grant**

High Performance Liquid Chromatography:
Principles and Methods in Biotechnology
Edited by **Elena D. Katz**

Tandem Techniques
by **Raymond P. W. Scott**

Tandem Techniques

by

Raymond P. W. Scott
Georgetown University, Washington DC, USA

and

Birbeck College, University of London, UK

JOHN WILEY & SONS
Chichester • New York • Weinheim • Brisbane • Singapore • Toronto

Copyright © 1997 by John Wiley & Sons Ltd,
Baffins Lane, Chichester
West Sussex PO19 1UD, England

National   01243 779777
International (+44) 1243 779777
e-mail (for orders and customer service enquiries): cs-books@wiley.co.uk
Visit our Home Page on http://www.wiley.co.uk
or http://www.wiley.com

All Rights Reserved. No part of this publication may be reproduced, stored in a retrieval system, or transmitted, in any form or by any means, electronic, mechanical, photocopying, recording, scanning or otherwise, except under the terms of the Copyright Designs and Patents Act 1988 or under the terms of a licence issued by the Copyright Licensing Agency, 90 Tottenham Court Road, London W1P 9HE, UK, without the permission in writing of the Publisher

Other Wiley Editorial Offices

John Wiley and Sons, Inc., 605 Third Avenue,
New York, NY 10158-0012, USA

VCH Verlagsgesellschaft mbH, Pappelallee 3,
D-69469 Weinheim, Germany

Jacaranda Wiley Ltd, 33 Park Road, Milton,
Queensland 4064, Australia

John Wiley & Sons (Asia) Pte Ltd, Clementi Loop #02-01,
Jin Xing Distripark, Singapore 129809

John Wiley & Sons (Canada) Ltd, 22 Worcester Road,
Rexdale, Ontario M9W 1L1, Canada

Library of Congress Cataloging-in-Publication Data

Scott, Raymond P. W. (Raymond Peter William), 1924–
  Tandem techniques / Raymond P. W. Scott.
    p.   cm. — (Separation science series)
  Includes bibiographical references and index.
  ISBN 0-471-96760-2
  1. Chromatographic analysis. I. Title. II. Series.
  QD79.C4S3837 1997   96-42339
  543'. 089 — dc20   CIP

British Library Cataloguing in Publication Data:

A catalogue record for this book is available from the British Library

ISBN 0 471 96760 2

Typeset by the author
Printed and bound in Great Britain by Bookcraft (Bath) Ltd.
This book is printed on acid-free paper responsibly manufactured from sustainable forestation, for which at least two trees are planted for each one used for paper production.

Contents

PART I INTRODUCTION TO TANDEM SYSTEMS 1

CHAPTER 1 INTRODUCTION TO TANDEM TECHNIQUES 3

The Evolution of Analytical Instruments 3
Separation Techniques for Tandem Systems 7
Chromatography 7
Gas Chromatography 9
Gas Chromatography Columns 13
Liquid Chromatography 18
Liquid Chromatography Columns 19
Thin Layer Chromatography 25
The Function of the Chromatographic Column 27
Peak Dispersion 30
Capillary Electrophoresis 35
Zone Electrophoresis 36
Isotachophoresis 37
Isoelectric Focusing 38
Capillary Electrophoresis Apparatus 40
Synopsis 41
References 43

CHAPTER 2 IDENTIFICATION TECHNIQUES FOR TANDEM USE 45

Introduction to Identification Techniques 45
Spectroscopy Involving the Absorption and Emission of Electromagnetic Waves 48
UV and Visible Spectroscopy 48
Fluorescence Spectroscopy 55
Infrared Spectroscopy 59
IR Spectroscopy Instrumentation 62
Raman Spectroscopy 66
Atomic Spectra 73
Atomic Absorption Spectroscopy 75
Chiro-Optical Measurements 76
Polarization Modulation 79
Practical Chiral Measuring Devices 80

Nuclear Magnetic Resonance Spectroscopy 82
The NMR Spectrometer 87
Mass Spectrometry 90
The Sector Mass Spectrometer. 90
The Quadrupole Mass Spectrometer 94
The Ion Trap Detector 97
The Time of Flight Mass Spectrometer 99
Synopsis 101
References 103

CHAPTER 3   INTERFACE CONDUITS 105

Dispersion in Open Tubes... 105
Gas Chromatography Columns and Connecting Tubes 105
Liquid Chromatography Columns and Connecting Tubes..................... 113
Low Dispersion Connecting Tubes 122
Serpentine Tubes 126
Synopsis 128
References 129

PART 2   GAS CHROMATOGRAPHY TANDEM SYSTEMS 131

CHAPTER 4 GAS CHROMATOGRAPHY IR SPECTROSCOPY (GC/IR) TANDEM SYSTEMS 133

The Early GC/IR/MS Triplet System 134
Modern GC/IR Systems 138
Light Pipe Interfaces 138
The Cryostatic Interface 147
Thermogravimetric Analysis (TGA) Coupled with GC/IR and GC/MS 153
Synopsis 162
References 163

CHAPTER 5 GAS CHROMATOGRAPHY / MASS SPECTROSCOPY (GC/MS) TANDEM SYSTEMS 165

Vapor Concentrators.. 165
The Ryhage Concentrator 166
The Bieman Concentrator 167
Ion Sources Used in GC/MS 168
Electron Impact Ionization 168
Chemical Ionization 171
The Inductively Coupled Plasma Mass Spectrometer Ion Source 174

Examples of Some Gas Chromatography/Mass Spectrometry
Applications... 177
Determination of Trace Concentrations of the 2,3,7,8-Chlorine-
Substituted Dibenzo-p-Dioxins and Furans in Beef Fat........................ 178
Analysis of Anabolic Steroids in Biological materials 179
Analysis of Waxes and Lipid Type Materials 183
Some General Applications of GC/MS tandem Systems........................ 187
Urine Analysis by GC/MS. 190
GC/MS Tandem Systems in Environmental Analysis 196
Derivatization Techniques 208
Esterification 208
Acylation 209
Some Special Applications of GC/MS... 212
Novel Interfaces for Specific Applications.. 216
Affinity Membrane Mass Spectrometry Interfaces................................. 216
A Combustion Interface for Isotope Ratio Monitoring with GC/MS 219
Pyrolysis GC/MS of Biological Particulates Collected During Space
Shuttle Missions.. 220
Synopsis 223
References 224

**CHAPTER 6   GAS CHROMATOGRAPHY/ATOMIC
SPECTROSCOPY (GC/AS) TANDEM SYSTEMS**...................... 227

The Microwave-induced Plasma (MIP) Atomic Emission
Spectrometer in GC/ES Systems.. 228
Inductively Coupled Plasma (ICP) GC/ES Systems.............................. 242
Synopsis 249
References 250

PART 3 LIQUID CHROMATOGRAPHY TANDEM SYSTEMS........ 253

**CHAPTER 7 LIQUID CHROMATOGRAPHY/UV SPECTROSCOPY /
FLUORESCENCE SPECTROSCOPY TANDEM SYSTEMS** 255

LC/UV Tandem Systems 255
LC/Fluorescence Spectroscopy Tandem Systems 267
The Hewlett-Packard Fluorescence Spectrometer Designed for Use
as a LC/FS Tandem System... 268
The Multiwave Fluorescence Detector 279
A Tandem Instrument that Monitors UV Absorption, Fluorescence
and Luminescence... 284
Laser-induced Fluorescence Detection Employing a LC/FL Tandem
Combination.. 286
Synopsis 288

References 288

CHAPTER 8 LC/IR TANDEM SYSTEMS 289

LC/IR Transport Interface 290
An Early LC/FTIR Interface 290
The Development of LC/IR Transport Interfaces 292
The Series 100 LC–Transform™ LC/FTIR Interface 308
Interface for the Combination of Liquid Chromatography and
Raman Spectroscopy 323
Synopsis 326
References 327

**CHAPTER 9 LIQUID CHROMATOGRAPHY/MASS
SPECTROSCOPY (LC/MS) TANDEM SYSTEMS** 329

Secondary Ion Mass Spectrometry (SIMS) 330
Fast Atom Bombardment (FAB) 331
Plasma Desorption Mass Spectrometry 332
The Inductively Coupled Plasma Ionization Source 333
Laser Desorption Mass Spectrometry 335
Matrix Assisted Laser Desorption/Ionization (MALDI) 337
Field Desorption Ionization 338
Liquid Chromatography/Mass Spectrometry (LC/MS) Techniques 340
Transport Interfaces
    The Wire Transport Detector 341
    The Belt Transport Detector 344
    Other Transport Systems 346
Direct Inlet LC/MS Interfaces 349
    The Thermospray Interface 350
    The Electrospray Interface 359
Applications of the Electrospray LC/MS Interface 364
Post Column Additives in LC/MS using Electrospray Interfaces 375
The Atmospheric Ionization Interface (API) 384
Inductively Coupled Plasma LC/MS Interfaces 389
The Particle Beam Interface 394
The Permeable Membrane Interface 398
Synopsis 400
References 402

**CHAPTER 10 LIQUID CHROMATOGRAPHY/ATOMIC
SPECTROSCOPY (LC/AS) TANDEM SYSTEMS** 405

Liquid Chromatography Flame Atomic Absorption Spectroscopy
(LC/AAS) Systems 406

Laser Enhanced Ionization for Measuring Organotin Compounds in
Liquid Chromatography Column Eluents 412
Laser Excited Atomic Fluorescence for Measuring Organo-
manganese and Organotin Compounds in Liquid Chromatography
Column Eluents 414
The LC/AAS Tandem Instrument Utilizing the Graphite Furnace
Interface 415
Inductively Coupled Plasma LC/AS Interfaces 417
A Liquid Chromatography Atomic Absorption Tandem System
Involving Thermochemical Hydride Generation 427
The Moving Wheel Liquid Chromatography Helium Microwave
Induced Plasma Interface 429
Synopsis 431
References 432

CHAPTER 11 LIQUID CHROMATOGRAPHY/NUCLEAR MAGNETIC RESONANCE SPECTROSCOPY (LC/NMR) TANDEM SYSTEMS 435

The Modern LC/NMR Tandem Instrument 439
The Basic LC/NMR Tandem System 440
    Capillary LC-NMR Coupling
    High Resolution 1H NMR Spectroscopy in the Nanoliter
    Scale 444
Optimization of LC and NMR Operating Conditions for Tandem
LC/NMR Systems 447
Employing 750 MHz Spectrometer in LC/NMR Tandem Systems 448
LC/NMR Detection of Peptides Employing Micro-Cells 449
Synopsis 453
References 454

PART 4 OTHER TANDEM SYSTEMS 455

CHAPTER 12 THIN LAYER CHROMATOGRAPHY / SPECTROSCOPY TANDEM SYSTEMS 457

Scanning Densitometry 458
The IR Scanning of TLC Plates 464
Diffuse Reflectance IR Fourier Transform Spectrometry (DRIFTS) 466
Scanning Thin Layer Plates by Photoacoustic Spectroscopy 470
Alternative TLC/IR Interfaces 472
Thin Layer Chromatography/Mass Spectroscopy 475
Synopsis 479
References 481

CHAPTER 13 ELECTROPHORESIS /SPECTROSCOPY TANDEM SYSTEMS 483

Sensitivity Enhancement Techniques...486
Concentration Enhancement Techniques...487
Capillary Electrophoresis in Tandem with the Ion Trap Mass Spectrometer..490
Capillary Electrophoresis Operated in Tandem with the Magnetic Sector Mass Spectrometer..492
Capillary Electrophoresis Operated in Tandem with the Time of Flight Mass Spectrometer 495
The Capillary Electrophoresis/Quadrupole Mass Spectrometer Tandem System Employing an Atmospheric Pressure Ionization Source...500
Capillary Electrochromatography Employing the Electrospray Interface...501
Off-line Matrix Assisted Laser Desorption Ionization Time of Flight Mass Spectrometric Monitoring of Capillary Electrophoresis Separations 504
Interface for Capillary Electrophoresis and Inductively Coupled Plasma Mass Spectrometry 507
Capillary Electrophoresis Coupled with Inductively Coupled Plasma Ionization and Atomic Emission Spectroscopy 509
NMR Spectroscopy on the Nanoliter Scale in Tandem with Capillary Electrophoresis 511
Synopsis 514
References...515

INDEX..517

Preface

Among the plethora of investigative procedures that are available to the contemporary scientist, it is clear that the most important and frequently used techniques are concerned with analytical chemistry. The majority of modern analyses embody two essential parts. The first isolates the substances of interest from the matrix of the sample, the second identifies and quantitatively estimates them. In the past the two stages have been carried out sequentially, as separate and independent procedures. Complex, high efficiency instruments are available to separate the components of interest from the matrix and each other, and equally complex spectroscopic devices are available to unambiguously identify them However, in order to cope with the high sample load experienced by many analytical laboratories today, the speed of analysis had to be increased by combining the separating device with the identifying spectrometer in a single *tandem* instrument.

The combination of two complex measuring systems must be achieved with neither instrument degrading the performance of the other. This combination is difficult but, nevertheless, has been successfully achieved by the use of some cleverly designed interfaces. There are many diverse separating procedures and many different identifying instruments, and each particular combination demands a unique interface. This book has been written to acquaint the reader with the many different tandem systems that have been devised and the details of the interfaces with which they have been employed. A wide range of spectroscopic techniques are discussed together with the many forms of chromatographic and electrophoretic separating procedures with which they can be used. Application examples of most tandem combinations are included. This book, is intended to introduce the reader to the wide range of tandem techniques that are now available, to provide a basic understanding of their function and operation, and to help in the selection of the appropriate instrumentation for any chosen application.

Raymond P. W. Scott   October 1996
Birkbeck College, London
Georgetown University, Washington, DC.

Acknowledgments

The information provided by Dr. Nicholson on LC/NMR techniques and also that provided Dr. A. F. Drake on chiral measuring systems, both of Birkbeck College London, are gratefully acknowledged.

Many instrument manufacturers have kindly provided detailed information on their products including photographs of devices and instruments and their help is greatly appreciated. Among the many, special thanks are due to ATI Unicam Ltd., Bruker Inc., CAMAG Scientific Inc., the ΔPACKARD Company, ESA Inc., the Hewlett-Packard Corporation, JM Science Inc., LabConections Inc., LDC Analytical, Nicolet Analytical, the Perkin Elmer Corporation, Polymer Laboratories Inc., Supelco Inc., VALCO Instruments Inc., Varian Instruments Inc. and VG Organic Inc. for information provided and permission to reproduce instrument details and application examples from their technical literature.

I would like to thank the Royal Society of Chemistry for permission to reproduce the following figures from papers (detailed references given) published in the *Analyst*. Figures 5.11 to 5.20 , figures 5.22 to 5.28 and figures 5.30 and 5. 31 in chapter 5: Figures 6.9 to 6.12 in chapter 6: Figures 9.19 and 9.21, figures 9.29 to 9.31, and figures 9.40, 9.53, 9.54 and 9.57 in chapter 9: Finally, figures 10.4, 10.5, 10.8, 10.9, 10.10 , 10.12, 10.14 to 10.16 in chapter 10.

I would also thank the American Chemical Society for permission to reproduce a number of diagrams from their journal, *Analytical Chemistry*. Appropriate recognition has been included with each diagram.

Figure 1.7 in Chapter 1 is reprinted from *J. Chromatogr.*, **125**(1976)251, Figures 1.5 and 1.6 in Chapter 1 are reprinted from *J. Chromatogr.*, **253**(1982)159; Figure 3.5, 3.8 and 3.9 in chapter 3 are reprinted from *J. Chromatogr.*, **268**(1982)169 with kind permission of Elsevier Science-NL, Sara Bugerhartstraat 25, 1055 KV Amsterdam, The Netherlands.

PART 1

INTRODUCTION TO TANDEM SYSTEMS

CHAPTER 1

INTRODUCTION TO TANDEM TECHNIQUES

The evolution of analytical instruments

In the early days of chemical analysis, analytical procedures were carried out sequentially, not concurrently, and consequently were very time consuming. Furthermore, the techniques to choose from were few in number, very crude in nature, and demanded considerable skill on the part of the analyst. At the beginning of the twentieth century, distillation, filtration and crystallization were the only separation techniques that were practical. Identification was achieved by employing such measurements as boiling points, melting points, vapor pressures and refractive indices; for quantitative evaluation, the analyst could only resort to gravimetric and volumetric procedures.

Fifty years ago analytical instruments as we know them today were virtually non-existent. The Abbe Refractometer, some crude pH meters, a manually operated UV spectrometer and a carbon–hydrogen–nitrogen analyzer (often constructed in the laboratory) comprised the limit of instrument availability. At the beginning of this century the structure of some fairly complex organic compounds had been elucidated, including such substances as morphine, vitamin C, thyroxin etc. These relatively complex structures were determined in the classical manner of that time, by analysis followed by synthesis, using the very primitive techniques already mentioned. Such was the caliber of the early twentieth century analysts who, with grossly inferior tools, laid the foundations of modern chemistry.

The analytical instrument industry became firmly established in the late 1940s and early 1950s, provoked by the technological advances that took place during the Second World War. Some of the first instruments to appear on the industrial scene were the polarograph, various types of pH meters and the manually operated UV spectrometer. This was the first time that reliable, and generally affordable scientific measuring equipment was available to universities and industry. In the mid 1950s the scanning UV spectrometer was introduced, and this was rapidly followed by the scanning IR dispersion spectrometer. At this time the mass spectrometer, the nuclear magnetic resonance spectrometer (NMR) and the X–ray crystallograph were still novelties in the laboratories of academia. In the mid 1950s the technique of gas chromatography (GC) was invented by James and Martin [1], and in the late 1950s the first isothermal gas chromatograph was produced. Excluding the rather massive and cumbersome Craig Machine, that was used largely in preparative LLC, the gas chromatograph was the first analytical separation instrument to be developed, despite the fact that liquid chromatography had been known since the latter part of the nineteenth century. The pre-eminence of GC was partly due to the nature of the technique, and partly due to the fact that it was at a more advanced stage of development, and was far easier to automate than LC.

In the early 1960s the first effective mass spectrometers began to appear in conjunction with the first low–resolution 60 kHz NMR machines. During this period, the GC instrument was further updated to include temperature programming, and the first primitive liquid chromatographs began to make their appearance. The 1970s were the heyday of analytical instrument development, spurred on by the introduction of the mini and then the micro computers. Fourier transform IR and NMR machines were introduced with consequent impressive increases in resolution and sensitivity. Liquid chromatographs were fitted with versatile gradient elution features, and both GC and LC instruments incorporated sophisticated automatic samplers, and computer data acquisition and processing systems. For the scientist involved with the development of analytical instrumentation, these were the 'golden years' of analytical chemistry. In the late 1970s and to the present time, a plethora of new analytical instruments were developed and manufactured, including the various types of spectrometer, the X-ray crystallograph, electrophoretic and isotachophoretic instruments and many others. Today the fully trained analyst has a wide range of techniques and instruments to choose from, but to exploit them fully, a knowledge of electronics, engineering and computer technology will be necessary. Above all, the modern analyst must have a sound understanding of the basic principles behind the function of any instrument or technique that is employed. Perhaps the complexity of modern analytical instrumentation, coupled with the necessary broad training of the contemporary analyst, accounts for the long-overdue acceptance of analytical chemistry as an established and important branch of chemistry.

Despite the speed and accuracy of the instrumental techniques that can now be employed, the use of more than one, separately and in sequence, can still be a very time-consuming procedure. To increase the speed of analysis, many of the techniques have been operated concurrently, so that two or more analytical procedures can be carried out simultaneously. The tandem operation of two different instruments greatly increases the efficiency of the analysis, but sometimes, due to unpredictable interactions between one technique and the other, the combination can become difficult in practice. These difficulties can become exacerbated if optimum performance is required from both instruments.

The purpose of this book is to describe the various techniques that can be combined in tandem form and discuss the difficulties associated with their interfacing, together with the procedures that have been developed to surmount them. Examples will be given that demonstrate the efficacy of the different tandem systems that are described, and will include a range of different separation techniques as well as different single and combined spectroscopic systems.

Today, unless techniques are used that are specific to the compound of interest, any successful analysis entails three distinctly different procedures. In the first instance the sample must be separated into its constituents so that the pertinent substances are isolated from the main bulk of the sample. This will probably involve the use of an appropriate separation technique such as chromatography. Secondly, the components that have been isolated will need to be identified, involving techniques such as spectroscopy or X-ray crystallography. Finally the amount of each component isolated must be quantitatively determined, ideally by the use of primary measurements such as those provided by gravimetric or volumetric techniques. A more practical alternative for quantitative assessment is to employ secondary methods of measurement, using devices that have been calibrated by primary procedures. An example of the latter is the chromatographic detector that has been calibrated for response and linearity against known concentrations of standard substances.

The three basic requirements, separation, identification and quantitative measurement, coupled with the availability of reliable and appropriate separating and identifying instrumentation evoked the combination of two types of analytical systems. The first efficient tandem system to be developed was the combination of the GC with the mass spectrometer (MS) and the first effective GC/MS instrument was assembled in the very early 1960s. This evolution was natural, as both the gas chromatograph and the mass spectrometer had reached a reasonably advanced stage of development and, although not particularly straightforward, interfacing the chromatograph with the mass spectrometer was one of the easier tandem systems to fabricate. GC can only separate volatile substances and so it was the needs of the analysts of the petroleum, essential oil and solvent industries that were the first to be satisfied by a separation/identification combination of instruments. It should be remembered that the gas chromatography detector gives a linear response with concentration and thus the GC/MS combination satisfies all the demands of the analyst. It follows that for the satisfactory use of tandem instruments, the analyst requires an armory of separation and identification techniques from which to choose. The choice of available

Introduction to Tandem Techniques

separation techniques that are presently available and satisfactory for use with combination systems will first be discussed.

Separation Techniques for Tandem Systems

The separation techniques employed with combination systems are, in most cases, some form of chromatography. There are some instances where classical electrophoresis has been used for preliminary separation before subsequent off–line spectroscopic examination. In fact electropherograms have been scanned directly by reflectance spectroscopy. However, a more appropriate electrophoretic technique suitable for tandem operation is capillary electrophoresis and so the basic principles involved in capillary electrophoretic separations will also be discussed. In the main, however, chromatography is much the favored separation procedure chosen for tandem systems. A word of caution might be added here. Isotachophoresis is at present receiving increased attention as it offers excellent possibilities for the separation of biological polymers and, with further development, might become an attractive alternative to chromatography for association with an appropriate identification technique.

It has been pointed out that to employ tandem techniques in an effective manner, the basic principles of both the separation process involved and the identifying technique must be clearly understood. Consequently, the essential characteristics of both chromatographic and electrophoretic separations will be discussed. Particular emphasis will be given to those properties of the separation technique that influences both the tandem interface and the efficient operation of the associated identifying technique.

Chromatography

Chromatography, by classical definition, is a separation process where resolution is achieved by the distribution of the components of a mixture between two phases, a stationary phase and a mobile phase. Those components held preferentially in the stationary phase are retained longer in the system than those that are distributed in the mobile phase. As a consequence solutes are eluted from the system in the order of their increasing distribution coefficients with respect to the stationary phase; *ipso facto* a separation is achieved. The mobile phase can be a gas or a liquid which gives rise to the two basic forms of chromatography, namely, gas chromatography (GC) and liquid chromatography (LC). The stationary phase can also take two forms, solid and liquid, which provides two subgroups of GC and LC, namely; gas–solid chromatography (GSC) and gas–liquid chromatography (GLC), together with liquid solid chromatography (LSC) and liquid chromatography (LLC). The different forms of chromatography are summarized in Table 1.

Table 1.1   Different Forms of Chromatography

MOBILE PHASE	STATIONARY PHASE
GAS Gas Chromatography **(GC)**	**LIQUID** Gas-Liquid Chromatography **(GLC)**
	SOLID Gas-Solid Chromatography **(GSC)**
LIQUID Liquid Chromatography **(LC)**	**LIQUID** Liquid-Liquid Chromatography **(LLC)**
	SOLID Liquid-Solid Chromatography **(LSC)**

It should be pointed out that the above classification is fundamental and all chromatographic separations will fit into one of the four classes. For sundry reasons, various other terms have been introduced by analysts in the field to describe certain subgroups of chromatography. Thin–layer chromatography (TLC) has been introduced to describe certain forms of lamina chromatography systems, size exclusion chromatography (SEC) for separations based on molecular size, reversed-phase chromatography (RPC) for chromatographic separations that are determined largely by dispersive interactions with the stationary phase, and there are many more. These terms can be confusing to those unfamiliar with the technique but,

Introduction to Tandem Techniques

from the point of view of tandem techniques, the important characteristics are defined by the four classifications given above.

GSC is used largely for the separation of permanent gases and very low boiling liquids and is rarely used in direct association with identification techniques, except occasionally with the quadruple mass spectrometer. GLC, on the other hand, is frequently used in conjunction with both the IR spectrometer and the mass spectrometer for the separation and identification of virtually all classes of volatile compounds. LLC systems can be extremely difficult to maintain in sufficiently stable conditions to permit an associated spectrometer or other identifying instrument to function in a consistent and reliable manner. This is due to the labile nature of the stationary liquid phase in contact with the flowing mobile phase. It follows that LLC is only rarely employed in tandem systems but, should some means of stabilizing the stationary phase be developed, this situation could rapidly change. On the other hand, LSC is the most popular form of liquid chromatography and is now being employed more extensively than even GLC; in fact LSC is now used in many applications that were originally in the realm of GLC. The stationary phases employed in LSC are either silica gel or, more commonly, a bonded phase and more recently certain polymeric materials. A bonded phase is formed when silica gel is treated with an appropriate silanizing reagent to link a hydrocarbon chain or other organic moiety on each of the hydroxyl groups situated on the silica surface. Such stationary phases are fairly stable, unless exposed to mobile phases of extreme pH, and thus are very satisfactory for the operation of instruments in tandem. The polymeric phases are mostly made from micro-reticulated polystyrene resin cross-linked with divinyl benzene. The two basic chromatographic techniques will now be discussed in some detail to allow the pertinent factors that effect the interfacing between the chromatograph and spectrometer or other identifying technique to be clearly understood.

Gas Chromatography

A block diagram of the basic gas chromatograph is shown Figure 1.1. It consists of a gas supply, the nature and complexity of which depends largely on the type of detector that is being used. If a katherometer detector (hot wire detector) is employed, a single gas supply of helium will be adequate. Conversely, if a flame ionization detector (FID) is used, then a three-component gas supply will be necessary comprised of hydrogen and air for the FID and helium (or occasionally nitrogen) for the mobile phase (commonly called the carrier gas). The different gases pass to appropriate flow controllers, one for each gas supply, and that from the carrier gas controller passes to a heated injection system, where the sample is introduced onto the column.

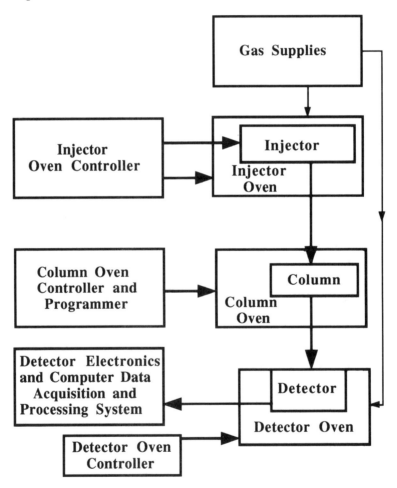

Figure 1.1 A Block Diagram of a Gas Chromatograph

Introduction to Tandem Techniques

There are a number of different types of injection devices available and these will be discussed in some detail later, in conjunction with the type of column with which they are used. The injector is connected directly to the column which is situated in an oven, and its temperature can be either controlled isothermally or programmed between chosen limits at selected rates. The carrier gas leaves the column and passes directly to the detector which is situated in its own thermostatically controlled oven. This is normally controlled isothermally at a temperature at least 20°C higher than the maximum temperature the column oven will reach during the development of the separation. This prevents any solute condensing in the detector or detector connecting tubes. If the chromatograph is connected to some other instrument, and the detector is destructive (*e.g.* the FID), the carrier gas is split into two streams between the column and the detector; one stream passes to the detector and the other to the associated instrument. If the detector is non–destructive, (*i.e.* the katherometer), then all the carrier gas can pass through the detector and part, or all, the exit gas fed to the second instrument. In some instances, the sensor of the associated equipment can be used as the chromatographic detector, in which case the whole column eluent can pass directly to the tandem instrument bypassing the GC detector. This procedure assumes that the response of the sensor in the associated instrument is linear and that the quantitative accuracy is not impaired.

It is not germane to the subject of this book to discuss detectors in detail, but some points will be made that are important with respect to tandem systems. The most practical and useful detectors for combined techniques employing gas chromatography as the separation process are the two previously mentioned, the katherometer and the FID. The katherometer has the advantage of being non-destructive and thus, all the eluent can pass to the second instrument, conserving the sample, and providing the maximum material for measurement. On the other hand, the katherometer has limited sensitivity relative to the FID, and thus requires more sample, which precludes its use with certain types of column. The katherometer has also a limited linear dynamic range but, within that range, can be used for accurate quantitative work. The destructive nature of the FID demands the use of a split system and thus the second instrument is partially deprived of sample. However, the FID is far more sensitive than the katherometer and consequently, only needs a very small proportion of the eluent for monitoring. Furthermore, it has a very wide linear dynamic range and thus, in general, the quantitative accuracy is not impaired.

The only two other detectors that are used to any significant extent with GC tandem systems are the Nitrogen–Phosphorous Detector (NPD) a modified form of the FID and the electron capture detector. Both these detectors are extremely sensitive and both are normally used with splitters, the former because it is destructive, the latter because its volume is relatively large and the resulting dispersion would impair the performance of the associated instrument. This is an important aspect of interfaces between the two instruments and will be discussed in considerable detail later. The NPD has a reasonable linear dynamic range and provides accurate quantitative analyses.

The electron capture detector, on the other hand, has a limited linear range and thus, the sample size must be carefully selected to satisfy both the quantitative accuracy required and the needs of the associated instrument. In some instances the associated instrument, *e.g.* a mass spectrometer, can be used as the detector. In such cases the total column eluent can be made to bypass the GC detector and enter the mass spectrometer directly. If this procedure is adopted, again care must be taken to ensure that the quantitative accuracy of the analysis is not impaired due to an inadequate linear response of the associated instrument.

The signal from the detector sensor is processed in a suitable manner by the detector amplifier and the output is usually passed to the chromatograph computer. The computer acquires the data and processes the results, and also controls the operation of the chromatograph and automatic sampler (if part of the equipment). The chromatograph can also communicate with the computer of the associated instrument to help synchronize measurements as the peaks are eluted. Interface design and construction will be discussed in detail in those chapters dealing with the

Introduction to Tandem Techniques   13

specific tandem techniques as they will differ significantly with each type of instrument employed.

Gas Chromatography Columns

There are basically two types of chromatographic column, the packed column and the capillary column. Both types are usually formed into coils or loops to make them sufficiently compact to fit inside small ovens. The packed column is usually 3 or 4 mm I.D. and 3 to 6 ft long, although, if designed correctly, and if the apparatus can cope with a sufficiently high inlet pressure, columns up to 100 ft long can be prepared [2]. The column case itself is usually made of glass but sometimes of metal and is packed with a diatomaceous earth such as celite coated with the stationary phase. The celite is ground and screened to about 100 µm particle diameter and de-activated by treatment with a suitable silane such as hexachloro-disilazane. The packing is then coated with stationary phase by wetting it with a solution of the stationary phase in a volatile solvent and evaporating the solvent in a rotary evaporator. The loading of stationary phase can vary (usually between 5% and 10% w/w), the higher the load, the larger the size of the sample that can be placed on the column. This can be advantageous when the associated instrument lacks a comparable sensitivity, *e.g.* the IR spectrometer. The column is packed by adding the packing and tapping until a maximum packing density is achieved.

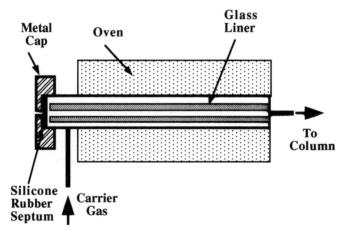

Figure 1.2   **The Septum Injector**

Injection systems for packed columns are simple, reliable and can be designed to give both accurate and precise quantitative results. A diagram of a septum injector for a packed column is shown in Figure 1.2.

The injector is situated in a small oven, often a simple metal block that is maintained at a temperature about 20°C above the maximum the column will attain during the development of the chromatogram. This is necessary to ensure that all the sample is volatilized and is called flash heating; the oven is sometimes referred to as the 'flash heater'. The injector itself consists of a tube with a septum cap that maintains the system pressure tight when the hypodermic syringe enters the injector. The carrier gas enters the side of the injector and the column is connected to the end by a suitable union. The injector usually contains a glass liner to prevent the sample coming in contact with the catalytically active metal surface of the injector body. The injection procedure involves drawing the sample, as a fluid, into a specially designed hypodermic syringe, inserting it through the septum via a hole in the metal cap and discharging the contents onto the glass liner. The sample instantly volatilizes and is swept onto the column. An alternative procedure that is, in fact, to be preferred as it eliminates any possibility of thermal decomposition is *on-column* injection. The same injector can be used, but the injection oven is turned off and a syringe with a long needle is used so that it penetrates to the column packing and the sample is discharged onto the column itself. Both methods, if carried out with due care, can provide precise and accurate quantitative results.

The capillary column was invented by Golay [3] and takes the form of an open tube with the stationary phase coated on the inside surface as a thin film. Originally the tube was constructed of metal but, due to surface activity of the metal, it was soon replaced by the soft glass capillary tubes invented by Desty [4]. The soft glass tubes were rigid and thus somewhat difficult to fit into the chromatograph and so they, in turn, were replaced by flexible fused silica tubes developed by Dandenau [5]. The flexible silica tubes were fabricated by coating the tube immediately after drawing with a polyimide resin. This prevented stress corrosion by sealing the external surface of the tube from contact with air and, in particular, from water vapor. The technique was extended to soft glass by Ogan *et al.* [6] who also developed a stable flexible soft glass tubing [7,8] that required no external polyimide coating. This was achieved by a careful annealing procedure immediately after drawing. It appears that flexible fused silica tubing has also been constructed without a polyimide coating [9]. Capillary columns can be fabricated having diameters between 50 and 1000 μm and lengths between 10 and 500 m. The film of stationary phase, usually between 0.1 and 1.0 μm thick, can be formed by either passing a solution of the stationary phase through the column and subsequently evaporating the solvent from the surface film so produced, or the stationary phase can be deposited by direct surface polymerization.

The injection of a sample onto a capillary column is far more difficult than placing it on a packed column. This is because the loading capacity of the capillary column is one to two orders of magnitude less than that of the packed column. Consequently, the sample size must be proportionally reduced and this renders on-column injection impractical for the smaller-bore high–efficiency columns. The reduced sample size is achieved by means of a split flow injection system as shown diagramatically in Figure 1.3.

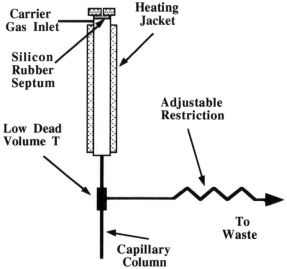

Figure 1.3   **A Split Flow Injection System**

It consists of the normal septum injection device, but of somewhat smaller proportions, and the exit carrier gas, containing the sample as vapor, splits into two streams; one stream passes to the column and the other through an adjustable restriction to waste. By adjusting the restriction, the proportion of the sample that passes onto the column can be selected. This would appear to be a satisfactory solution to the problem of injecting samples onto a capillary column. Unfortunately, however, the split proportion varies with the diffusivity of the solute and consequently its molecular weight. Thus the sample placed onto the column may not be truly representative of the original sample injected, and the precision and accuracy of the quantitative aspect of the analysis can be impaired. Nevertheless, the reduced quantitative accuracy is sometimes tolerated for the sake of improved resolution and reduced analysis time.

In an attempt to effect a compromise between the packed and capillary columns, the large bore capillary column was introduced (0.053 in. I.D.). The large–bore column carries a much larger load of stationary phase and is wide enough to allow a hypodermic needle to enter and allow on-column injection. Unfortunately, the wider bore significantly reduces the resolution obtained from a given length of column and also introduces other injection problems. On injection into an open tube, a sample is often split into two zones, which results in single solutes providing double peaks on elution from the column. Solutions to this problem have been suggested that tend to complicate the normally simple injection procedure. These include the 'retention gap' method [10], where the first few centimeters of column are stripped of stationary phase so that the sample is injected into an empty tube. This method can reduce the peak splitting but a more successful, though more complicated procedure is to employ the solute focusing method [11]. This procedure requires two ovens that can be heated and cooled independently. The sample is injected onto the first part of the column, which is hot, and the sample vapor subsequently condensed onto a second cooled portion of the main column. As a consequence, all the solutes are focused at a point on the front portion of the cold column. The column is subsequently heated to develop the separation in the normal way. This procedure reduces the chance of peak splitting, but is more involved and demands more complicated and expensive equipment. A

Introduction to Tandem Techniques

better solution for some samples might be, where possible, to employ a packed column.

Other types of columns have been developed, such as the SCOT (Solid Coated Open Tubes) columns. These are open tubular columns that have a coating of celite deposited on the internal surface on which the stationary phase has been coated. This type of column attempts to retain the low flow impedance of the open tube, while increasing the stationary phase load to recover some of the advantages of the high stationary phase loading of the packed column. The basic difference between the capillary and the packed column is their flow impedance. Due to the continual change in direction of the flow of gas in a packed column, as it winds its way between the particles of packing, considerable energy is utilized and this is obtained at the expense of a greater pressure drop across the packed column. Thus, the flow impedance of a packed column is much greater than that of the open tube. This is reflected in the relative magnitudes of the constants in the d'Arcy equation, for the flow of fluids through packed beds, compared with that of the Poiseuille equation, for fluid flow through open tubes. It follows that for a given inlet gas pressure, a capillary column can be much longer than a packed column, perhaps by as much as one to two orders of magnitude, and thus the capillary column has a far greater potential for high column efficiencies. There are, however, in addition to injection difficulties, certain disadvantages to capillary columns so that they are not always optimum for tandem systems; these will be discussed a little later.

The characteristics of the GC system, that are important for tandem operation, are the solute concentration leaving the column and the volume flow rate. The concentration of solute in the eluting gas from a GC column can range from 1×10^{-6} g/ml to 5×10^{-12} g/ml. In any particular GC system, the actual concentration will depend both on the mass of sample injected and the dimensions of the column. The flow rate, depending on the column diameter and the carrier gas employed, will range from about 0.5 ml/min. for a capillary column 0.25 mm I.D. to about 25 ml/min. for a standard packed column 4.6 mm I.D.. It is clear that the interface will need to be carefully designed, on the one hand, to accommodate the elution characteristics of the gas chromatograph and, on the other, to meet the inlet requirements of the tandem instrument.

Liquid Chromatography

A block diagram of the basic liquid chromatograph is shown in Figure 1.4. The liquid chromatograph consists of a solvent supply system that provides at least dual solvent capability and, in some instruments, four solvents may be selectively available. The different solvents are blended by means of a solvent programmer that can change the solvent composition according to a selectable range of time functions, and also provide isocratic development for chosen periods. The blended solvent then passes to the injection valve that places the sample onto the column. When used with tandem systems, the sample volume can range from 0.1 to 20 µl and is chosen to be appropriate for the particular column employed and the type of instrument with which the chromatograph is to be associated. There are also a number of different types of column employed in LC, but virtually all those used with tandem instruments utilize the standard sample valve for injection purposes. The valve and the attached column should be situated in a thermostatted oven.

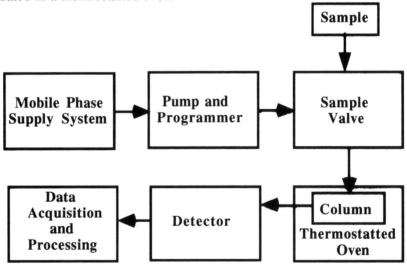

Figure 1.4   **The Basic Liquid Chromatograph**

Introduction to Tandem Techniques

The exit flow from the column passes to the detector. Column detector connections are designed to minimize any extra column dispersion that might impair the chromatographic resolution. The most commonly used detectors, in order of their popularity and versatility, are the UV detectors (single or variable wavelength and the diode array detector), the refractive index detector, the fluorescence detector and the electrical conductivity detector. With the exception of the refractive index detector, all the detectors are fairly sensitive and reasonably linear in response. The most sensitive is the fluorescence detector which unfortunately has the most restricted linear dynamic range, *i.e.* little more than two orders of magnitude. The UV and electrical conductivity detectors are the next most sensitive and have linear dynamic ranges in excess of three orders of magnitude. The refractive index detector, although the least sensitive, has the advantage of being the closest to a universal detector and is, therefore, frequently used for detecting the many substances that are not conducting, do not adsorb light in the UV range of wavelengths, and do not fluoresce. All the detectors are non-destructive and are flow-through devices and so the exit flow can be interfaced directly with another instrument. In a similar manner to the gas chromatograph, the output of the LC detector usually passes to a computer for data processing. In addition, the same computer will control the operation of the chromatograph as well communicating with any tandem instrument to synchronize their operation.

Liquid Chromatography Columns

There are also two types of liquid chromatography columns, the packed column and the capillary column. The latter, however, at the time of writing this book, is an academic novelty rather than a practical analytical system. LC capillary columns are still in a state of development. They are used to a small extent in supercritical liquid chromatography where they are sometimes associated with the mass spectrometer in the tandem LC/MS technique. Unfortunately the loading capacity of the LC capillary column is far too small for the sensitivities available from present–day LC detectors and consequently they must be overloaded for practical use. Overloaded columns give relatively poor efficiencies and thus, due to the inherent limitations of the detector, the high resolution theoretically possible from such columns cannot be realized.

The packed LC column, on the other hand, is well developed and is the 'work horse' of the separation chemist. Generally, the diameter of the analytical column can range from 0.5 mm to 10 mm and its length from 2 cm to 2000 cm although columns up to 10 m long have been fabricated [2]. The packing is usually silica or silanized silica (reverse phase) but polystyrene cross-linked with divinyl benzene has become more popular particularly for ion exchange media. This is because bonded silica ion-exchange media are unstable at high salt concentrations or outside the pH range of about 5–8. LC packings are normally available having particle diameters of 3, 5, 10 and 20 µm. Low-resolution, high-speed columns should be short, with relatively wide diameters and are packed with very small particles. An example of a high-speed separation [13] is shown in Figure 1.5.

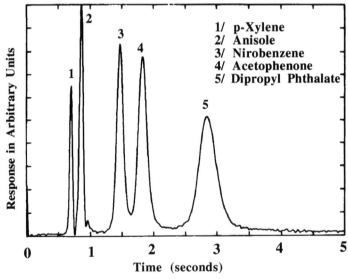

Packing, Hypersil, (silica) particle size 3 µm., Column Length 2.5 cm, Column Diameter 2.6 mm, Mobile phase 2.2% methyl acetate in n-pentane, Mobile Phase Velocity 3.3 cm/sec

Figure 1.5 The Separation of a Five–Component Mixture in 3.5 Seconds

Introduction to Tandem Techniques

The separation was carried out on a column 2.5 cm long, 2.6 mm I.D., packed with silica gel particles, 3 μm in diameter. It is seen that the first pair of solutes are separated and eluted in less than 1 sec. This type of separation is not practical for most tandem techniques, except possibly for LC/MS, as the response of most spectrometers and associated instruments would not be fast enough to obtain the necessary data during the elution of each peak. Nevertheless, it demonstrates that extremely fast separations can be obtained from contemporary columns if so desired. In contrast, very high efficiencies and consequently high resolution, are achieved by using very long columns of very small diameter packed with particles of relatively large diameter [14]. As the columns must be packed, the minimum diameter that appears to be practical is about 0.5 mm. An example of a chromatogram from a high resolving power column is shown in Figure 1.6.

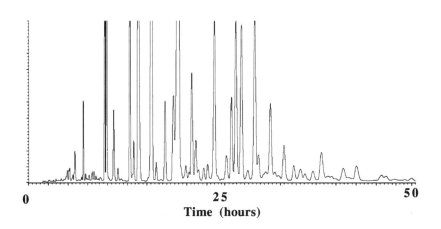

Figure 1.6 The Separation of a Coal Extract on a High-Efficiency Column

The column was 2 m long, 1 mm I.D., and packed with a reverse phase having a particle diameter of 10 μm. The development was isocratic, at a flow-rate of 20 μl/ min. It should be noted that the chromatogram takes over 50 hours to develop. Unfortunately, the 'cost' of a chromatographic separation is counted in units of column inlet pressure and time. Once the maximum available pressure is employed, then time must be sacrificed if very high resolution is required.

The commonly used stationary phases in LC are silica, bonded silica's (*e.g.* the reverse phases) and micro–reticulated cross-linked polystyrene resins. Silica is manufactured in a wide variety of pore sizes, the magnitude of which can significantly affect its chromatographic performance and the properties of any bonded phases made from it. A graph showing the distribution of pore size with pore volume is shown in Figure 1.7. The curves show that the distribution of pore diameters can differ greatly between different silica gels. In fact, silica gel can be used as an exclusion medium for separation on the basis of molecular size [15].

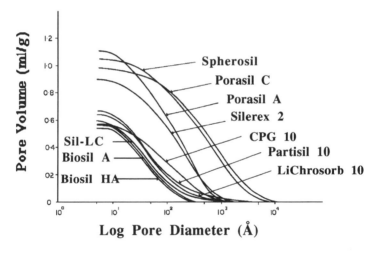

Figure 1.7 Graph of Pore Volume per Gram of Silica Gel against Log Pore Diameter

The range of pore sizes that can occur in a silica gel, including bonded silica, render LC a poor technique for thermodynamic measurements and solute identification from retention data. This is because each solute has a unique *dead volume* and, depending on its molecular size, interacts with a unique fraction of the available stationary phase [15]. This molecular size dependence has a very strong effect on the capacity ratio (k') of a solute which is inversely proportional to the dead volume of the solute. For the want of a better chromatographic measurement (k') values are often employed for solute identification. The uncertainty of solute identification from (k') data, and other measurements based on solute retention, is another reason why the use of tandem techniques is so important and becoming so popular.

When silica gel is treated with an appropriate silane reagent, organic groups can be bonded to the surface. Depending on the nature of the bonded moiety, the character of the surface can range from strongly polar to dispersive or even ionic. The silanizing process will not be discussed in detail, but it should be said that there are basically three types of reverse phase. The use of a silane with a single functional group, such as octyl-dimethyl-chlorosilane, will react only with the hydroxyl groups on the surface and produce the 'brush' type reverse phase. Many chromatographers consider that the two methyl groups on the silicon atom of the reagent stearically hinders any further reaction with a directly adjacent hydroxyl group. In fact, many consider that all attached organic groups are separated from their neighbors by a silicon atom with a hydroxyl group attached (as depicted below). Nevertheless, experimental proof of this remains forthcoming. stationary phases: brush type

A 'Brush' Type Bonded Phase

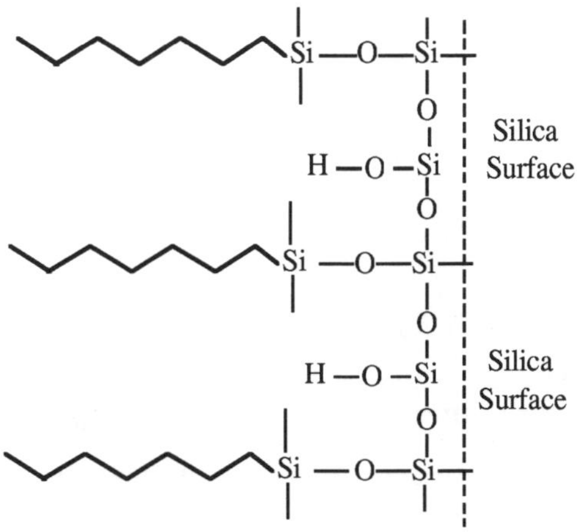

If the silica is treated alternately, with a bifunctional silane such as octyl-dichloro-silane and then water, an oligomeric phase is built up on the

surface. Each oligomeric chain consists of series of organic moieties, joined by silicon atoms to a silanol group on the surface. After the last oligomer has been added, the end of the oligomer is capped with trimethylchlorosilane or hexamethyldisilazane. The oligomeric phases are difficult to manufacture as their synthesis involves a series of steps that are best carried out in a fluidized bed reactor. As a consequence they are not, at present, commercially available. They do offer greater stability to aqueous solvents and higher salt solutions than do the other bonded phases [16] and could, therefore, find application in the separation of materials of biological origin. The third type of bonded phase, the 'bulk' type phase, results from the use of a trifunctional silane reagent in the bonded phase synthesis.

An Oligomeric Bonded Phase

$$\begin{array}{c}
CH_3 \\
| \\
CH_3-Si-CH_3 \\
| \\
O \\
| \\
\sim\sim\sim\sim Si-CH_3 \quad + HCl \\
| \\
O \\
| \\
\sim\sim\sim\sim Si-CH_3 \\
| \\
O \\
| \\
\sim\sim\sim\sim Si-CH_3 \\
| \\
O \\
| \\
-O-Si-O- \text{ Silica Surface} \\
|
\end{array}$$

The bulk phase is polymeric in form, and is produced, for example, by the use of trichlorosilyl reagents with water added to the silica, prior to reaction. If the silica surface is saturated with water, the octyltrichlorosilane reacts with both the hydroxyls of the silica surface *and* the adsorbed water, causing cross-linking and an octylsilyl polymer is built up on the surface. Due to the polymerization process, the stationary phase has a multi-layer character and consequently, has been given the term 'bulk' phase. The bulk phase can also be synthesized by alternately treating

Introduction to Tandem Techniques

the silica with the trifunctional reagent and then with water, in a similar manner to the synthesis of the oligomeric phase. However, with the trifunctional silane the polymer is produced as opposed to an oligomeric chain.

All three types of phases behave differently, particularly when in contact with solvent mixtures of very high water content [17–19] but have no specific influence on any associated tandem instrument. Reverse phases are used in over 75% of all LC analyses. However, these stationary phases are almost always employed with mobile phases consisting of aqueous solvent mixtures. Such solvents can pose certain difficulties when used with tandem instruments such as the IR spectrometer or the NMR spectrometer. These difficulties and their solution will be discussed under the appropriate spectroscopic technique.

The characteristics of the LC system that are important for tandem operation are, again, the solute concentration leaving the column and the volume flow rate. The concentration of solute in the eluting solvent from a LC column can range from about 5×10^{-9} g/ml to 1×10^{-4} g/ml. The eluent concentrations that can be used with LC detectors is seen to be several orders greater than that used with GC detectors. The actual concentration will again depend both on the mass of sample injected and the dimensions of the column. The column flow rate will range from about 5 µl/min., for a microbore column about 0.5 mm I.D., to about 2 ml/min., for a standard packed column about 4.6 mm I.D. Again it is seen that any interface used for a particular tandem instrument will need to be carefully designed to accommodate both the elution characteristics of the LC column and the inlet requirements enjoined by the tandem instrument.

Thin Layer Chromatography

Thin layer chromatography (TLC) is basically a form of LSC, but with two distinct differences. The separation is achieved on a lamina sheet of stationary phase, as opposed to a cylindrical column, and the mobile phase is not pumped through the system, but allowed to percolate through the stationary phase layer by surface tension forces. The stationary phase, usually either silica or silanized silica, is spread on the surface of a glass plate which is spotted at one end with the sample. As a rule, the end of the plate where the sample was placed is then dipped in a trough of mobile phase and the separation developed. As the solvent rises up the plate, the solutes are eluted to different distances along the plate, and a separation is achieved. Inherently the technique has limited resolving power and restricted sample size, although it is very inexpensive to operate. The latter advantage is its main cause for survival. Perhaps the last comment is a little severe, as there are certain types of sample, and certain monitoring procedures, where the simplicity and economy of TLC make it the separation technique of choice.

Various modifications of the basic system have been developed such as the circular plate, where the sample is placed in the center and the mobile phase is introduced by a siphoning procedure at the same point. The separation results in the individual solutes forming concentric rings, which can be made visible by the usual procedures (*e.g.* charring and staining etc.). Another modification of the technique is two-dimensional TLC development. After initial development along one axis of a rectangular plate, the plate is dried and then the complementary edge of the plate immersed in a different solvent and the separation developed perpendicular to the previous axis. This procedure increases not only the resolution obtainable, but also the component capacity, and thus allows the technique to be used for the separation of more complex mixtures.

TLC is not normally employed as a separation technique for tandem operation with spectroscopic instruments. The reason for this is fairly obvious. Unless the plate itself is spectroscopically scanned, the solute bands must be located by a non-destructive means and the solutes removed by scraping the plate and subsequent extraction. Only then can the material separated be examined. UV and fluorescence scanning of TLC plates has been successfully carried out [20–21]. However, as will be discussed later, UV and fluorescence spectra, although helping to confirm the identity of a suspected compound, provide very limited information if the solutes of interest are completely unknown. Mass spectra, NMR spectra and even IR

Introduction to Tandem Techniques

spectra are usually obtained by the somewhat tedious 'spot' recovery process.

An excellent treatment of TLC separation techniques and the spectroscopic examination of TLC plates has been given by Poole [22] and is strongly recommended to those readers interested in TLC or tandem techniques employing TLC as the primary separation process.

The Function of the Chromatographic Column

During the development of a chromatogram, two processes are active in the column that proceed concurrently and more or less independently of each other. These processes are common to both GC and LC columns and have two-dimensional equivalents in TLC.

First, the individual solutes contained in the sample move through the column at different rates, due to their different distribution coefficients with respect to the stationary phase. This results in the individual solutes moving away from each other as they move along the column and are thus separated.

Second the individual solute bands begin to spread, due to the various dispersion processes that occur. It follows that the column must be designed to contain this dispersion so that each solute peak can be eluted discretely. The spreading of the peak, both in the column and in any interface between the column and a second instrument, is extremely important in tandem systems. This is so because just as the peak dispersion must be contained in the column so that the peaks do not merge, so must the dispersion in the interface between the two instruments be constrained, so that the integrity of the separation is maintained in the tandem instrument. Furthermore, the dispersion in the column controls the total volume of the eluted peak.

Thus the sensing cell in the associated instrument will have a maximum permissible volume to ensure that, at any time, only one solute peak can be present in the cell and be monitored. More than one sample in the sensing cell of the associated instrument would provide confused data and make identification very difficult, if not impossible.

The retention volume, $V_{R(A)}$, of solute A is given by

$$V_{R(A)} = V_0 + K_{(A)}V_L \qquad (1.1)$$

where (V_0) is the dead volume, that is the volume of mobile phase between the injection point and the detector,
($K_{(A)}$) is the distribution coefficient of solute (A) between the two phases,
and (V_L) is the volume of stationary phase in the is the volume of stationary phase in the

and for solute B,

$$V_{R(B)} = V_0 + K_{(B)}V_L \qquad (1.2)$$

where ($K_{(B)}$) is the distribution coefficient of solute (B) between the two phases.

It is seen that the retention of a solute depends only on the volume of mobile phase and stationary phase in the column, and the distribution coefficient of the solute between the two phases. It is independent of the mobile phase velocity and the particle size of the packing.

The volume between the peaks, which is one measure that describes the magnitude of the resolving power of the column, is given by

$$V_{R(B)} - V_{R(A)} = V_0 + K_{(B)}V_L - V_0 - K_{(A)}V_L$$

$$= (K_{(B)} - K_{(A)})V_L \qquad (1.3)$$

It is seen that for any given pair of solutes the volume between the peaks increases with the volume of stationary phase in the column. Thus, as a packed column contains one to two orders of magnitude more stationary phase than the capillary column then, for columns of equal length, the packed column will give the greater *separation*. However, it must also be

Introduction to Tandem Techniques

pointed out that *resolution* depends not only on the degree of separation but also on the extent to which the peaks are dispersed (the more narrow the peaks, the higher the efficiency of the column). The capillary column can be made much longer than the packed column and provide much higher efficiencies (more narrow peaks) thus the highest resolution achievable for a given inlet pressure will be obtained from the capillary column. The standard deviation of the eluted peak is given by the following equation [23],

$$\sigma = \frac{V_R}{\sqrt{n}} \qquad (1.4)$$

where (σ) is the standard deviation of the peak,
and (n) is the column efficiency in theoretical plates.

Substituting for (V_R) from equation (1.1),

$$\sigma = \frac{(V_o + KV_L)}{\sqrt{n}} \qquad (1.5)$$

Now $\quad V_o + KV_L = V_o(1 + K\frac{V_L}{V_o}) = V_o(1 + k') \qquad (1.6)$

and as (V_o) the dead volume of the column is about 60% of the total column volume,

$$V_o = 0.6 \pi r^2 l$$

where (r) is the radius of the column,
and (l) is the length of the column

Substituting for (V_o) in equation (6), and for (V_o+KV_L) in equation (1.5),

$$\sigma = \frac{0.6 \pi r^2 l (1+k')}{\sqrt{n}} \qquad (1.7)$$

It is seen that the standard deviation of a peak is directly related to the column volume and the capacity ratio of the respective solute. The peak volume can be taken as equivalent to four volume standard deviations, and thus the peak volume is also related to column volume and hence its dimensions and the capacity ratio of the eluted peak. It follows that the sensing volume of any associated spectroscopic instrument, *i.e.* the sample cell, can be much larger for a packed column than a capillary column before it is in danger of holding more than one solute peak. However, for an extremely complex mixture, the capillary column must be employed in GC to realize the necessary resolution. The peak volume increases as the retention time increases for all types of column. Consequently, as each peak must be ascribed equal importance, the interface system and sensor must be designed to accommodate the peaks of smallest volume, *i.e.* the first peaks eluted. This can place even more stringent demands on the sensor-interface system when a capillary column is employed.

There is a choice of two distinctly different types of GC columns, as already discussed, and advantage should be taken of this choice when designing tandem systems. Capillary columns, despite being nearly thirty years old, have the image of being the 'state of the art' in column technology. This has arisen partly because they are easier to fabricate commercially in an optimum form, and partly because of the glamour of the exceptionally high efficiencies that are available from them. However, they can sometimes place almost impossible demands on any associated technique and, in many cases, the very high efficiencies may not be required. There is no 'Philosopher's Stone' for the analyst, the column chosen should always be on the basis of efficacy and not popularity and, if used in conjunction with a tandem instrument, it must be in accordance with the unique requirements of the associated equipment.

Peak Dispersion

The second process that occurs in the column is the progressive dispersion of each solute band during its passage through the column. In contrast to solute retention or solute selectivity, peak dispersion is strongly dependent on the particle size of the packing, the physical properties of the solute and

Introduction to Tandem Techniques

phase system and the linear velocity of the mobile phase. The dispersive capacity of a column is measured as the variance per unit length of the column, (H), which can be calculated from the ratio of the column length, (l), to the number of theoretical plates in the column, (n).

Thus
$$H = \frac{l}{n} \tag{1.8}$$

Due to the nature of equation (1.8), (H) has also been given the term 'height of the theoretical plate'. The first equation for (H) was derived for packed columns in gas chromatography by Van Deemter [23], and took the following form:-

$$H = A + \frac{B}{u} + Cu \tag{1.9}$$

where (A),(B),(C) are constants for a given system,
and (u) is the linear mobile phase velocity.

Each expression in equation (1.9) pertains to a specific dispersion process each of which will now be described.

The Multipath Process

During progress along a packed column, the solute molecules describe a tortuous path through the interstices between the particles of the stationary phase support. It is fairly obvious that some molecules will randomly travel shorter paths than others. Those that on the average, pass along shorter paths will move ahead of the maximum of the concentration profile, while those molecules that pass along paths of greater length will lag behind the maximum of the concentration profile. This will result in dispersion of the solute band and its contribution to (H) was deduced by Van Deemter as independent of the linear velocity and equal to a constant (A), where,

$$A = 2\lambda dp$$

where (λ) is a constant,
and (dp) is the particle diameter.

Longitudinal Diffusion

Dispersion due to longitudinal diffusion ensues from the fact that, while the solute is in the mobile phase, the band will spread as a result of the normal diffusion processes driven by concentration gradients that occur in any fluid. Obviously the longer the solute remains in the column, the more the solute will diffuse, and thus the variance of the band due to this effect will be inversely proportional to the linear mobile phase velocity. Van Deemter proposed the following function to describe the contribution to variance from longitudinal diffusion:

$$\frac{B}{u} = \frac{2\gamma D_m}{u}$$

where (D_M) is the diffusivity of the solute and (γ) is a constant.

Resistance to Mass Transfer

The third dispersion process that Van Deemter considered to contribute to (H) and which, at high mobile phase velocities, becomes the major contribution to peak variance, is the resistance to mass transfer between the two phases. This dispersion process results from the finite time necessary for a solute molecule to pass through the mobile phase to enter the stationary phase and conversely to pass through the stationary phase and enter the mobile phase. It follows that some molecules are stationary in the column during the period that the transfer processes are taking place, while those molecules in the mobile phase continue to move along the column. As a consequence band dispersion takes place. The two functions proposed by Van Deemter to describe the resistance to mass transfer between the two phase were as follows:-

$$Cu = f_1(k')\frac{dp^2}{D_m}u + f_2(k')\frac{df^2}{D_S}u$$

where (k') is the capacity factor of the solute,
(df) is the film thickness of the stationary phase,
and (D_S) is the diffusivity of the solute in the stationary phase.

Introduction to Tandem Techniques

Thus equation (9) can be expanded to the form:-

$$H = 2\lambda dp + \frac{2\gamma D_m}{u} + f_1(k')\frac{dp^2}{D_m}u + f_2(k')\frac{df^2}{D_S}u \quad (1.10)$$

Unfortunately equation (1.10) is not precise as it assumes a constant linear mobile phase velocity throughout the column. Due to the fact that in gas chromatography the mobile phase is compressible, the mean mobile phase velocity can not be employed, unless the pressure drop across the column is extremely small. Ogan *et al.* [24] extended the Van Deemter equation further, to take into account the compressibility of the carrier gas and produced the following equation:

$$H = 2\lambda dp + \frac{2\gamma D_o}{u_o} + f_1(k')\frac{dp^2}{D_o}u_o + f_2(k')\frac{df^2}{D_S}(G-1)u_o \quad (1.11)$$

where (G) is the inlet/outlet pressure ratio of the column,
  (u_o) is the mobile phase velocity at the column exit,
and (D_o) is the diffusivity of the solute in the mobile phase at atmospheric pressure.

An explicit equation for the variance per unit length of a chromatographic column is important in the design of interfaces for tandem systems, as it allows the factors that control the dispersion to be identified, and the actual dispersion to be calculated if required. Equation (1.11) is the first example and others will now be considered for capillary columns and LC columns. In 1958 Golay [25] derived the following equation for the variance per unit length of a capillary column:

$$H = \frac{2D_m}{u} + \frac{1 + 6k' + 11k'^2}{24(1+k')^2}\frac{r^2}{D_m}u + \frac{k'}{6(1+k')^2}\frac{df^2}{K^2 D_S}u \quad (1.12)$$

where (r) is the radius of the capillary column.

However, Golay also assumed a constant mobile phase velocity throughout the column and consequently Ogan *et al.* introduced a modified equation to account for the compressibility of the carrier gas, *i.e*:

$$H=\frac{2D_o}{u_o}+\frac{1+6k'+11k'^2}{24(1+k')^2}\frac{r^2}{D_o}u_o+\frac{k'}{6(1+k')^2}\frac{df^2}{K^2D_S}(G-1)u_o \quad (1.13)$$

It should be noted that as there is no packing present in the open tubular column consequently there is no multipath term. The dispersion that takes place in an LC column has been studied by a number of workers [26–29], each providing an alternative equation for (H).

HETP Curve from Experimental Data of Katz *et al*

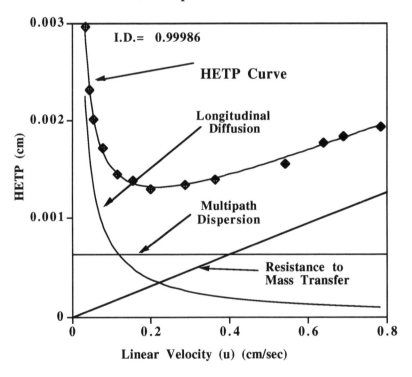

Figure 1.8   **Experimental Van Deemter Curve.**

Each equation was tested against an extensive set of data by Katz *et al.* [30], who identified the Van Deemter equation as the one that best fitted the experimental data. An example of a fit of their data to the Van Deemter equations is shown in Figure 1.8. From the curve fit, the individual contributions to the column variance per unit length depicted in the Van Deemter equation could be isolated. The horizontal line, independent of mobile phase velocity, represents the Multipath Effect, the reciprocal function, the Longitudinal Diffusion contribution and the linear curve the Resistance to Mass Transfer effect. It is also seen that the index of determination is very close to unity, indicating very good correlation between the experimental data and the Van Deemter equation. It should also be noted that in LC, as the liquid mobile phase is virtually incompressible relative to that of a gas, the mean linear velocity can be employed in the equation and a correction involving the inlet/outlet pressure ratio is unnecessary.

The demand for high resolution from the column, and the need for low dispersion in the interface of a tandem system, are in direct conflict. Low dispersion is required in the column to ensure adequate resolution, which means the peaks must be narrow and their volume small. In contrast, the interface requires the peaks to have as large a volume as possible, to ensure that the dispersion that must take place in the interface has a minimum impact on the resolution. It follows that column and interface design must always be a compromise. In some cases the ideal column that allows the most efficient use of the tandem spectroscopic instrument cannot be employed. The design of interfaces will be discussed in a subsequent chapter.

Capillary Electrophoresis

To date, capillary electrophoresis has not been used extensively as a separation method in conjunction with spectroscopic or other identifying techniques, but a number of examples have been reported, so this situation may well change in the future. For this reason capillary electrophoresis, the most likely electrophoretic method that would be used in tandem systems, will be briefly described.

Electrically charged compounds can be transported in a gel or liquid under the influence of an electric field and this process has been termed electrophoresis. There are basically three different electrophoretic methods of separation zone electrophoresis, isotachophoresis and isoelectric focusing.

Zone Electrophoresis

Zone electrophoresis is carried out in an electrolyte, across which is applied an electric field. The samples to be separated are placed in the center of the electrolyte, which could be a electrophoretic plate or capillary tube depending on the system being used. Those substances carrying a negative charge migrate to the anode, while those carrying a positive charge migrate to the cathode. Eventually the substances separate into individual bands, their relative positions depending on their individual mobilities. This type of separation carried out in a capillary tube is the most amenable to tandem operation. A zone electrophoretic separation is depicted, diagramatically, in Figure 1.9.

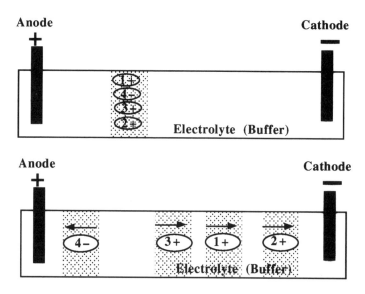

Figure 1.9   **Zone Electrophoresis**

Introduction to Tandem Techniques

Isotachophoresis

Although based on the same principle of electrophoretic migration, isotachophoresis is carried out in a different way. The sample is placed at the junction between a leading and terminating electrolyte contained in a capillary tube. The leading electrolyte must have a higher mobility than any of the sample components and the terminating electrolyte must have a mobility that is less than any of the sample components. In addition, the leading electrolyte should have a buffering capability at the pH at which the samples are to be separated. On the application of the electric field, the compound with the highest mobility, (3), will migrate faster leaving those moving at a slower rate, (1) and (2), behind. This results initially in a mixed zone being formed before the leading electrolyte and after the terminating electrolyte. The sample components can never enter the leading electrolyte because their mobility is less than that of the electrolyte. In a similar manner, the terminating electrolyte can never enter the sample mixture as its mobility is less than those of the sample components.

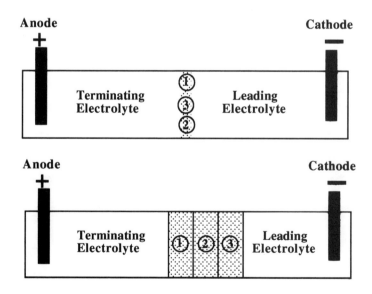

Figure 1.10   **An Isotachophoresis Separation**

Eventually each sample component is separated from its neighbor in order of their increasing mobilities, the one with the least mobility being situated next to the terminating electrolyte and that with the greatest mobility next to the leading electrolyte. The separation is depicted in Figure 1.10. The bands in isotachophoresis are not dispersed by diffusion in the normal way, as any dispersion that does occur must result in the solute entering its neighboring bands. However, the solute will immediately be driven back as a result of their differential mobilities and thus the solute bands are *self-sharpening*. This system, in its present form, would be difficult to use in tandem with another instrument, but with some development, tandems systems employing isotachophoresis as the separating technique may well become a practical and useful combination in the future.

Isoelectric Focusing

Isoelectric focusing is used to separate amphoteric substances and, as a consequence, the separation is based not on their differential mobilities but on their different isoelectric points (pI). The isoelectric point of an ampholyte is that (pH) at which it has no net charge.

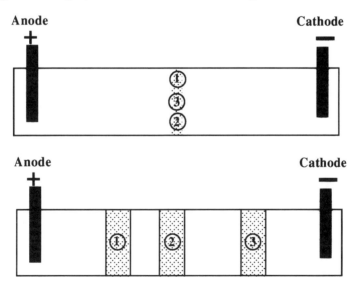

Figure 1.11   A Separation by Isoelectric Focusing

At the (pI) the ampholyte will not migrate in an electric field as it will be existing probably as a zwitterion (an internal salt) with no charge. A separation due to isoelectric focusing is depicted in Figure 1.11. Using a mixture of ampholytes (usually polyamino polycarboxylic acids) contained in the capillary tube, the anode vessel containing and acid solution, and the cathode vessel filled with an alkaline solution, a (pH) gradient will be formed along the capillary tube on the application of an electric field. When this gradient is stabilized, the sample is introduced into the center of the electrically arranged ampholytes. Each substance will migrate under the applied field until it reaches the position where the (pH) is equal to its (pI) and, as at that point it will no longer be charged, it will come to a halt. If any solute tends to move out of the isoelectric point, it will immediately ionize and become charged again and under the electric field forced back into its isoelectric position; *i.e.* it is *focused*. As a result, a series of bands of solutes are formed in the capillary tube in the order of their increasing (pI) values. It is clear that the resolution of such a system can be extremely high and separation will depend on all the substances having sufficiently different isoelectric points.

Electro-osmotic Flow (Electro-endosmosis)

The movement of a liquid, when in contact with a charged surface, situated in a strong electric field is called electro-endosmosis. The flow of liquid through a silica tube under electro-endosmosis is of 'plug' form and does not exhibit the parabolic velocity profile that normally occurs in Newtonian flow. It is interesting to note that there is very little band dispersion due to resistance to mass transfer when the flow is electro-osmotically driven.

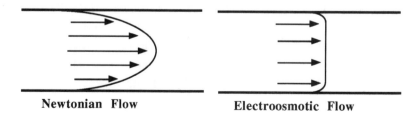

Figure 1.12   **Newtonian Flow and Electro-endosmosis**

The different types of flows are illustrated in Figure 1.12. In electrophoretic separations, electro-osmotic flow can often affect the separation adversely, but it can also be used to advantage in placing samples onto the capillary electrophoretic system.

Capillary Electrophoresis Apparatus

A diagram of the basic instrument used for capillary electrophoresis is shown in Figure 1.13. It consists of two reservoirs, one carrying the anode and anode electrolyte, and the other the cathode and cathode electrolyte.

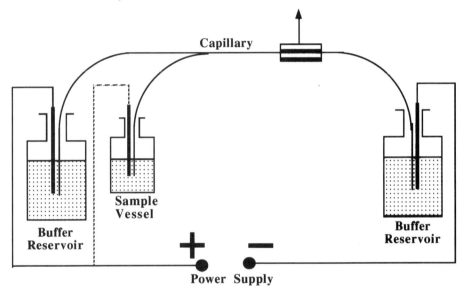

Figure 1.13   Capillary Electrophoresis Apparatus

Each end of a fused quartz capillary dips into two reservoirs, thus joining them electrically. At one end of the tube, there is a 'T' join connecting to another reservoir containing the sample solution and, at the other end, a detector. The capillary is usually a polymer–coated fused silica tube and the detector consists of a small section of a similar tube, from which the polymer coating has been removed. Detection is achieved by a variety of procedures, the most common being the adsorption of UV light by the solutes as they pass the aperture in the polymer coating, usually employing

Introduction to Tandem Techniques

a fiber optical system. Alternatively the section of tube is irradiated by an appropriate laser, and the intensity of the fluorescent light monitored with an appropriate light-sensing cell. The sample is introduced into the system using electro-osmotic flow by connecting a high potential to the electrode in the sample reservoir. When sufficient sample has passed into the tube, the high potential is again connected to the anode in the anode reservoir and the separation developed. Whereas tandem systems using GC and LC as the basic separation techniques are now commonplace, this is not so for tandem systems based on capillary electrophoretic separations. However, as already suggested, this may well change in the near future.

Synopsis

Early analytical procedures were carried out sequentially, not concurrently, and consequently were very time consuming. Analytical instruments were first introduced in the late 1940s and started to become established in the 1950s. The first instruments to be introduced were UV and visible spectrometers followed by the gas chromatograph and IR spectrometer. The development of sophisticated analytical instruments opened up the possibility of combining a separation technique with an identification technique and thus permit concurrent analyses. There are two basic separation techniques employed with tandem systems, GC and LC; these techniques can be sub-divided, depending on the nature of the stationary phase, to GLC, GSC, LLC, and LSC. A gas chromatograph consists of a gas supply, gas controllers, injector, column oven and detector. The three most common detectors are the FID, the katherometer and the electron capture detector. There are two types of column employed in the gas chromatograph, the packed column and the capillary column. The former provides higher loading capacities, larger peak volumes but lower resolution, the latter has a very low loading capacity, very small peak volumes but can provide very high resolution. The packed column requires a simple septum injector for sampling whereas the capillary column needs more involved systems such as the split injector. The basic liquid chromatograph consists of a solvent programmer, pump, injection valve, column and column oven, and a detector. Packed columns are almost exclusively employed in LC, although some have very small diameters. The capillary column has not, at the time of writing this book, been developed sufficiently for general analytical use in LC. There are two basic types of LC stationary phases: silica which separates substances on the basis of polarity, and the reverse phase, that separates on the basis of dispersive interactions. The latter is the most commonly used. Very fast separations can be achieved by LC and efficiencies of three quarters of a million plates have been realized. Silica and to a lesser extent reverse phases have a range of pore sizes and thus exhibit exclusion properties. Thin layer chromatography is a lamina form of LSC, but has only very limited use in tandem systems. In any column, the solutes are both retained and individually dispersed. Retention depends primarily on the thermodynamic properties of the distribution system and the quantity of stationary phase in the column. Dispersion, on the other hand, depends on the kinetic properties of the system, such as the mobile phase velocity, the particle diameter of the support in a packed column, or the radius of a capillary column, together with the physical properties of the solute and the two phases. There are three major contributions to band variance: the multipath effect, longitudinal diffusion and the resistance to mass transfer of the solute between the two phases. All three need to be minimized to produce an efficient column. However, a high–efficiency column producing very narrow solute bands can make the design of the interface between the column and the tandem spectroscopic system more difficult.

Capillary electrophoresis has only been used to a limited extent in tandem systems, but holds good possibilities for the future. There are three electrophoretic methods of separation: zone electrophoresis, where the components are separated on a basis of relative mobilities; isotachophoresis, where the separation is again based on relative mobilities but where the solutes are sandwiched between leading and terminating electrolytes; and finally, isoelectric focusing, where the solutes are separated according to their isoelectric points. In all electrophoretic systems, movement of charged species is always accompanied to a greater or lesser extent by electro-osmotic flow. Electro-osmotic flow is the movement of a liquid when in contact with a charged surface under the influence of an electric field. The flow is 'plug' form and is not accompanied by the parabolic velocity profile associated with Newtonian flow. In 'plug' flow there is virtually no band dispersion resulting from resistance to mass transfer effects. The capillary electrophoretic apparatus consists of two reservoirs, each containing electrolyte, and fitted with electrodes. Each reservoir is inter-connected by fused quartz capillary tubing dipping into each reservoir. A third reservoir, containing another electrode, is connected by a 'T' to one end of the capillary, through which the sample is applied to the tube. At the other end of the tube is the detector. The most common detector is a UV absorption system measuring the intensity of the light passing through a cleaned section of the tube using fiber optics. Using the same physical system detection by fluorescence measurement has also been found satisfactory.

References

1. A. T. James and A. J. P. Martin, *Biochemist. J.*, **50**(1952)679.
2. R. P. W. Scott in *Gas Chromatography 1958*, D. H. Desty (Ed.), Butterworth London (1958)189.
3. M. J. E. Golay, *idem*, 36.
4. D. H. Desty, A. Goldup and B. F. Wyman, *J. Inst. Petrol.*, **45**(1959)287.
5. R. D. Dandenau and E. M. Zenner. *J. High Res. Chromatogr.*, **2**(1979)351.
6. K. L. Ogan, C. Reese and R. P. W. Scott, *J. Chromatogr. Sci.*, **20**(1982)425.
7. K. L. Ogan, U. S. A. Patent Application (pending).
8. J. Stuff, M. Wojtas and P. Uden, Analyst (in press).
9. D. R. Biswas, ITT, Virginia. Private Communication.
10. K. Grob Jr., *J. Chromatogr.*, **237**(1982)15.
11. J. V. Hinshaw, Jr. and F. J. Yang, *J. High Resol. Chromatogr.*, **6**(1983)554.
12. E. Katz and R. P. W. Scott, *J. Chromatogr.*, **253**(1982)159.
13. R. P. W. Scott and P. Kucera, *J. Chromatogr.* **169**(1979)51.
14. R. P. W. Scott and P. Kucera, *J. Chromatogr.* **125**(1976)251.
15. A. Alhedai, D. E. Martire and R. P. W. Scott, *Analyst,* **114**(1989)869.
16. H. Zhou, C. F. Poole, J. Triska and A. Zlatkis, *J. High Resolution Chromatog.*, **3**(1980)440.
16. S. O. Akapo, R. P. W. Scott and C. F. Simpson, *J. Chromatogr. Sci.*, **14**(1991)217.
17. C. H. Lochmuller and D. R. Wilder, *J. Chromatogr. Sci.*, **17**(1979)574.
18. R. K. Gilpin and J. A. Squires, *J. Chromatogr. Sci.* **20**(1982)345.
19. R. P. W. Scott and C. F. Simpson, *J. Chromatogr.*, **197**(1980)11.

20. H. T. Butler, M. E. Coddens and C. F. Poole, *J. Chromatogr.*, **290**(1984)113.
21. H. T. Butler, M. E. Coddens, S. Khatib and C. F. Poole , *J. Chromatogr. Sci.,* **23**(1985)20.
22. C. F. Poole, H. T. Butler, M. .E. Coddens and S. A. Schuette, in *Analytical and Chromatographic Techniques in Radiopharmaceutical Chemistry*, D. M. Wieland, T. J. Manger and M. C. Tobes (Eds.), Springer-Verlag, New York, (1985)3.
23. J. J. Van Deemter, F. J. Zuiderweg and A. Klinkenberg, *C. Eng. Sci.*, **5**(1956)271.
24. K. Ogan and R. P. W. Scott, *J. High. Res. Chromatog.*, **7**(1984)382.
25. M. J. E. Golay, in Gas *Chromatography 1958* D.M.Desty (Ed.), Butterworth, London (1958)26.
26. J. Calvin Giddings, *J. Chromatog.*, **5**(1961)46.
27. J. F. K. Huber and J. A. R. J. Hulsman, *Anal. Chem. Acta*, **38**(1967)305.
28. G. J. Kenedy and J. H. Knox, *J. Chromatogr. Sci.*, **10**(1972)606.
29. C. Horvath and H. Lin, *J. Chromatog.*, **126**(1976)401.
30. E. Katz, K. Ogan and R. P. W. Scott, *J. Chromatogr.*, **270**(1983)51.

CHAPTER 2

IDENTIFICATION TECHNIQUES FOR TANDEM USE

Introduction to Identification Techniques

This chapter will give a brief introduction to the different spectroscopic techniques that have been used in tandem systems. The intent is to provide those completely unfamiliar with spectroscopy, sufficient understanding to help them operate tandem systems and perhaps identify eluted components by spectra matching. The information given here is quite inadequate to permit structure elucidation of an unknown substance from spectroscopic data, which would require the services of an experienced spectroscopist.

Most of the identification methods employed in tandem systems are spectroscopic in nature. The word, spectrum, was originally given to the colored bands formed by passing visible light through a prism and the study of these colored bands was given the term *spectroscopy*. The study of visible light was, in due course, extended to shorter wavelength and higher frequencies and thus, as the shortest wavelength of visible light is violet, the term *ultra violet* spectroscopy was introduced. At the other end of the visible spectrum (red), light of longer wavelength and lower frequencies are to be found and this range was termed the *infrared* region. Below the ultraviolet region are the X-rays and at even shorter wavelengths and higher frequencies are the γ–rays. Beyond the infrared region is the *far infrared* and *microwave* region, followed by the range of frequencies used for radio and television transmission. Intermediate in the radio frequency range is the waveband that encompasses the frequencies employed in NMR spectroscopy *i.e.* 60,100, 300 and 1000 MHz.

A diagram showing the distribution of the different wave bands of interest is shown in Figure 2.1. It is seen that the spectrum has been divided into a number of different regions. The γray region occurs below 0.5 Å and the X-ray region extends between 0.5 Å and 50 Å. The regions of interest, and pertinent to the use of tandem techniques, is part of the UV region between about 150 nm and 400 nm, the visible region between 400 nm and 700 nm and the infrared region between about 2.5 m and 25 m.

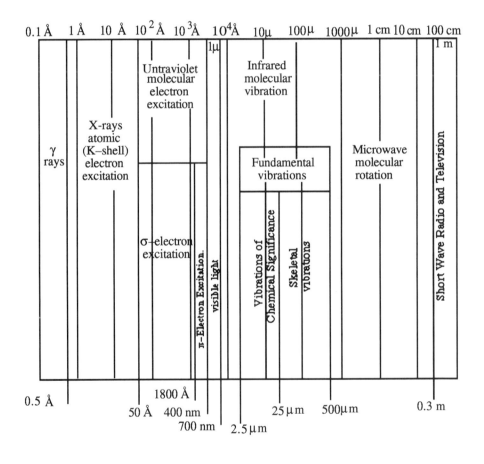

Figure 2.1   The Electromagnetic Spectrum

The wavelength of the electromagnetic radiation employed in NMR spectroscopy ranges from 3 m to 15 m which extends into the radio wave region.

So far, all the spectroscopic techniques mentioned are based on the measurement of the adsorption of electromagnetic radiation similar in nature to that of visible light, but having different wavelengths and frequencies. Mass spectrometry, however, is the exception and this technique should not, on a rational basis, be termed a spectroscopic technique. Nevertheless, partly because mass spectra have some similarities to other types of spectra, and partly for convenience, it has been included with other forms of spectroscopy in the group of analytical methods known as spectroscopic techniques.

All spectroscopic techniques (with the exception of mass spectrometry) depend on the adsorption of waves similar to those of visible light. Electro-magnetic waves differ only in frequency and wavelength. They all travel at the same speed in a vacuum (that of light) and thus the product of the frequency and wavelength, for any given light wave, is a constant and equal to the speed of light. Physically, an electromagnetic wave could be considered to consist of a sinusoidal electric field, acting at right angles and in phase with a sinusoidal magnetic field. It is the electric vector of the electromagnetic wave form that provides the energy that is measured by spectroscopic instruments. The electric and magnetic fields interact with matter in ways that are unique to the chemical nature of the substance through which they pass. Consequently, the magnitude and nature of this interaction can provide a basis for identifying the chemical structure of the interacting substance.

Electromagnetic radiation of different frequencies interact with different parts of a molecule to differing extents. Thus each of the spectroscopic techniques that operate over different ranges of frequency can provide information that is unique to some specific part of the molecule of which the interacting material is composed. Of course, radiation of some wavelengths may not interact at all with the medium through which it is passing, in which case the substance is said to be transparent to radiation of that wavelength.

As already stated, the frequency (ν) and the wavelength (λ) are related to the velocity of light (c) in the following way:-

$$\nu \lambda = c \tag{2.1}$$

Wave frequencies are measured in Herz (cycles/second), kHz (10^3 cycles/second), MHz (10^6 cycles/second), or GHz (10^9 cycles/second). As seen in Figure 2.1, wavelength can be measured in a number of different units depending on the magnitude. The units commonly employed are Angstroms (10^{-10} m), milli-micron or nanometers (10^{-9} m), microns (10^{-6} m), centimeters or meters.

Spectroscopy Involving the Absorption and Emission of Electromagnetic Waves

The manner in which electromagnetic radiation interacts with a substance, and the type of information that is provided to help determine its chemical structure, will be briefly discussed. The basic principles involved in spectroscopic measurement will be considered with particular reference to tandem systems. In particular the sample requirements (sample size, sample volume etc.) will be considered to evaluate their effect on any associated separation apparatus.

UV and Visible Spectroscopy

The practical wavelength range for UV/visible spectroscopy, useful in tandem techniques, is between about 180 nm and 700 nm, the wavelength range between 180 nm and 400 nm, being part of the UV spectrum and 400 nm to 700 nm the whole of the visible spectrum. In general it is difficult to operate at wavelengths below 180 nm as most substances adsorb in this region. This means that, in tandem systems, the carrier fluid from the separation technique employed will dominate the measurement and the solute properties will not be discernible from those of the carrier.

Identification Techniques for Tandem Use

This problem is particularly difficult when liquid chromatography is employed as the separation technique, but not such a problem with gas chromatography, as many gases are transparent to light of a wavelength significantly below 180 nm. However, there are other practical difficulties that are met when working in this wavelength range. The walls of the vessel or device in which the UV light is generated may also be opaque to light of such wavelengths and transmission can be very poor. In practice, the lower wavelength limit for tandem techniques is at present about 150 nm and most UV spectroscopic measurements are carried out at wavelengths between 180 nm and 400 nm. The energy carried by an electromagnetic wave is not continuous, but propagated in finite parcels called quanta. The relationship between the radiation energy and wavelength is given by:

$$E = \frac{hc}{\lambda} \qquad (2.2)$$

Where (E) is the energy of the radiation,
(λ) is the wavelength of the radiation,
(c) is the velocity of light,
and (h) is Planks constant.

Radiation is only adsorbed by a substance when the energy of the radiation (as given by equation (2.2)) corresponds to the that needed to increase the potential energy of the substance by one or more increments. The transfer of energy is achieved by the interaction of its electric vector with the substance. Adsorption of UV/visible radiation changes the electronic state of a molecule and can, for example, raise an electron from the ground state to one of its excited states. The electronic ground state is one in which all of the electrons of the species are in their most stable orbitals. An electronic excited state is one in which at least one of the electrons occupies a orbital of higher energy than that of the ground state.

It would appear that if the radiation passing through a substance was programmed with respect to wavelength, and the transmitted light monitored by an appropriate sensor, then the curve resulting from the output of the sensor being plotted against wavelength would show a series of sharp adsorption lines. These lines would occur at frequencies where the radiation energy was equal to that of specific electronic transitions in the molecules of the substance.

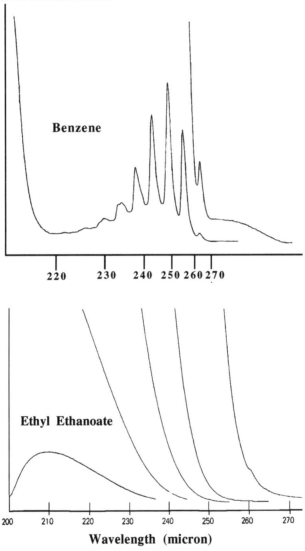

Figure 2.2 UV Adsorption Spectra of Benzene and Ethyl Ethanoate

Identification Techniques for Tandem Use 51

This, in fact, is observed in atomic spectroscopy (which will be discussed later) but in tandem systems, most spectroscopic measurements are made in the presence of the mobile phase which complicates the basic simple adsorption spectrum. In solution, a given molecule may exhibit numerous adsorption levels that have energies very close to one another. The bands are so close that they cannot be observed individually and, as a result, they occur under one envelope giving a broad band in the UV adsorption spectrum. The breadth of the band may extend from 50 to 300 nm. Examples of the adsorption spectra of two different compounds are shown in Figure 2.2.

It is seen that the aromatic structure of benzene gives a fairly complex spectrum that could easily be used for identification purposes. In contrast, the spectrum for ethyl ethanoate is a very simple spectrum containing no fine structure and would be of little use for solute identification. Unfortunately, there are many other compounds, particularly those containing the ester and acid groups, that would give very similar spectra to that of the example ester and would only be identified from their UV spectrum with considerable difficulty. In addition, the spectra for ethyl ethanoate is generally more typical of the vast majority of the UV spectra of organic compounds.

UV spectroscopy is the least helpful of all the spectroscopic techniques from the point of view of structure confirmation or structure identification. It is, however, the most sensitive and the easiest to employ in tandem systems. In addition, the UV spectrometer is relatively inexpensive and consequently, despite its technical limitations, it is one of the more common spectroscopic techniques to be employed in conjunction with separation instruments.

The UV spectrometer can take two basic forms, the dispersive spectrometer, and the diode array spectrometer. There are normally two light sources in a UV/Visible Spectrometer. The source of UV is usually a low-pressure deuterium lamp that emits light over the wavelength range of 185 nm to 380 nm, although xenon lamps that emit over a similar wavelength range are also used. The visible–light source is usually a tungsten filament lamp that can be made to either manually or automatically replace the deuterium lamp when the wavelength programmer has reached the appropriate transmission wavelength.

The UV/Visible Dispersive Spectrometer

A diagram of the optical system of a UV/Visible Dispersive Spectrometer is shown in Figure 2.3

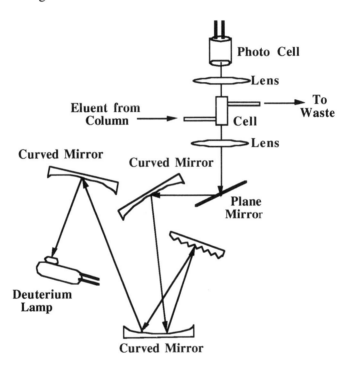

Figure 2.3 The Multiwave Length Dispersive Spectrometer-Detector

Light from the deuterium discharge lamp is collimated by two curved mirrors onto a holographic diffraction grating. The dispersed light is focused, by means of a curved mirror, onto a plane mirror and the light of the required specific wavelength selected by adjusting the angle of the plane mirror. Light of the selected wavelength is then focused by means of

Identification Techniques for Tandem Use

a lens through the flow cell. The exit beam from the cell is focused onto a photo-cell which gives a response that is some function of the intensity of the transmitted light. The wavelength is scanned by rotating the plane mirror. The output from the photocell, electronically modified and presented on a chart recorder or computer printer, will provide an adsorption curve relating adsorption to wavelength (or frequency). The adsorption curve is known as the adsorption spectrum and its shape will be characteristic for the substance being examined.

The Diode Array UV/Visible Spectrometer

The diode array spectrometer functions in an entirely different way from the dispersive instrument. A diagram of a diode array detector is shown in Figure 4. Light from a deuterium lamp is collimated by an achromatic lens system so that the *total light* passes through the detector cell onto a holographic grating. In this way the sample is subjected to light of all wavelengths generated by the lamp.

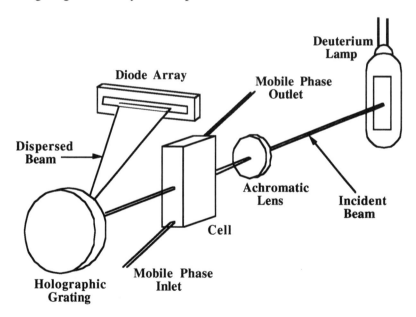

Figure 2.4 The Diode Array Detector

The dispersed light from the grating is then focused onto a diode array. The array may contain many hundreds of diodes and the output from each diode is regularly sampled by a computer and stored on a hard disk. The spectrum of the solute can be obtained by recalling from memory the output of each of the diodes, *i.e.* a curve relating adsorption to wavelength. The only disadvantage of this type of spectrometer is that its resolution is limited by the number of diodes in the diode array.

Although, the two systems described appear to be satisfactory, both suffer from certain disadvantages due to second-order effects. When light having a narrow band wavelengths is passed through the cell (as in the dispersive spectrometer) and the whole of the transmitted light allowed to fall on the sensor, then the light received will not only be that light of the same wavelength that was transmitted through the cell, but also any fluorescent light that the incident light induced. Thus the light measured is not solely transmitted light of the selected wavelength but will also contain any fluorescent light that was induced in the sample.

In a similar way, when light containing the whole range of wavelengths is passed through the cell (as in the diode array spectrometer), then the light sensed at a selected wavelength will not only contain the transmitted light of that wavelength but also any fluorescent light of that wavelength that was induced in the sample by incident light of other wavelengths. It is seen that neither spectrometer system can be certain of measuring true light absorption at a specific wavelength. In fact, true light absorption can only be measured by selecting the incident wavelength and then passing the transmitted light to another monochromator and then selecting the same wavelength to monitor.

An excellent text on the absorption of light and ultraviolet radiation has been written by Schenk [1]. This book is probably now out of print but is likely to be available in many libraries. A more recent publication on the subject (1992) is that by H. H. Perkampus [2] which deals not only with the theory of the subject but also includes many applications. UV spectroscopy is the simplest to use in tandem with a chromatograph but, unfortunately, provides limited information about the structure of the solute being eluted.

Fluorescence Spectroscopy

Fluorescence spectroscopy has two unique advantages over UV spectroscopy for solute identification in tandem systems. Firstly, the spectrometer has an order of magnitude greater sensitivity. Secondly florescence spectra often show greater detail than UV spectra and are consequently more reliable for identification purposes by spectra matching. In addition, for a given compound, fluorescence spectra can be obtained over a large number of different excitation wavelengths, each providing a unique spectrum that improves the confidence of identification.

Fluorescence is a specific type of luminescence. When molecules are excited by electromagnetic radiation to produce luminescence, the emitted light is called photoluminescence. If the release of electromagnetic energy is immediate, or stops on the removal of the exciting radiation, the substance is said to be fluorescent. If the release of energy is delayed, or persists after the removal of the exciting radiation, then the substance is said to be phosphorescent. When light is adsorbed by a molecule, and a transition to a higher electronic state occurs, the wavelength at which this happens will be determined by the structure of the particular substance. When electrons are raised to an upper excited single state as a result of the adsorption of light energy, the transitions produce the characteristic UV or visible adsorption spectrum of the respective compound. If the excess energy of the molecule in the excited state is not dissipated rapidly by collisions with other molecules, or by other means, the electron will return to the ground state with the emission of electromagnetic radiation in the form of fluorescence. As some energy is inevitably lost before the emission occurs, the emitted fluorescent light is always of a longer wavelength than that absorbed on excitation. In due course Raman spectroscopy will be discussed and it will be seen the two processes have considerable similarity. However, the wavelength of the emitted light is different from that produced by fluorescence.

Discussions on the theoretical basis of fluorescence have been given by Guilbault [3], Udenfriend [4] and Rhys Williams [5].

The fluorescent signal (I_F) is given by

$$I_F = \phi I_0 \left(1 - e^{-kcl}\right)$$

where is the quantum yield (the ratio of the number of photons
(I_F) emitted to the number of photons absorbed),
(I_0) is the intensity of the incident light,
(c) is the concentration of the solute,
(k) is the molar absorbance,
and (l) is the path length of the cell.

There are two basic types of fluorescent spectra that can be taken:

1. The fluorescence intensity taken at a fixed wavelength while programming the excitation wavelength.

2. The fluorescence intensity taken over a range of wavelengths while employing a fixed excitation wavelength.

This flexibility can produce a very large number of fluorescence spectra from a given compound either monitored at a range of fixed emission wavelengths or over a range of fixed excitation wavelengths, all of which will be significantly different. Examples of a pair of such spectra are given in Figure 2.5. The spectrum on the left was obtained by monitoring the fluorescence at 405 nm and programming the excitation light from 230 nm to 410 nm which provides an excitation spectrum. The spectrum on the right is obtained by fixing the excitation light at 292 nm and monitoring the fluorescent light from 360 nm to 480 nm which provides an emission spectrum. Fluorescence spectra can be used for identifying substances by comparing them with reference spectra. However, they have very limited use for the structure elucidation of a completely unknown substance.

Figure 2.6 shows the basic fluorescence spectrometer detector. The excitation source that emits UV light over a wide range of wavelengths (usually a deuterium lamp) is situated at the focal point of an ellipsoidal mirror shown at the top left hand corner of the diagram.

Identification Techniques for Tandem Use

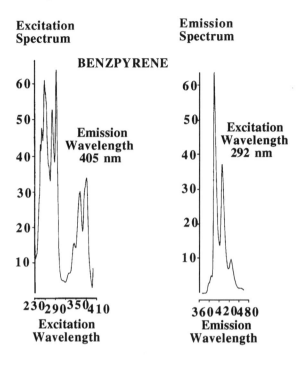

Courtesy of the Perkin Elmer Corporation

Figure 2.5   Two Fluorescence Spectra one Monitored at 405 nm and the Other Excited at 292 nm

The parallel beam of light is arranged to fall onto a toroidal mirror that focuses it onto a grating on the left-hand side of the diagram. This grating allows the frequency of the excitation light to be selected, or the whole spectrum scanned providing excitation spectra. The selected wavelength then passes to a spherical mirror and then to an ellipsoidal mirror, at the base of the diagram, which focuses it onto the sample. Between the spherical mirror and the ellipsoidal mirror, in the center of the diagram, is a beam splitter that reflects a portion of the incident light onto another toroidal mirror. This toroidal mirror focuses the portion of incident light onto the reference photocell providing an output that is proportional to the strength of the incident light. Fluorescent light from the cell is focused by an ellipsoidal mirror onto a spherical mirror at the top right-hand side of the diagram.

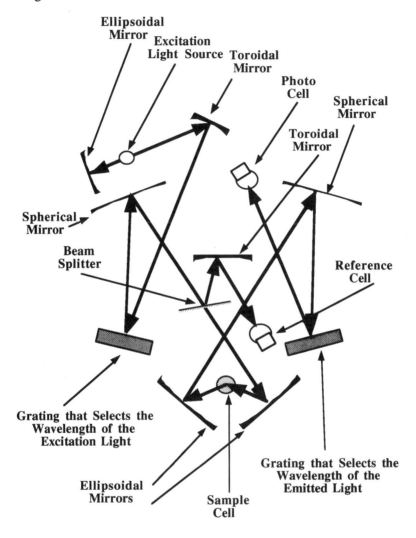

Courtesy of the Perkin Elmer Corporation

Figure 2.6 **The Fluorescence Spectrometer Detector**

This mirror focuses the light onto a grating situated at about the center-right of the Figure. This grating can select a specific wavelength of the fluorescent light to monitor, or scan the fluorescent light and provide an emission fluorescent spectrum. Light from the grating passes to another photocell which monitors its intensity. The instrument is rather complex and, as a result, rather expensive. However, from the point of view of measuring fluorescence spectra it is extremely versatile. Much less sample is required to produce useful fluorescence spectra compared to UV spectra and, in general, fluorescence spectra contain more fine detail and are thus more useful for solute identification where sample size is limited. It is seen that as both the wavelength of the excitation light and that of the fluorescent light are exclusively selected, consequently the spectra are not liable to same errors as those of from the diode array discussed previously.

Infrared Spectroscopy

IR spectroscopy involves the absorption of light having a wavelength longer than the visible spectrum, *i.e.* between 2 and 15 micron. In the practice of IR spectroscopy, spectra are usually displayed using wavenumbers as the independent variable, as opposed to wavelength. Both forms of presentation are as shown in Figure 1.7. The wavenumber is the number of waves per centimeter taken as $(1/\lambda)$ and expressed in reciprocal centimeters (cm^{-1}). Optically, the IR spectroscopic system is very similar to that used to measure UV absorption except that the materials of construction must be different. The optical components must be transparent to IR light of the pertinent wavelength as opposed to being transparent to UV light. As already stated, UV absorption occurs at frequencies where the radiation energy is equal to that of specific electronic transitions in the molecules of the substance. In contrast, IR adsorption occurs at frequencies where the radiation energy is equal to that of changes in vibrational and rotational energy of the molecule. A molecule can be regarded as an assembly of balls (atomic nuclei) and springs (chemical bonds). Such a system can vibrate in a very complex manner and for most compounds, a molecule containing (n) atoms will have (3n-6) modes of vibration. Consequently a characteristic fundamental frequency and an adsorption band will be associated with each vibration mode. Both UV and IR absorption provide spectra that are characteristic of the molecule and both can be used for compound identification. IR spectra, however, because of the relatively large number of possible absorption bands, show considerable differences between diverse molecules and contain much fine structure. In contrast, and as already discussed, the majority of UV spectra are very similar even though the structure of the molecules may differ considerably.

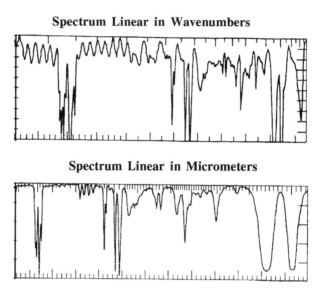

Figure 2.7 IR Spectra Presented in Wavenumbers and Microns

Consequently, IR spectra can be far more useful for confirming compound identity than UV spectra. Unfortunately, the measurement of an IR spectrum requires considerably more sample than that required to obtain a UV spectrum and thus, although the IR spectrum is more informative, the technique is not as sensitive. For further details on IR spectroscopy the books by Conley [6] and Alpert [7] are recommended.

In tandem systems, the IR spectrometer is commonly associated with either a gas or a liquid chromatograph. Consequently the spectra in a GC tandem system will be taken as a vapor in the carrier gas, whereas in an LC tandem system the spectra would normally need to be taken as a solution in a liquid (the mobile phase). In general, spectra of the same materials taken in a liquid are very similar to that taken as a vapor sample in a gas. In fact, Welti [8] suggests vapor spectra might be considered as liquid spectra taken at infinite dilution. This is only true, however, if there is no interaction between the sample and solvent. The vapor and liquid film spectra of n-hexanol are shown in Figure 2.8.

It is seen that they are indeed very similar and that vapor spectra can be confidently used for the confirmation of sample identity, providing a reference vapor sample spectrum was available. It should be noted, that the disperse peak at about 3400 wavenumbers, shown in the liquid sample spectrum, demonstrates the effect of intra-hydrogen bonding between the OH groups of the n-hexanol (and possibly the presence of water in the sample or solvent), which is not present in the vapor spectrum.

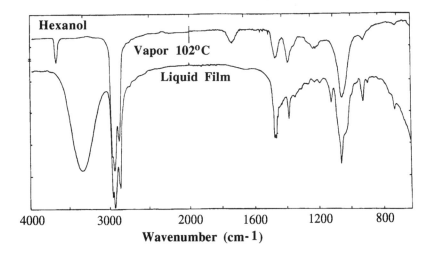

Figure 2.8   Vapor and Liquid Spectra of n–Hexanol

Correlation charts can be constructed to help assign specific absorption bands to certain chemical bonds or groups present in an unknown molecule. An example of such a correlation chart, after Stuart [9], is shown in Figure 2.9. The presence of certain bands at specific wavenumbers helps identify the major groups present in the molecule but gives very little evidence on the size of the molecule or the manner in which the actual groups are joined. The interpretation of spectra requires considerable skill and, unless the analyst is also an experienced spectroscopist, verification should be sought to confirm any conclusions that are drawn from the spectra of an unknown substance.

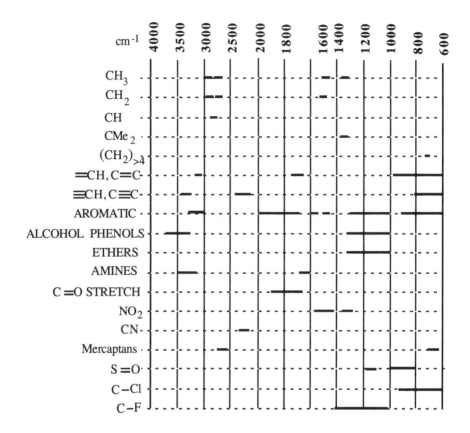

Figure 2.9 An IR Correlation Chart for Interpreting IR Spectra

IR Spectroscopy Instrumentation

There are two basic types of IR spectrometer in use today: the simple grating dispersion IR spectrometer and the Fourier transform IR spectrometer (FTIR). The former is rapidly being replaced by the latter which is much faster and it is likely that in a few years *only* the FTIR instrument will be in use with tandem systems. However, as both are still in use at this time, both systems will be described. The two spectrometers are quite different in design and function.

The Simple Grating Infrared Spectrometer

A diagram of the grating IR spectrometer is shown in Figure 2.10. Light from the IR source is reflected from a mirror, through the reference cell, through an attenuator system and onto a second mirror. The beam is attenuated by imposing a device that removes a continuous but controllable fraction of the light from the reference beam. The attenuator usually takes the form of a comb, the teeth of which are cut so that the amount of light attenuated is linearly related to the lateral movement of the comb through the beam. After passing through the attenuator to the second mirror, the attenuated light passes through a chopper to a grating, where light of a specific wavelength is selected and passes to an IR sensor.

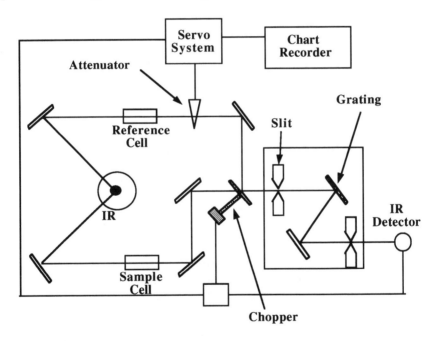

Figure 2.10 The Simple Grating IR Spectrometer

A second beam of light from the source is reflected through the sample cell by a plane mirror and then, by means of another mirror, also passes through the chopper to the grating and then to the IR sensor.

The chopper arranges for the sensor to alternately receive light that has been transmitted through the cell, and light that has passed through the attenuator. The attenuator is controlled by a servo mechanism that adjusts the light transmission until both beams have the same intensity. The amount of light that is absorbed is indicated by the position of the attenuator. In other words, the amount of light, of a particular wavelength, which is absorbed by the sample, is measured by attenuating the reference beam until its intensity is equivalent to that of the beam transmitted through the sample. The resolution is controlled by the width of the slit which is adjustable. In the older versions of this type of IR spectrometer, an analog plotter, mechanically associated with the attenuator, recorded the spectrum. Even if modified to provide an output that is proportional to absorption, the big disadvantage of this type of spectrometer for use in tandem systems is its very slow rate of scanning.

The Fourier Transform IR Spectrometer.

The Fourier transform IR (FTIR) spectrometer works on an entirely different principle, and involves much simpler instrumentation but more complicated data processing. The FTIR spectrometer can scan a sample far more rapidly than the dispersive instrument. The faster scan speed, obviously, makes it more suitable for tandem operation in conjunction with a chromatograph. The basic difference is that the dispersion instrument scans the sample one wavelength at a time, whereas the FTIR spectrometer examines the sample using all the wavelengths coincidentally. A diagram of the basic system is shown in Figure 2.11. Collimated light from a broad band infrared source passes into the optical system and impinges on a beam splitter that comprises a very thin film of germanium. Approximately 50% of the light passes through the film and is reflected back along its path by a fixed mirror, where half of the light intensity (25% of the original light intensity) is reflected by the same beam splitter, through the sample cell, to the infrared sensor. The other 50% fraction of the incident light is reflected at right angles to its incident path onto a moving mirror. Light from the moving mirror returns along its original path and half of the light intensity is transmitted through the beam splitter, through the sample cell, to the infrared sensor. Thus 25% of the incident

Identification Techniques for Tandem Use   65

collimated light from the source reaches the sensor from the fixed mirror and 25% from the movable mirror. As the path length of the two light beams striking the sensor will differ, there will be destructive and constructive interference and, in fact, the system constitutes a Michelson interferometer.

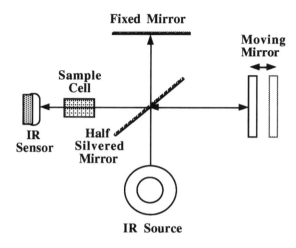

Courtesy of Nicolet Inc.

Figure 2.11   The Basic FT-IR System Concept

As the movable mirror traverses its programmed path, it will produce a series of maxima and minima that will be recorded by the sensor, as all the different wavelengths generated by the source pass through conditions of constructive and destructive interference. In addition, the frequency of this wave form will be determined by the velocity of the moving mirror which is controllable. In fact, the interferometer is actually taking a Fourier transform of the incoming signal. An example of an interferogram obtained from the FTIR is shown in Figure 2.11. It might appear that the resolution of the interferogram is not very high, and an inverse Fourier transform might not provide a conventional IR spectrum with very good resolution. In fact one scan from an FTIR instrument, taking about a second, would give an conventional IR spectrum with resolution equivalent to that obtained by a dispersive instrument scanning, over a period of 10 to 15 minutes. Nevertheless, the resolution can be improved by taking a number of scans and adding them together and processing the sum. Such procedures improve the resolution proportionally to the square root of the number accumulated scans. For example, 16 accumulated scans would increase the resolution by a factor of four. This, of course, is achieved at the expense of time.

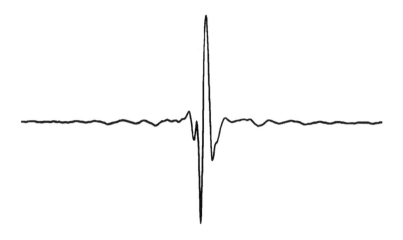

Courtesy of Nicolet Inc.

Figure 2.12 A Typical Interferogram

The accumulation of spectra also increases the sensitivity of the device, which is a useful advantage in tandem systems, where the available sample from the chromatograph may be limited. Furthermore, as a single scan may only take about a second to run, in order to double the sensitivity, the total scanning time would only require to be increased to four seconds.

In general, tandem systems involving IR measurements have not been the most successful and, without the introduction of the Fourier transform IR spectrometer, it is doubtful if IR tandem instruments would have survived after the introduction of the faster and more efficient chromatography columns.

Raman Spectroscopy

If any substance, beitmay gaseous, liquid or even solid, is exposed to radiation of a defined frequency, then the light scattered at right angles contains frequencies that differ from that of the incident light and which will be characteristic of the substance that causes the scattering. The accepted mechanism of the Raman effect, is that a molecule absorbs the incident radiation and as a consequence, is raised to a higher level of energy. The molecule then emits light at the Raman frequency and falls to a new level, usually somewhere between the initial and final states. If the frequency of the incident light is (v_i) and that of the scattered light (v_s) then the Raman frequency (Δv) is given by.

$$\Delta v = v_i - v_s$$

The situation is depicted in Figure 2.13.

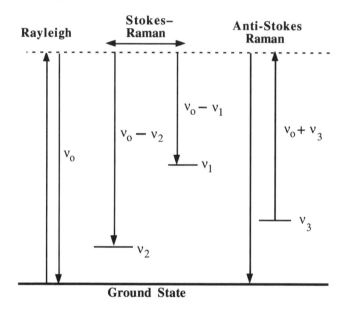

Figure 2.13. Different Forms of Light Scattering

If the excitation energy is (hv_0), and the molecule is raised from ground state to an excited level and then falls back to the ground state, the frequency of the light emitted will be the same as that of the incident light and this phenomenon is called *Rayleigh scattering*. If, however, the light excites the molecule from ground level and then falls back to an energy level of (hv_1) then the frequency of the emitted light will be ($v_0 - v_1$), and this is called *Stokes–Raman* scattering. In Stokes–Raman scattering, the scattered light always has a frequency *lower* than that of the incident light. Finally, if a molecule that already exits at a raised energy level of ($h\nu_3$), is raised further to an excited state by the incident light and then falls back to the ground state, the frequency of the emitted light will now be ($\nu_0 + \nu_3$). This is called *Anti-Stokes-Raman* scattering, and the light scattered by this process always has a frequency *greater* than that of the incident light.

Analysis of the Raman spectra from a wide range of compounds has shown that ($hc\Delta\nu$) is almost invariably equal to the change in rotational or vibrational energy of the molecule. Raman spectroscopy discloses the presence of non-polar bonds and aromatic rings. It also discloses changes in polarization and the 'shape' of the electron distribution in the molecules as it vibrates. In contrast, infrared spectroscopy detects changes in the dipole moment of the molecule as it vibrates and is more sensitive to polar functional groups. The information provided by the two techniques, Raman spectroscopy and infrared spectroscopy, is often claimed to be complementary. When a particular bond between atoms produces a strong infrared signal, it is less likely to produce a strong Raman signal, and *vice versa*. The study of Raman spectra has a number of advantages as, by the appropriate choice of the incident radiation, the scattered lines can be brought into a convenient region of the spectrum where they can be easily observed. The energy of the incident radiation determines the spectroscopic region where the Raman scattering is observed. Originally, both incident and scattered radiation were measured in the visible region of the spectrum but, for various reasons, one of which is discussed below, the near-infra-red radiation is now the most frequently employed.

In the early days, the practice of Raman spectroscopy was experimentally more difficult than today, as the intensity of the scattered light is only 0.0001% of that of the incident light (one part in a million). The difficulties were greatly reduced with the introduction of the laser light sources of high energy in the 1960s and, in particular, the argon laser with its intense blue and green emission. Nevertheless, although the high-intensity laser light sources has aided in Raman spectroscopic techniques, it has also led to other problems, such as photochemical reaction in the sample and sample heating with resultant black-body radiation. One serious problem associated with Raman spectroscopy is the fluorescence that can accompany the Raman scattering, and can be as much as six to eight orders of magnitude stronger then the Raman light. Often, when trying to examine Raman scattering, fluorescence is the only phenomenon observed. The fluorescence can come from a variety of sources. It can be caused by trace impurities, coatings on polymers, additives etc., that provide so much background fluorescence that the Raman spectrum of the major component cannot be discerned. The use of near infrared excitation can help solve this problem. It has been found, that the use of light having a wavelength around 1 mµ to irradiate the sample virtually eliminates the fluorescent problem.

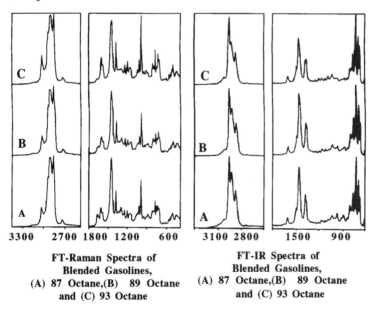

FT-Raman Spectra of Blended Gasolines, (A) 87 Octane,(B) 89 Octane and (C) 93 Octane

FT-IR Spectra of Blended Gasolines, (A) 87 Octane,(B) 89 Octane and (C) 93 Octane

Courtesy of Nicolet Inc.

Figure 2.14 Infrared and Raman Spectra of Some Gasoline's of Different Octane Ratings

However, other problems remain such as photochemical changes in the sample and black-body radiation produced by local heating. These problems must be carefully distinguished from fluorescence, as the experimental procedure necessary to correct for these effects differ greatly from the precautions that need to be taken against fluorescence.

A given substance will give a characteristic Raman spectrum under controlled conditions and the spectrum can be used to confirm the identity of the substance. An example of the relative absorption curves for infrared and Raman spectroscopy is shown in Figure 2.14. It is seen that the spectra for the gasoline's having different octane rating are very similar but, although the IR spectra show minimal differences between the samples, there are clear and unambiguous differences in the Raman spectra. In Figure 2.15 spectra are shown that have been taken from aspirin powder by diffuse reflectance infrared and Raman spectroscopy.

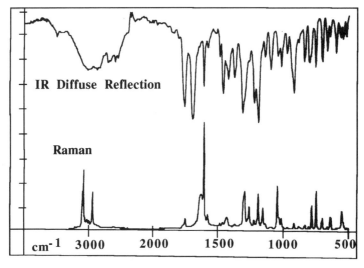

Courtesy of the Perkin Elmer Corporation

Figure 2.15 IR Diffuse Reflectance and Raman Spectra of Aspirin Powder

It is seen that absorbance and Raman scattering takes place at very similar wavelengths and there is little to choose between the two spectra for substance identification. However if the sample is in another form, *e.g.* as the unprepared tablets themselves, the spectra differ considerably, as shown in Figure 2.16.

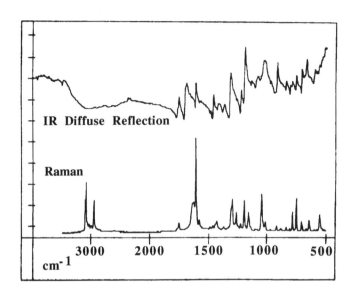

Courtesy of the Perkin Elmer Corporation

Figure 2.16 IR Diffuse Reflectance and Raman Spectra of an Aspirin Tablet

It is seen that the situation has now quite changed, the infrared spectrum is virtually useless whereas the Raman spectrum remains similar to that obtained from the powder. One of the great advantages of Raman spectroscopy is that it is virtually independent of the form that the sample takes. This could be a benefit that should be taken into account when a tandem system incorporating the Raman spectrometer is being considered.

Unfortunately, the technique is even less sensitive than IR spectroscopy so its value as an identifying technique in a tandem system appears to be rather limited. Nevertheless, the use of a laser light source in conjunction with FT/IR will help in this respect, particularly if employed with the LC/IR interfaces presently available. Such a combination might render an LC/Raman tandem system quite practical and such possibilities will be discussed when dealing with LC/IR interfaces.

Raman Spectroscopy Instrumentation

The modern Raman spectrometer is based on the use of a laser light source and a FTIR spectrometer. The laser light source must generate light of a suitable wavelength that does not produce significant fluorescence and consequently, the excitation wavelength is chosen to be about 1 micron. A diagram of the Raman attachment that is designed to fit a standard FTIR spectrometer is shown in Figure 2.17.

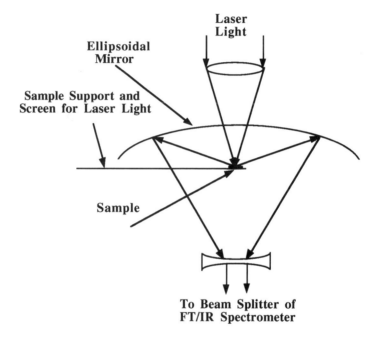

Figure 2.17 The Optical Arrangement for Measuring Raman Spectra

Laser light from an appropriate source is focused onto the solid sample that is held on a support which also acts as a screen to prevent laser light directly entering the optical system of the FTIR spectrometer. The scattered light is then focused by means of a large ellipsoidal mirror and a collimating lens, so that it enters the FTIR optical system and strikes the beam splitter. From there on, the optical system is very similar to that of the FTIR spectrometer as shown in Figure 2.11. It can be seen that if the sample support took the form of a suitable sample transport system then the Raman spectrometer could, indeed, be associated with a liquid chromatograph as a tandem instrument.

Atomic Spectra

When a solid is heated to incandescence, it emits a more or less continuous spectrum over a wide range of wavelengths. However, when gases or vapors are heated under the same conditions, spectroscopic examination of the light emitted discloses a series of lines, often very complicated in structure, at those specific wavelengths that are characteristic of the elements present. These bands, or lines of emitted light, represent energy changes that occur when electrons orbiting the nucleus of the respective atom change energy levels. Atomic emission spectroscopy is commonly used to identify the presence of certain elements in a sample; the procedure can be very sensitive and, at the same time, provide completely unambiguous identification. Atomic emission spectrometers are in common use in many analytical laboratories, and they are fairly easy to incorporate with a chromatograph in the form of a tandem system. The older types of spectrometer were a little cumbersome, and not very sensitive, but with the advent of simple and inexpensive ways of producing inert gas plasma the situation has changed radically. The system is now very sensitive and, in conjunction with the gas chromatograph, has been used successfully for a number of years as a tandem combination for specific element detection.

The Atomic Emission Spectrometer

The atomic emission spectrometer is extremely versatile, with a very high sensitivity and selectivity. The model described here will be that manufactured by the Hewlett-Packard Corporation. Basically, atomic emission is achieved by means of a helium plasma, and the light emitted is analyzed by a diode array spectrometer. A diagram showing the basic principles of the helium plasma atomic emission spectrometer is shown in Figure 2.18. The plasma is microwave induced into a helium stream employing a water-cooled transducer. The sample, mixed with the pure helium make-up gas, enters the plasma and the elements present in the solute emit light, the wavelength of which is characteristic for each element. The sample residue subsequently passes to waste. The light emitted is transmitted through a quartz window, and is then focused by a quartz lens and spherical mirror onto a diffraction grating.

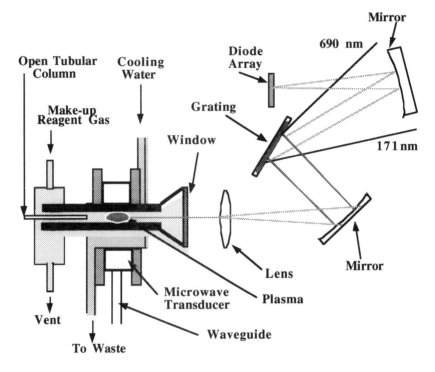

Courtesy of the Hewlett-Packard Corporation

Figure 2.18 The Helium Plasma Atomic Emission Spectrometer

The dispersed light from the grating is then focused by a second mirror onto the diode array. The diode array can have any number of diodes, but the contemporary instrument manufactured by Hewlett-Packard has 211, and its position is adjustable. Consequently, different wavelength ranges can be selected for monitoring, from the full spectrum provided by the grating. The total wavelength range covered by the instrument is from 170 nm to 800 nm. Each diode element is 61 μm wide and the entrance slit is 50 μm wide. In tandem operation, the diode array would be scanned continuously during the development of the chromatogram and the data wavelength pertinent to a particular element, a chromatogram could be constructed that monitors that specific element. This could be considered as the atomic emission equivalent to single-ion-monitoring in chromatography/mass spectrometry tandem systems. Employing data for carbon, hydrogen, oxygen and nitrogen, the empirical formula for an organic compound can also be approximately determined.

Atomic Absorption Spectroscopy

Atomic absorption spectroscopy is another element specific spectroscopic monitoring system that can determine the presence of specific elements when they exist at high temperature in a flame or in a graphite furnace. The device is, in fact, the complement of the atomic emission spectrometer, in that the *absorption* of light specific to a particular element is measured, as opposed to the light emitted. The amount of light absorbed is proportional to the amount of the element present which, in turn, is proportional to the amount of the element that is continuously fed into the flame or furnace.

The Flame Atomic Absorption Spectrometer

A diagram of a flame atomic absorption spectrometer is shown in Figure 2.19. The light source is a cold cathode lamp that produces, almost exclusively, the light that would be naturally emitted by the element to be measured. A whole range of such lamps are available that includes the vast majority of the elements of general analytical interest. Consequently, the light will contain specifically those wavelengths that the element in the flame will selectively absorb. The light passes through the flame, which is usually rectangular in shape so as to provide an adequate path length of flame for the light to be absorbed, and then into the optical system of the spectrometer. The flame is fed with a combustible gas, customarily air/acetylene, nitrous oxide/acetylene or air/propane or butane. The sample, dissolved in a suitable solvent, is nebulized and fed into the gas stream at the base of the burner. The light, having passed through the flame, can be focused directly onto a photo cell or onto a diffraction grating by means of a spherical mirror. The diffraction grating is

grating by means of a spherical mirror. The diffraction grating is movable, and so it can be set to monitor a particular wavelength that is characteristic of the element being measured, or it can be scanned to produce a complete absorption spectrum of the sample.

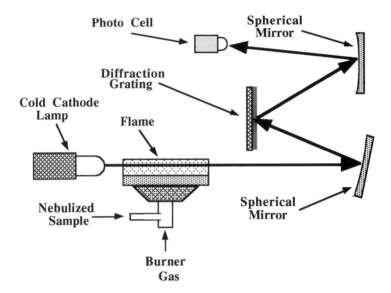

Figure 2.19 The Flame Atomic Absorption Spectrometer

After leaving the grating, light of a selected wavelength, or range of wavelengths, is focused onto the photo cell. The position of the diffraction grating determines the wavelength of the light that is to be monitored. The flame absorption spectrometer is fairly sensitive, and can be readily used as a tandem instrument combined with a chromatograph, providing that an appropriate interface is employed.

Chiro-Optical Measurements

Over the last decade there has been a growing interest in *chiral chromatography,* a term given to the separation of optically active compounds by chromatographic techniques. This interest has arisen largely as a result of the recognition that the majority of physiologically active substances exist in chiral form. Furthermore, in many countries, it is now mandatory that if a drug can exist with different chiral structures,

Identification Techniques for Tandem Use

is now mandatory that if a drug can exist with different chiral structures, the relative physiological activity of the different enantiomers must be determined. It follows, that the separation and identification of enantiomers is now a very important analytical problem. The chirality of a substance is measured by its capacity for rotating polarized light

As discussed earlier, light consists of a sinusoidally changing electric field normal to, and in phase with, a sinusoidally changing magnetic field. The plane of the electric vector in normal light, takes no particular orientation, but in plane polarized light, the electric vector is either vertically or horizontally polarized. If the electric vector transcribes an helical path, either to the right or left, the light is said to be *circularly polarized*. A linearly polarized beam of light, can be considered to be the resultant of two equal-intensity, in-phase components, one left and the other right circularly polarized, or of two orthogonal linear components at $\pm 45°$.

The differential absorption of these two $\pm 45°$ linear components in a medium is known as *linear dichroism*; if there is a differential velocity between the two $\pm 45°$ linear components, when passing through a medium (*i.e.* the refractive index of the medium to the two light components differ), then this is known as *linear birefringence*. In an analogous manner, the difference in the adsorption characteristics of a medium to left and right circularly polarized light is termed *circular dichroism* (CD) and it follows, that the difference in refractive index of a substance to the two light components is called *optical rotary dispersion* (ORD), sometimes reported as *specific optical rotation*.

CD spectra are usually measured as the differential absorption of left and right circularly polarized light, *i.e.* (A_L-A_R) and is usually reported as the differential molar extinction coefficient ($\Delta\varepsilon$), where

$$\Delta\varepsilon = (\varepsilon_L - \varepsilon_R) = \frac{(A_L - A_R)}{cl}$$

where (l) is the length of the cell,
and (c) is the morality of solute

It is apparent that a tandem system incorporating some chiral measurement, or even a chiral spectrometer in conjunction with a suitable separation technique, could be extremely useful in drug analysis or in the separation and identification of physiologically active materials.

The basic apparatus that is used for measuring circular dichroism is shown in Figure 2.20. It consists of a light source, a linear polarizer, a Fresnel rhomb that converts the linear polarized light to circularly polarized light, a sample cell, and finally an appropriate light intensity measuring device.

The rotation of the linear polarizer ± 45°, to the appropriate Fresnel rhomb axis, induces the generation of left or right polarized light.

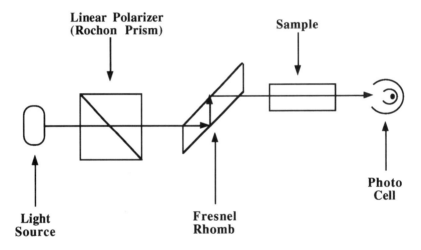

Figure 2.20 The Basic Apparatus for Measuring Circular Dichroism.

The modern form of the CD spectrometer is shown in Figure 2.21, which would be the type appropriate for tandem operation. Light from a broad emission source passes through a chopper and then to a monochromator that allows light of a selected wavelength to be passed through a filter to the polarizer. The polarizer can be a Rochon prism and the polarized light would then be passed through a photoelastic modulator (Pockels cell), the function of which will be described below. The selected left or right

Identification Techniques for Tandem Use

circularly polarized light is then passed through the cell and the intensity of the transmitted light monitored by the sensor.

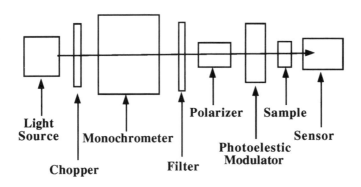

Figure 2.21 A Modern CD Spectrometer

Polarization Modulation.

The alternate production of left or right circularly polarized light, which is called polarization modulation, is an essential process for CD measurement. Now, a linearly polarized light beam can be said to be the resultant of two orthogonal, in phase, linear, light beams. Consider a block of isotropic fused silica that is rendered birefringent by pressure exerted along the (x) or (y) axis as shown in Figure 2.22.

Under these condition the refractive indexes (n_x and n_y) will differ, and

$$n_x \neq n_y$$

If the light beam passing through the block is oriented with its resultant axis at 45° to the pressure axis, one of the components will travel faster through the medium than the other. If $n_x > n_y$, then the (x) component of the light beam will travel more slowly. If the retardation is exactly $(\frac{\lambda}{4})$ the emergent light beam will be right circularly polarized.

If the retardation is $(-\frac{\lambda}{4}$ or $-\frac{3\lambda}{4}, -\frac{7\lambda}{4}$, etc.) then the emergent light beam will be left circularly polarized.

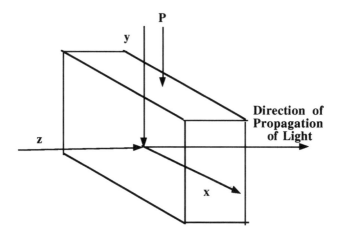

Figure 2 22 Polarization Modulation

This an example of a general principle which was first described by Brewster and given the term photo elasticity. Electro-optic Modulizers, that are used in some modern instruments, operate in a similar way.

Practical Chiral Measuring Devices

The successful development of a Chiral monitor, based on optical rotation measurement, hinges on the Faraday effect. If a plane polarized beam of light passes through a medium that is subjected to a strong magnetic field, the plane of the polarized beam is rotated a small angle (α), where (α) is given by,

$$\alpha = VdH$$

where, (V) is the Verdet Constant,
(d) is the path length
and (H) is the magnetic field strength.

The magnetic field is generated in an air core coil inside of which, is a rod made from glass having a high Verdet constant. Now,

$$H = iN$$

where, (i) is the current through the coil, and (N) is the number of turns in the coil. Thus,

$$\alpha = VdiN$$

Identification Techniques for Tandem Use 81

Thus (α) can be controlled by varying (i). The rotational resolution ($\Delta\alpha$) for a controlled change in current through the coils of (Δi) can be as little as 10^{-5} but due to heat losses in the coil, the maximum value of ($\Delta\alpha$) is limited to about ± 2°. A diagram of the detector is shown in Figure 22.

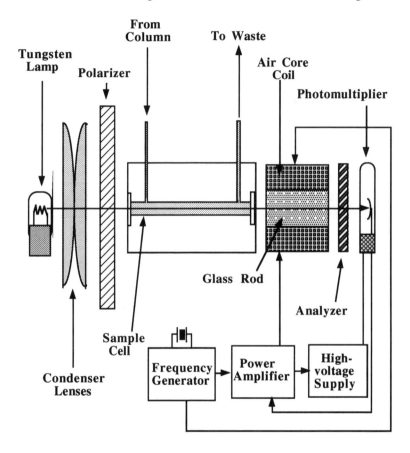

Courtesy of JM Science Inc.

Figure 2.23 A Chiral Detector Monitoring Optical Rotation

Light from a tungsten lamp passes through two condenser lenses to a polarizer. Plane polarized light from the polarizer passes through a temperature controlled cell, to a Faraday modulator and thence, through an analyzer onto a photomultiplier. The modulator is supplied with a crystal-controlled AC component. If an optically active sample is present, the intensity of the light falling onto the photomultiplier changes. By the use of a phase sensitive amplifier, and electrical feed-back, the current though the modulator is automatically adjusted until no AC component appears on the photomultiplier output. The feed-back current is then taken as a measure of the optical rotation of the sample.

Nuclear Magnetic Resonance Spectroscopy

In order to understand the technique of nuclear magnetic resonance (NMR), it is necessary to consider the magnetic properties of atomic nuclei. The nucleus of an atom spins and thus, if the charge is not symmetrically placed on the atom, the spinning charge will constitute a circular current which will produce an associated magnetic field like a small bar magnet. Not all nuclei possess an asymmetric charge but among those that do are the hydrogen and the ^{13}C nuclei, which are the nuclei of major importance for NMR spectroscopy.

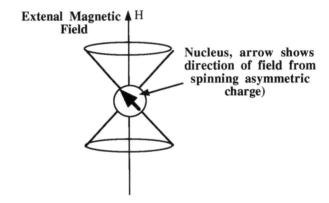

Figure 2.24 A Diagram of a Precessing Nucleus

If the spinning nucleus, with its associated magnetic field, is placed in a strong external magnetic field the external field will act upon the spinning nucleus to try to change its spinning axis to be in line with the magnetic field. Now according to Newton's law, *when a force acts upon a spinning body to change its axis of rotation then, to conserve the angular momentum, the spinning body will precess*. The precessing nucleus is depicted diagramatically in Figure 2.24.

Identification Techniques for Tandem Use

From quantum rules, the precessing nucleus has only two possible orientations. Consequently, if energy is supplied to the spinning nucleus, by way of electromagnetic radiation of the necessary frequency, energy can be absorbed and the precessing nucleus changed from one orientation to the other. The magnetic quantum number for the proton is ± 1/2 and consequently, the frequency (v) at which this transition can occur can be calculated from the equation:

$$v = \frac{\mu H}{h\left(\frac{1}{2}\right)}$$

where (μ) is the nuclear magnet moment,
(H) is the external magnetic field strength,
and (h) is planks constant

Now the nuclear magnet moment for the proton is 2.793 Bohr magnetons (1 magnetron = 5.093 × 10^{-24} erg/gauss) and thus, if it is situated in a field of 15,000 Gauss, the necessary frequency to make the transition is given by,

$$v = \frac{2.793\left(5.049 \times 10^{-23}\right)}{h\left(\frac{1}{2}\right)} = 63.9 \text{ Mhz}$$

In a similar manner, taking a range of transition frequencies, the different field strengths can be calculated and are shown in Table 2.1.

Table 2.1 NMR Field Strengths and Frequencies for Some Nuclei

Nucleus	20 MHz	60 MHz	100 MHz	250 MHz	750 MHz
^{1}H	4,700	14,000	23,500	58,750	176,250
^{2}H	30,600	91,800	153,000	382,500	1,147,500
^{13}C	18,700	56,00	93,400	233,750	701,250
^{14}N	65,000	195,000	325,000	812,500	2,437,500

Consider a sample containing protons, situated in a strong magnetic field and irradiated at the transition frequency. If the sample is then scanned by a second, low intensity magnetic field, when transition actually occurs energy will be absorbed and can be detected.

The manner in which the experimental measurements are made will be discussed later, but consider the results of an experiment using alcohol as the sample. Assume that the device is examined on a low-resolution spectrometer (a concept which will also be discussed in due course). The sample is irradiated with electromagnetic waves of the calculated frequency. The magnetic field intensity is then scanned over a narrow range close to that where the absorption of energy is expected to take place. The energy of absorption, which is sensed electronically, is plotted against field strength to provide the NMR spectrum. The spectrum for ethyl alcohol would look something like that depicted in Figure 2.25.

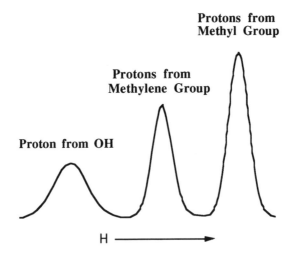

Figure 2.25 A Low-Resolution NMR Spectrum of Ethyl Alcohol

There are several points of interest arising from the simple spectrum. First, the three peaks have areas in the proportions of 1:2:3 indicating they are from the proton on the oxygen, the methylene protons and the methyl protons. Thus the spectrum discloses the number of protons associated with each peak from the relative peak areas. Second, the proton peaks appear at

Identification Techniques for Tandem Use

different values of the scanned magnetic field. This is because, due to their specific environment, they experience shielding from the electron clouds from neighboring atoms and thus, although they all absorb energy at the same frequency (because the frequency in this experiment is fixed) the applied field must be different to compensate for their unique atomic environment. Now,

$$H = H_F + H_S - H_C$$

where (H) is the net magnetic field experienced by the proton,
(H_F) is the high intensity applied magnetic field,
(H_S) is the small applied scanning magnetic field,
and (H_C) is the shielding effect provided by the atomic environment of the proton

and
$$H_C = \alpha(H_F + H_S)$$

Where (α) is the shielding effect of the electron environment of the proton.

In fact it is the *chemical environment* of the proton that will affect the diamagnetic shielding constant (α). Consequently, the relative positions of the absorption peaks, determined by the magnitude of (α), will disclose the nature of the chemical environment and contribute information with regard to the overall structure of the molecule.

One further point. If the resolution of the NMR machine is increased (which, as in chromatography, means that the widths of the peaks are reduced relative to their movement apart, *viz.* chemical shift) then the proton peaks show a well-defined and predictable fine structure. An example of the spectrum of ethyl alcohol that would be obtained on a NMR spectrometer having greater resolution is shown in Figure 2.26. It is now seen that the magnetic field experienced by a proton is also influenced by protons on the adjacent carbon atoms. For example, the methylene protons can contribute magnetic influence at three different levels to the field experienced by the methyl protons. The magnetic fields due to the two methylene protons can act in opposite directions or both in one direction or the other. Thus, as there are three different contributions, the methyl protons will display three peaks. Furthermore, as the probability of both protons acting in the same direction is half that of them acting in opposition, the center peak will be twice the height of the side peaks.

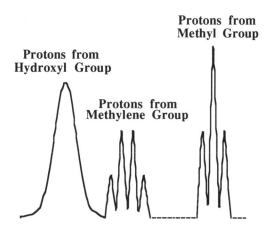

Figure 2.26 A High Resolution NMR Spectrum of Ethyl Alcohol

In a similar way, the three protons of the methyl group can contribute fields at four different levels to the methylene protons. There are two possibilities for them all to act in one direction and two possibilities where two are acting in one direction and the other in opposition. Thus the methylene protons will display four peaks. As the probability of the two protons acting in one direction and the other in opposition is twice as great as all the protons acting in one direction or the other, the two center peaks will be twice the height of the outside peaks.

Thus the position of the peaks, or what is known as the chemical shift, indicates the chemical nature of the neighboring groups and the fine structure of the proton peaks provides information on the degree of saturation of the neighboring atoms. In addition, the area of the peaks provides quantitative information on the distribution of the protons throughout the molecule. Even with this brief and somewhat superficial treatment of NMR spectroscopy, the value of the technique for structure

Identification Techniques for Tandem Use

elucidation becomes quite obvious. Unfortunately, the combination of NMR with chromatographic systems is one of the most difficult tandem combinations to operate and, although significant advances have been made over the past few years, much remains to be accomplished. For those requiring more information on NMR, the books by Banwell and McCash (11) and Paudler (12) are strongly recommended.

The NMR Spectrometer

The early NMR spectrometers operated at about 60 MHz and thus the necessary field of 14 kilogauss could be produced by conventional iron cored electromagnets or even by ceramic permanent magnets.

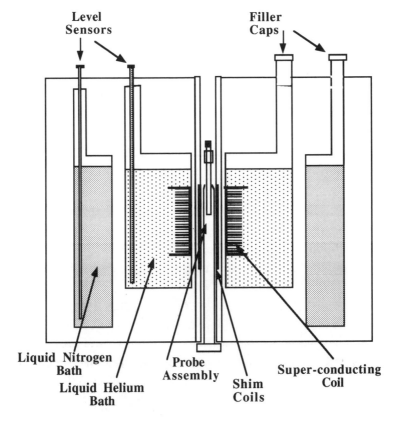

Figure 2.27 The NMR Spectrometer with Super-conducting Magnet

Today, however, modern high-resolution NMR spectrometers operate at a frequencies of 250 or even 750 MHz and, as seen from Table 2.1, for proton spectroscopy, this requires magnetic fields of about 60 and 180 kilogauss respectively. Such fields cannot be obtained from conventional electromagnets and consequently, superconducting magnets must be used.

Superconducting magnets require a continuous supply of current and, unfortunately, consume large quantities of cryoscopic fluids. The magnet consists of a main field superconducting coil with a number of other smaller coils that can control field gradients in different directions with respect to the main field. These smaller coils are called shim coils and are used to improve the homogeneity of the field.

A modern NMR machine fitted with a superconducting magnet is depicted in Figure 2.27. The superconducting coils must remain submerged in liquid helium during use, with the current in each, established during installation. Outside the liquid helium bath is a liquid nitrogen bath, which reduces the heat transfer to the helium bath, and thus conserves helium. There are liquid level sensors that actuate warning devices in both the helium and nitrogen baths to ensure they do not become exhausted. There is also a number of shim coils associated with the probe inside the magnet that operate at room temperature. These shim coils provide the final adjustments to field homogeneity which, in modern instruments, are usually under computer control. An air supply, provided through appropriate conduits to the probe, actuates the turbine that spins the sample and provide energy for any automatic sample handling devices.

A diagram of an NMR probe is shown in Figure 2.28. Inside the probe is a Dewar vessel which holds the sample tube, the various sensor coils and the conduits to the system. The Dewar is also fitted with a heater to control the probe and sample at a prescribed temperature. The Dewar contains two coils; the rf lock-coil that is usually tuned to deuterium as the reference nucleus which, in effect, provides the calibrating scale for the spectrum, and the rf coil for the nucleus under examination. The total rf circuit is not included in the diagram to avoid confusion. There are two trimming capacitors situated in the probe, one for each of the two coils mentioned, and are adjustable from outside the probe.

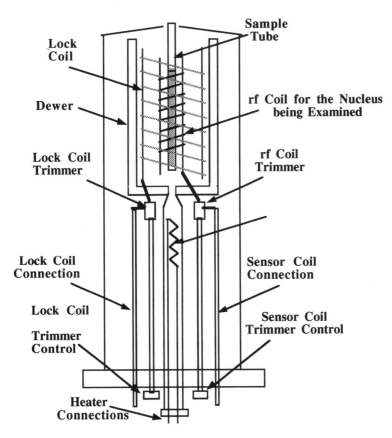

Figure 2.28 The Probe of a NMR Spectrometer with a Superconducting Magnet

The probe shown is the standard type of NMR probe and cannot be used for flow through samples. Consequently, it would not be appropriate if the NMR probe was to be used in a tandem configuration. Although the electrical connections remain the same, the physical shape must be changed. In the standard instrument the probe is not accessible from both the top and the base of the magnet but usually only from the base. This means that the sample cannot flow though the cell vertically downwards exiting at the base of the instrument but must take the form of a U. This obviously requires considerable modification to the probe. However, the probe design for tandem use will be discussed later with other interfaces.

Mass Spectrometry

Mass spectrometry, as opposed to the other spectrometric methods that have been discussed so far, is not involved with the absorption of electromagnetic radiation and thus is an entirely different type of analytical technique. Basically, mass spectrometry involves first, the production of ions from the sample and these ions can be molecular ions, ion fragments or ion complexes, depending on the ionization process that is used. Second, the ions are then accelerated to high velocities in a vacuum and, by applying a range of different physical and electrical techniques, the ions are separated into their individual masses and each mass-group sensed and identified. The advantages of this type of analytical approach and its value as a tandem instrument are very obvious.

There are three basic types of mass spectrometer, the sector mass spectrometer, the quadrupole mass spectrometer (which includes the mass analyzer) and the time-of-flight mass spectrometer. All three have been used (and indeed are still used) in tandem configuration, and consequently the basic principles of all three will be described. However, the quadrupole mass spectrometer in one of its forms is by far the most popular mass spectrometer in tandem use.

The Sector Mass Spectrometer.

The sector mass spectrometer functions on the combined mass selection of an electrostatic field and a magnetic field. An instrument that utilizes both an electric and a magnetic field to analyze the ions is often called a double focusing sector mass analyzer. A diagram of the basic double focusing mass analyzer is shown in Figure 2. 29.

Consider an ion, mass (m) and charge (z) entering the electric field (E) at velocity (v). The ion will describe an arc of radius (R_1) and then enter the magnetic sector. Now the electrostatic deflecting force will equal the centrifugal force of any ion that will enter the magnetic sector entry slit at

(R_1). The centrifugal force and the electrostatic force can be equated algebraically, to determine the conditions under which an ion will enter the magnetic sector slit.

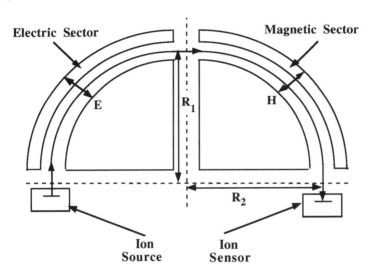

Figure 2.29 The Double Focusing Sector Mass Spectrometer.

Thus

$$\frac{mv^2}{R_1} = ezE \qquad (2.3)$$

where (e) is the charge on the electron.

Thus solving for (v)

$$v = \sqrt{\frac{eEzR_1}{m}} \qquad (2.4)$$

Now having entered the magnetic sector the only ions that reach the detector at radius (R_2) will be those where the centrifugal force will equal the deflecting force from the magnetic field.

Thus bearing in mind that an electron of charge (e) traveling at a velocity (v) constitutes a current of (ev) then,

$$\frac{mv^2}{R_2} = Hezv \quad \text{or} \quad mv = HeZR_2$$

Substituting for (v) from equation (2.4),

$$m\sqrt{\frac{eEzR_1}{m}} = HeZR_2$$

Rearranging,

$$\frac{m}{z} = \frac{H^2 e R_2^2}{E R_1} \qquad (2.5)$$

It is seen that by scanning (H) over an appropriate range, the ions with increasing (m/z) ratios will be sensed by the detector. This is a very simple treatment of the double focusing mass spectrometer and, in fact, if the slit entering the magnetic sector was made sufficiently narrow to produce well-resolved mass peaks, then the number of ions collected would be very small and the system relatively insensitive. Nevertheless, the explanation given illustrates the function of the sector instruments and how it is operated. The low sensitivity can be avoided by the choice of a suitable combination of electrostatic and magnetic sectors such that the velocity dispersion in the two sectors is equal and opposite. Consequently, a larger slit can be used and the resulting high sensitivity recovered.

Ionization Methods

There are number of ways of producing sample ions, many of which will be discussed when the actual tandem interfaces are described. The most common form of ionization is by electron impact.

Electron impact ionization is a fairly harsh method of ionization and usually produces a range of molecular fragments that in most cases helps to elucidate the structure of the molecule. However, although molecular ions are often produced, which is important for structure elucidation, sometimes only small fragments of the molecule are observed, with no molecular ion. Under such circumstances alternative ionizing procedures must be used. A diagram of an electron impact ion source is shown in Figure 2.30.

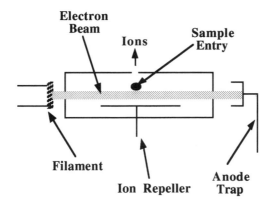

Figure 2.30 An Electron Impact Ionization Source

Electrons are generated by a heated filament which then pass across the ion source to the anode trap. The sample vapor is introduced in the center of the source and the molecules drift, by diffusion, into the path of the electron beam. Collision with the electrons produce a molecular ion and ionized molecular fragments, the size of which is determined by the energy of the electrons. The electron energy is controlled by the accelerating potential applied to the anode trap. The ions that are produced are driven by a potential applied to the ion-repeller electrode into the accelerating region of the mass spectrometer.

Chemical Ionization If a large excess of a reagent gas is employed together with the sample, an entirely different type of ionization takes place. As there is an excess of the reagent gas, the reagent molecules are preferentially ionized and the reagent ions then collide with the sample molecules and produce sample + reagent ions or in some cases protonated ions. This type of ionization is called *chemical ionization* and is a very gentle form of ionization. Very little fragmentation takes place and parent ions + a proton or a molecule of the reagent gas is produced. The molecular weight of the parent ion are thus easily obtained. Little modification to the normal electron impact source is required and a conduit for supplying the reagent gas is all that is necessary.

The Quadrupole Mass Spectrometer

The most popular mass spectrometer used in tandem systems is the quadrupole mass spectrometer, either as a single quadrupole or as a triple quadrupole which can provide MS/MS spectra (at technique that will be discussed later. A diagram of a Quadrupole mass spectrometer is shown in Figure 2.31.

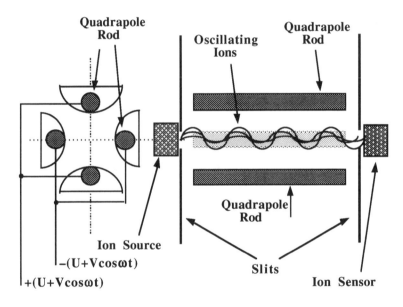

Figure 2.31 The Quadrupole Mass Spectrometer.

The operation of the quadrupole mass spectrometer is quite different from that of the sector instrument. The quadrupole spectrometer consists of four rods which must be precisely straight and parallel and so arranged that the beam of ions is directed axially between them. Ideally the rods should be hyperbolic in cross section but in practice less expensive cylindrical rods are nearly as satisfactory. A voltage comprising a DC component (U) and a radio frequency component ($V\cos\omega t$) is applied between adjacent rods, opposite rods being electrically connected, as rods, with a relatively small potential ranging from 10 to 20 volts. Once inside the quadrupole, the ions oscillate in the (x) and (y) dimensions as a result of the high-frequency electric field.

The stability of the oscillating ions are determined by the magnitude of two parameters (a) and (q) which are defined by the following equations:

$$a = \frac{8eU}{mr_0^2\omega^2} \quad \text{and} \quad q = \frac{4eV_0}{mr_0^2\omega^2}$$

where, (r_0) is half the distance between opposite rods of the quadrupole system and the other symbols have the meaning previously ascribed to them.

The oscillations of the ions will only remain stable for certain combined values of (a) and (q). Outside these values the oscillations become infinite and will strike the rods and become dissipated. The relationship between (a) and (q) is shown in Figure 2.32

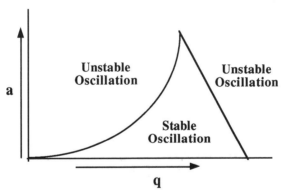

Figure 2.32 Conditions for Stable Oscillation in a Quadrupole Mass Spectrometer

It is seen that there is a very restricted range of values of (a) and (q) that can permit the mass spectrometer to operate in a stable mode. The mass range is scanned by changing U and V_0 while keeping the ratio U/V_0 constant.

The quadrupole mass spectrometer is compact, rugged and easy to operate and consequently is a popular instrument to use in tandem combinations. It is easily interfaced with a wide range of chromatographic systems. Unfortunately, the mass range of the quadrupole spectrometer does not extend to very high values but, as will be seen when dealing with different interfaces, under certain circumstances multiple charged ions can be generated and identified by the mass spectrometer. This, in effect, significantly increases the mass range of the device.

The quadrupole mass spectrometer can also constructed in such a way as to be able to provide a MS/MS performance. This is achieved by combining three quadrupole units in series. A diagram of a triple quadrupole mass spectrometer is shown in Figure 2.33. The sample enters the ion source and is usually fragmented by either an electron impact or chemical ionization process. In the first analyzer the various charged fragments are separated in the usual way, which then pass into the second quadrupole section sometimes called the collision cell. The first quadrupole behaves as a straightforward mass spectrometer.

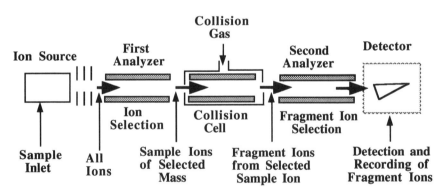

Figure 2.33 The Triple Quadrupole Mass Spectrometer.

Instead of the ions passing to a sensor, the ions pass into a second mass spectrometer. In this way a specific ion can be selected for further study. In the center quadrupole section the selected ion is further fragmented by collision ionization and the new fragments then pass into the third quadrupole which functions as a second analyzer. The second analyzer collision ionization and the new fragments then pass into the third quadrupole which functions as a second analyzer. The second analyzer segregates the new fragments into their individual masses which are detected by the sensor, producing the mass spectrum. In this way, the exclusive mass spectrum of a particular molecular or fragment ion can be obtained from the myriad of ions that may be produced from the sample in the first analyzer. It is seen that this is an extremely powerful analytical system that can handle exceedingly complex mixtures and very involved molecular structures. The system has a adequate resolving power and is valuable for structure elucidation. A photograph of the triple quadrupole mass spectrometer manufactured by VG Organic Inc. is shown in Figure 2.34.

Courtesy of VG Organic Inc.

Figure 2.34 The Triple Quadrupole Mass Spectrometer Manufactured by VG Instruments

The Ion Trap Detector

The ion trap detector is a modified form of the quadrupole mass spectrometer, but was designed more specifically as a chromatography detector than for use as a tandem instrument for structure elucidation and solute identification. The electrode orientation of the quadrupole ion trap mass spectrometer is shown in Figure 2.35.

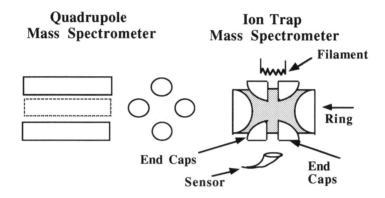

Figure 2.35 Pole Arrangement for the Quadrupole and Ion Trap Mass Spectrometers

It was shown in Figure 30, that the quadrupole spectrometer contains four rod electrodes. The ion trap mass spectrometer has a quite different electrode arrangement and consists of three cylindrically symmetrical electrodes comprised of two end caps and a ring. The device is small, the opposite internal electrode faces being only 2 cm apart. Each electrode has accurately machined hyperbolic internal faces. In a similar manner to the quadrupole spectrometer, an rf voltage together with an additional DC voltage is applied to the ring and the end caps are grounded. In the same way as the quadrupole mass spectrometer, the rf voltage causes rapid reversals of field direction, so any ions are alternately accelerated and decelerated in the axial direction and *vice versa* in the radial direction. There are operating parameters, (a), and (q), that define the conditions of oscillation which are analogous to those for the quadrupole mass spectrometer but, in this case, (r_0) is the internal radius of the ring electrode. The ion trap is small and (r_0) is typically about 1 cm. At a given voltage, ions of a specific mass range are held oscillating in the trap. Initially, the electron beam is used to produce ions and after a given time the beam is turned off. All the ions, except those selected by the magnitude of the applied rf voltage, are lost to the walls of the trap, and the remainder continue oscillating in the trap. The potential of the applied rf voltage is then increased, and the ions sequentially assume unstable trajectories and leave the trap via the aperture to the sensor. The ions exit the trap in order of their increasing m/z values. The first ion trap mass spectrometers were not very efficient, but it was found that the introduction traces of helium to the ion trap significantly improved the quality of the spectra. The improvement appeared to result from ion–helium collisions that reduced the energy of the ions and allow them to concentrate in the center of the trap. The spectra produced are quite satisfactory for solute identification by comparison with reference spectra. However, the spectrum produced for a given substance will probably differ considerably from that produced by the normal quadrupole mass spectrometer.

The Time of Flight Mass Spectrometer

The time of flight mass spectrometer was invented many years ago but, due to the factors controlling resolution not being clearly recognized and also due to certain design defects that occurred in the first models, it exhibited limited performance and was rapidly eclipsed by other developing mass spectrometer techniques. However, with improved design, modern fabrication methods and the introduction of Fourier transform techniques, the performance has been vastly improved. As a result, there has been a resurgence of interest in this particular form of mass spectrometry. A diagram of the time of flight mass spectrometer is shown in Figure 2.36.

In a time of flight mass spectrometer the following relationship holds,

$$t = \left(\frac{m}{2zeV}\right)^{\frac{1}{2}} L$$

where (t) is the time taken for the ion to travel a distance (L)
(V) is the accelerating voltage applied to the ion,
and (L) is the distance traveled by the ion to the ion sensor.

It follows that for a given system, the mass of the ion is directly proportional to the square of the transit time to the sensor. The sample is volatilized into the space between the first and second electrodes and a burst of electrons, over about a microsecond period, is allowed to produce ions. The extraction voltage, (E) is then applied for another short time period which, as those further from the second electrode will experience a greater force than those closer to the second electrode, will focus the ions.

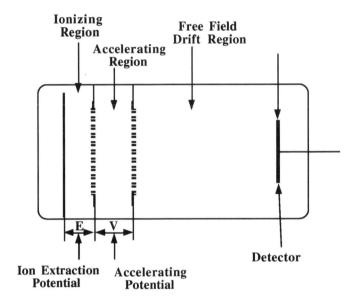

Figure 3.36 The Time of Flight Mass Spectrometer

After focusing, the accelerating potential (V) is applied for a much shorter period than that used for ion production (*ca* 100 nsec) so that all the ions in the source are accelerated almost simultaneously. The ions then pass through the third electrode into the drift zone and are then collected by the sensor electrode. The particular advantage of the time of flight mass spectrometer from the point of view of tandem systems lies in the fact that it is directly and simply compatible with direct desorption from a surface. Consequently, it can be employed with laser desorption and plasma desorption techniques which will be discussed in a later chapter.

An excellent discussion on general organic mass spectrometry is given in *Practical Organic Mass Spectrometry* edited by Chapman [13].

Synopsis

Most tandem systems involve the association of the separation technique with some form of spectroscopy, the most common being, ultraviolet, fluorescence, infrared, Raman, atomic, chiral, nuclear magnetic resonance, and mass spectroscopy. All except the last depends on the absorption or emission of electromagnetic radiation. UV spectroscopy provides information on electron transitions and is useful for identifying certain classes of compounds, such as aromatic and unsaturated compounds; although one of the more sensitive, it is the least useful of all the spectroscopic techniques for tandem operation. Fluorescence spectroscopy depends on the emission of light that occurs as an excited electron falls back to its ground state. Fluorescence spectra provide more information then UV spectra as they usually contain more fine structure. Furthermore, a large number of fluorescence spectra can be obtained from a single substance, by employing light selected from a range of different wavelengths. Infrared spectroscopy is more informative than UV or fluorescence spectroscopy. IR spectra demonstrate light absorption at different wavelengths resulting from the change in rotational and vibrational energy of the molecule and contain considerable fine structure. Consequently, they can be used to identify a substance from reference spectra, and can also demonstrate the presence of certain chemical groups such as carbonyl, hydroxy, aromatic rings etc. The technique is inherently less sensitive than UV spectroscopy but, with the use of Fourier transform techniques, the sensitivity can be significantly enhanced. Raman spectroscopy involves the measurement of light that is emitted when the rotational and vibration energy of a molecule is raised by the absorption of excitation light and then falls back to a different energy level. It is a relatively insensitive technique and special precautions must be taken to avoid fluorescent light, which can provide overwhelming background signal and noise. The information Raman spectra provide is very similar to that of IR spectra to which, they are in fact, complementary. Atomic emission spectrometry involves measuring either the light emitted by excited atoms of an element, or the light absorbed by an element in vapor form. This technique is particularly useful for trace elements analysis and for obtaining the element ratios of a substance. The characteristic emission of an element is usually obtained by passing the sample through a plasma and monitoring the spectrum produced. Atomic absorption spectra are usually obtained by passing light, having exclusively the wavelengths that will be absorbed by the element, through the vaporized sample and measuring the intensity of the light transmitted. Chiral substances rotate the plane of polarized light to the right, or to the left, depending on the molecular structure of the molecules through which the light is passing. Chiral spectroscopy will determine the extent of this rotation, and this information can be used either to identify the presence of the optical isomer, or help confirm its identity. The technique is relatively insensitive, but its use in tandem instruments is becoming more important because of the selective biological activity of the different optical isomers of many drugs and biological materials. Nuclear magnetic resonance spectroscopy is the most powerful technique for structure elucidation, unfortunately it is also one of the least sensitive. Certain spinning nuclei have asymmetrical charges and thus constitute nuclear magnets having associated magnetic fields. Because of the conservation of angular momentum, when such nuclei are situated in a strong magnetic field, they precess at a frequency that depends on the strength of the field. If exposed to electromagnetic waves of the precessing frequency, the nuclei can absorb energy and this energy absorption can be detected. The precise frequency that the nuclei precess will depend on the their magnetic environment which, in turn, will depend on the adjacent atoms, and the protons associated with the adjacent atoms. Thus the proton NMR spectra (the position of the absorption peaks and the fine structure) will provide information on the elements present in the molecule, the distribution of the protons in the molecule, and from the areas of the peaks in the spectrum, the relative proportion of the protons throughout the molecule. Mass spectrometry has a much higher sensitivity than NMR, and the mass spectrum is useful for both confirming the identity of a substance and for elucidating the details of an unknown molecular structure. The sample is ionized by appropriate procedures, to produce molecular ions, molecular ion fragments, or both. The ions are then subjected to either electrostatic or magnetic forces, in a manner that will discriminate between the masses of the ions present. There are three basic ion discriminating instruments: the sector instrument, the quadrupole instrument or the time of flight spectrometer. The quadrupole spectrometer is the most commonly used analyzer, but the time of flight mass spectrometer has been improved to the point of becoming a worthy competitor. Other than NMR, mass spectrometry is the best compromise between being the most informative identifying technique, and the most sensitive for use in tandem with a separating instrument.

References

1. G. H. Schenk, *Absorption of Light and Ultraviolet Radiation*, Allyn and Bacon Inc., Boston(1973).
2. H. H. Perkampus, *UV-VIS Spectroscopy and Its Applications*, Springer-Verlag, (1992).
3. G. G. Guilbault, *Practical Fluorescence*, Marcel Dekker, New York (1973).
4. S. Udenfriend, *Fluorescence Assay in Biology and Medicine*, Academeic Press, New York(1962).
5. A. T. Rhys Williams. *Fluoresence Detection in Liquid Chromatography*, Perkin Elmer Corporation, Beaconsfield, England (1980).
6. R. T. Conley, *Inrfrared Spectroscopy,* Allyn and Bacon Inc., Boston,(1972).
7. N. L. Alpert, IR; Theory and Practice of Infrared Spectroscopy, (1973).
8. D. Welti, *InfraredVapor Spectra*, Heyden and Sons, New York (1970).
9. B. Stuart, *Modern Infrared Spectroscopy,* John Wiley, Chichester(1996)
9. D. A. Long, *Raman Spectroscopy,* McGraw Hill, New York(1977).
10. C. N. Banwell and M. McCash, *Fundamentals of Molecular Spectroscopy*, McGraw-Hill, New York(1994).
11. W. W. Paudler, *Nuclear Magnetic Resonance*, Allyn and Bacon Inc., Boston (1971).
12. *Practical Organic Mass Spectrometry,* J. R. Chapman (Ed.) John Wiley,Chichester (1994)

CHAPTER 3

INTERFACE CONDUITS

All tandem interfaces contain conduits, through which the column eluent from the separation system (the chromatograph) passes, in order to enter the identifying device (the spectrometer). The interface can range in complexity from a short length of small diameter tubing to an elaborate device that concentrates the solute in the mobile phase, prior to it entering the spectrometer. However, irrespective of the complexity of the interface, it must be designed in such a manner that it causes little, or preferably no, loss in chromatographic resolution, and delivers the necessary amount of solute to the spectrometer. The solute bands must not be allowed to spread during passage through the interface, and cause closely eluted peaks to merge into one another, before being analyzed in the spectrometer. The most likely and certain sources of peak dispersion in any interface will be the connecting tubes. Furthermore, as lengths of open tubing are commonly included as conduits in most interfaces, the dispersion that takes place in open tubes will be considered in some detail.

Dispersion in Open Tubes

The magnitude of the dispersion that takes place in open tubes depends strongly on the nature of the fluid passing through it. It follows that dispersion in GC connecting tubes will differ considerably from those in LC, and consequently each system will be considered separately.

Gas Chromatography Columns and Connecting Tubes

The dispersion in simple open tubes results from the parabolic velocity profile that exists across the radii of such tubes, and is depicted in Figure 3.1. The length of the arrows represent the relative velocity of the mobile phase at different positions across the tube. This velocity profile causes the solute contained in the mobile phase close to the wall to move very slowly and that at the center to move more rapidly.

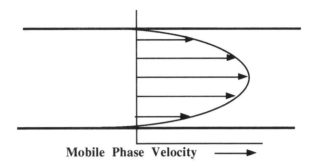

Figure 3.1 The Parabolic Velocity Profile of a Fluid Passing Through an Open Tube

As a consequence, any solute molecules in the center of the tube will be carried rapidly down the center of the conduit, and away from those situated close to the walls. This process will obviously result in the dispersion of the solute band. Furthermore, the dispersion will occur irrespective of there being stationary phase on the walls, as in a chromatographic column, or if the walls are bare, as in a simple tubular chromatography conduit. It will be seen that there is another dispersion process that takes place in an open tube, but the contribution is usually not significant at the flow rates normally employed in chromatographic development.

The band spreading that takes place in conduits containing local concentrations of solute vapor, carried by a flow of inert gas, was critically examined by Golay in the late 1950s. Golay employed electrical analogies to develop an equation that quantitatively described the dispersion of a band of solute vapor in an open tube, the internal wall of which was coated with stationary phase. The Golay equation [1], as it is called, expressed the variance contribution per unit length of tube, in terms of the velocity of the mobile phase through the tube and certain

Interface Conduits

physical properties of the solute and mobile phase. Golay's' original equation assumed an average velocity through the entire length of tube but, due to the compressibility of the gaseous mobile phase, this assumption leads to significant error when practical data was compared with that predicted theoretically. The equation, as already discussed in chapter 1, was modified by Ogan and Scott [2] to take into account the compressibility of the mobile phase, and their modified Golay equation is given as follows:

$$H = \frac{2 D_{m(o)}}{u_o} + \frac{\left(1 + 6k' + 11 k'^2\right) r^2}{24 (1 + k')^2 D_{m(o)}} u_o + \frac{2k' df^2}{3 (1+k')^2 D_s (\gamma+1)} u_o \quad (3.1)$$

where (H) is the variance per unit length of the column or the height equivalent to a theoretical plate.
(u_o) is the velocity of the mobile phase at the column outlet,
(D_m) is the diffusivity of the solute in the mobile phase measured at atmospheric pressure and the column temperature,
(D_s) is the diffusivity of the solute in the stationary phase, at the column temperature,
(r) is the radius of the column,
(k') is the capacity factor of the eluted solute,
(df) is the film thickness of the stationary phase,
and (γ) is the inlet/outlet pressure ratio of the column.

Now, for all practical columns, (r) will take values close to 0.25 mm and (df) values close to 0.25 μm. In addition, D_m will be approximately 0.1 cm^2/sec and D_s about 0.000015 cm^2/sec. It follows that the function involving (r^2) will be one to two orders of magnitude greater than the function involving (df^2) and thus, to a first approximation,

$$H = \frac{2 D_{m(o)}}{u_o} + \frac{\left(1 + 6k' + 11 k'^2\right) r^2}{24 (1 + k')^2 D_{m(o)}} u_o \quad (3.2)$$

Equation (3.2) will apply to both connecting tubes (where k'= 0, that is when there is no stationary phase coating on the internal walls of the tube) and to GC coated capillary columns If equation (3.2) is differentiated and equated to zero, expressions for the optimum exit velocity ($u_{o\,(opt)}$) and the minimum value for the height of the theoretical plate ($H_{(min.)}$) can be obtained and is given by the following two equations. ($H_{(min.)}$) is the plate height obtained by operating the column at the optimum velocity).

$$u_{o(opt.)} = \frac{2D_{m(o)}}{r}\left(\frac{12(1+k')^2}{(1+6k'+11k'^2)}\right)^{0.5} \quad (3.3)$$

and

$$H_{min.} = r\left(\frac{(1+6k'+11k'^2)}{3(1+k')^2}\right)^{0.5} \quad (3.4)$$

Now, from the Plate Theory [3], the peak variance ($\sigma^2_{c(v)}$) in volume units of an open tubular column of length (l_c) and radius (r_c) is given by,

$$\sigma^2_{c(v)} = \frac{(\text{Tube Volume})^2}{n} = \frac{(\pi r_c^2 l_c)^2}{n_c} = \frac{\pi^2 r_c^4 l_c^2}{n_c} \quad (5)$$

where (n_c) is the number of theoretical plates in the column. Now, by definition, $\dfrac{l_c}{n_c} = H_c$.

Thus, replacing ($\dfrac{l_c}{n_c}$) by (H_c), and substituting for (H_c) from equation (3.4.)

$$\sigma^2_{c(v)} = \pi^2 r_c^5 l_c \left(\frac{((1+6k'+11k'^2))}{3(1+k')^2}\right)^{0.5} \quad (3.6)$$

and for an empty tube or an unretained peak, $k'= 0$, thus

$$\sigma^2_{c(v)} = 0.577\pi^2 r_c^5 l_c \quad (3.7)$$

Interface Conduits

It should be pointed out that equation (3.6) and (3.7) apply to a capillary column *operated at its optimum velocity* [equation (3.3)], and is only applicable to a connecting tube if its *radius is the same as that of the column.*

If the connecting tube has a different radius (r_t) to that of the column and a different length (l_t) the equations become a little more complex.

The flow rate, (Q_o) leaving the column will be given by,

$$Q_{o(opt)} = \pi r_c^2 u_{o(opt.)} = \frac{2\pi r_c^2 D_{m(o)}}{r_c} \left(\frac{12(1+k')^2}{(1+6k'+11k'^2)} \right)^{0.5}$$

$$= 2\pi r_c D_{m(o)} \left(\frac{12(1+k')^2}{(1+6k'+11k'^2)} \right)^{0.5}$$

Thus the velocity of the mobile phase (u_t) flowing through a connecting tube of radius (r_t) will be given by

$$u_t = \frac{Q_o}{\pi r_t^2} = \frac{2 r_c D_{m(o)}}{r_t^2} \left(\frac{12(1+k')^2}{(1+6k'+11k'^2)} \right)^{0.5}$$

Consequently, from equation (3.2) the variance per unit length, or the height of the theoretical plate (H_t), of the connecting tube will be

$$H_t = \frac{2 D_{m(o)}}{\dfrac{2 r_c D_{m(o)}}{r_t^2} \left(\dfrac{12(1+k')^2}{(1+6k'+11k'^2)} \right)^{0.5}} + \frac{(1+6k'+11k'^2) r_t^2}{24(1+k')^2 D_{m(o)}} \frac{2 r_c D_{m(o)}}{r_t^2} \left(\frac{12(1+k')^2}{(1+6k'+11k'^2)} \right)^{0.5}$$

and if the connecting tube is empty (no stationary phase coating),

$$H_t = \frac{r_t^2}{r_c}\left(\frac{1+6k'+11k'^2}{12(1+k')^2}\right)^{0.5} + r_c\left(\frac{1+6k'+11k'^2}{12(1+k')^2}\right)^{0.5}$$

$$H_t = \left(\frac{r_t^2 + r_c^2}{r_c}\right)\left(\frac{1+6k'+11k'^2}{12(1+k')^2}\right)^{0.5} \quad (3.8)$$

Now, again from the Plate Theory [3], the peak variance ($\sigma_{t(v)}^2$) in volume units of a connecting tube of length (l_t) is given by

$$\sigma_{t(v)}^2 = \frac{(\text{Tube Volume})^2}{n_t} = \frac{(\pi r_t^2 l_t)^2}{n_t} = \frac{\pi^2 r_t^4 l_t^2}{n_t} \quad (3.9)$$

where (n_t) is the number of theoretical plates in the connecting tube. Now, again by definition, $\left(\frac{l_t}{n_t}\right) = H_t$. Thus, replacing $\left(\frac{l_t}{n_t}\right)$ by (H_t) from equation (3.9), and substituting for (H_t) from equation (3.8),

$$\sigma_{t(v)}^2 = \pi^2 r_t^4 l_t \left(\frac{r_t^2 + r_c^2}{r_c}\right)\left(\frac{1+6k'+11k'^2}{12(1+k')^2}\right)^{0.5} \quad (3.10)$$

and for an empty *column* or a solute eluted at the dead volume, i.e. k'= 0, thus

$$\sigma_{t(v)}^2 = 0.288\,\pi^2 r_t^4 l_t \left(\frac{r_t^2 + r_c^2}{r_c}\right) \quad (3.11)$$

Interface Conduits

Now having derived equations that describe the relative contributions from the column and the connecting tube, the magnitude of these effects can be examined. The data for the three most commonly used capillary columns in GC are summarized in Table 3.1, where ($u_{o(opt.)}$) and ($H_{min.}$) were calculated from equations (3.3) and (3.4).

Table 3.1 The Chromatographic Properties of Three GC Capillary Columns Having the Most Popular Dimensions.

Column Parameter	Column 1	Column 2	Column 3
l	30 m	50 m	30 m
d=2r	0.025 cm	0.025 cm	0.053 cm
V_0	1.47 ml	2.45 ml	6.62 ml
$V_{r(k'=1)}$	2.94 ml	4.90 ml	13.24 ml
$V_{r(k'=2)}$	4.41 ml	7.35 ml	19.86 ml
$u_{(0)(opt.)}$ k'=0	55.4 cm/sec	55.4 cm/sec	26.14 cm/sec
$u_{(0)(opt.)}$ k'=1	26.13 cm/sec	25.12 cm/sec	12.32 cm/sec
$u_{(0)(opt.)}$ k'=2	22.02 cm/s	22.02 cm/sec	10.38 cm/sec
H_{min} k'=0	0.0072 cm	0.0072 cm	0.0153 cm
H_{min} k'=1	0.0153 cm	0.0153 cm	0.0325 cm
H_{min} k'=2	0.0182 cm	0.0182 cm	0.0385 cm
n k'=0	416,700	694,400	196,100
n k'=1	196,100	326,800	92,300
n k'=2	164,800	274,700	77,900
σ_v^2 k'=0	5.22×10^{-6} ml^2	8.70×10^{-6} ml^2	2.23×10^{-4} ml^2
σ_v^2 k'=1	1.11×10^{-5} ml^2	1.84×10^{-5} ml^2	4.74×10^{-4} ml^2
σ_v^2 k'=2	1.31×10^{-4}	2.19×10^{-4} ml^2	5.62×10^{-4} ml^2
σ_v k'=0	2.28 µl	2.95 µl	14.9 µl
σ_v k'=1	3.33 µl	4.29 µl	21.8 µl
σ_v k'=2	3.62 µl	4.68 µl	23.7 µl

Table 3.1 provides some important information with respect to the general properties of capillary columns and in particular *a propos* to their use in tandem with other instruments. It should be noted, that the variances and widths of the eluted peaks increase gradually with elution volume ((k') value), but tend to an approximately constant value when k'>5. This

indicates that the early peaks are the most narrow, and any dispersion that takes place in the connecting tube will have the maximum deleterious effect on peaks eluted at k'< 1. It is also seen that the increase in peak width with (k') has become fairly small, even when k'=2. It follows, that if the chromatographic conditions can be adjusted to ensure that the peaks of interest are all eluted at a (k') of two or more, then the detrimental effect of extra column dispersion will be minimized. It is also seen that the optimum linear velocity has an unique value for each (k') value of the solute. It follows that the choice of flow rate will be optimum for only those solutes that are eluted at and around a specific (k') value. It is clear that the optimum velocity should be chosen at the (k') value of the closest eluted solutes where separation is the most difficult.

According to Klinkenberg [4], the variance due to extra column dispersion, ($\sigma^2_{t(v)}$) should be restrained to be no greater than 10% of the peak variance ($\sigma^2_{c(v)}$) to ensure its effect on column resolution is not serious. That is,

$$\sigma^2 = 1.1\sigma^2_{c(v)} = \sigma^2_{c(v)} + \sigma^2_{t(v)} \quad \text{or} \quad \sigma^2_{t(v)} = 0.1\sigma^2_{c(v)}$$

$$\text{i.e.} \quad \sigma_{t(v)} = 0.333\,\sigma_{c(v)}$$

It follows that the dispersion from the connecting tube must not be more than one third of the dispersion produced by the column. The dispersion from tubes 0.025 cm I.D. (*ca* .010 I.D.), and 0.0125 cm I.D. (*ca* 0.005 I.D.), when used in conjunction with GC columns 30 m long and 0.025 cm I.D., and 0.053 cm I.D., was calculated for the dead volume peak, *i.e.* the worst case scenario. The results are shown in Figure 3.2.

It is seen that for the GC columns the contribution of the connecting tube to the column dispersion is negligible, and far below the advised limit of one third the column dispersion. It is also seen that the small diameter tubing need not be used and at least 50 cm of the 0.025 cm I.D. tubing can be tolerated, without the inherent dispersion having any serious effect on the column resolution.

Interface Conduits

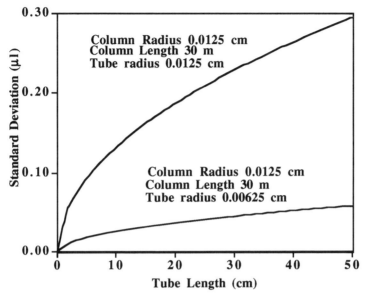

Figure 3.2 Graph of Standard Deviation of Dispersion from Connecting Tubes against Tube Length when Used with Different Columns

However, the same freedom of choice is not available when using LC columns, as the plate height attainable from an LC packed column is several orders of magnitude less than that available from GC capillary columns.

Liquid Chromatography Columns and Connecting Tubes

The equation describing the variance per unit length, or the height of the theoretical plate, (H_c), in a packed LC column is that of Van Deemter [5]:

$$H_c = 2\lambda d_p + \frac{2\gamma D_m}{u_c} + \frac{1+6k'+11k'^2}{24(1+k')^2}\frac{d_p^2}{D_m}u_c + \frac{8k'}{\pi^2(1+k')^2}\frac{d_f^2}{D_s}u_c \quad (3.12)$$

where (λ) is a dimensionless constant *ca.* 0.5,
(γ) is a dimensionless constant *ca.* 0.6,
and (dp) is the average particle diameter of the packing.
The other symbols have the meaning previously ascribed to them.

By differentiating equation (3.12) and equating to zero, expressions for the optimum velocity ($u_{c(opt)}$), and the minimum value for the height of the theoretical plate ($H_{c(min)}$), can be obtained.

$$u_{c(opt)t} = \left[\frac{2\gamma D_m}{\left(\frac{1 + 6k' + 11k'^2}{24(1+k')^2} \frac{d_p^2}{D_m} + \frac{8k'}{\pi^2(1+k')^2} \frac{d_f^2}{D_s} \right)} \right]^{0.5} \quad (3.13)$$

$$H_{c(min)} = 2\lambda d_p + 2\left(2\gamma D_m \left(\frac{1+6k'+11k'^2}{24(1+k')^2} \frac{d_p^2}{D_m} + \frac{8k'}{\pi^2(1+k')^2} \frac{d_f^2}{D_s} \right) \right)^{0.5} \quad (3.14)$$

Now in liquid chromatography, $D_m \approx D_s$, and $d_p \gg d_f$, thus

$$u_{c(opt)} = \frac{2D_m}{d_p} \left[\frac{12\gamma\left((1+k')^2\right)}{1 + 6k' + 11k'^2} \right]^{0.5} \quad (3.15)$$

and

$$H_{c(min)} = 2d_p \left(\lambda + \left(\gamma \left(\frac{1+6k'+11k'^2}{12(1+k')^2} \right) \right)^{0.5} \right) \quad (3.16)$$

Employing the equations that describe the relative contributions from the column (3.5), (3.15), and (3.16), the data for the three most commonly used packed LC columns can be calculated and the results are summarized in Table 3.2.

Table 3.2 The Chromatographic Properties of Three LC Packed Columns Having the Most Popular Dimensions

(The standard columns have diameters of 4.6 mm and the microbore column 1.0 mm. The mobile phase fraction of column volume (ε) is taken as 0.6 and $D_m = D_s$ as 1.5×10^{-5} cm^2 sec^{-1}))

Column Parameter	Column 1	Column 2	Column 3
l	3 cm	10 cm	25 cm
dp	3 µm	5 µm	5 µm
V_0	0.299 ml	1.0 ml	0.118 ml
$V_{r(k'=1)}$	0.598 ml	2.0 ml	0.236 ml
$V_{r(k'=2)}$	0.897 ml	3.0 ml	0.353 ml
u_{opt} k'=0	0.268 cm/sec	0.161 cm/sec	0.161 cm/sec
u_{opt} k'=1	0.126 cm/sec	0.0759 cm/sec	0.0759
u_{opt} k'=2	0.107 cm/sec	0.0640 cm/sec	0.0640
H_{min} k'=0	0.000434 cm	0.000724 cm	0.000724 cm
H_{min} k'=1	0.000585 cm	0.000975 cm	0.000975 cm
H_{min} k'=2	0.000635	0.001058	0.001063
n k'=0	6912	13810	34530
n k'=1	5128	10260	25641
n k'=2	4709	9450	23518
σ_v^2 k'=0	1.29×10^{-5} ml^2	7.25×10^{-5} ml^2	4.03×10^{-7} ml^2
σ_v^2 k'=1	6.97×10^{-5} ml^2	3.90×10^{-4} ml^2	2.17×10^{-6} ml^2
σ_v^2 k'=2	7.59×10^{-5} ml^2	9.52×10^{-4} ml^2	5.29×10^{-6} ml^2
σ_v k=0	3.59 µl	8.5 µl	0.64 µl
σ_v k=1	8.35 µl	19.7 µl	1.47 µl
σ_v k=2	8.71 µl	30.9 µl	2.30 µl

It is seen from table 3.2, that the GC capillary column and the LC packed columns behave somewhat similarly. The efficiencies decrease and consequently, the peaks get broader using larger particles or larger diameter capillaries. For both types of column, the efficiency falls with increasing (k') values over the range examined. The peak widths from the standard columns are not greatly different from the GC columns, although the GC columns are two orders of magnitude longer than the LC columns.

It is also clear that the microbore column gives peaks of much smaller volume, and thus extra column dispersion would be far more damaging to any separation from such columns. It is now interesting to examine the contribution to the peak dispersion made by connecting tubes.

The flow rate, $(Q_{o(opt)})$ leaving the column will be given by,

$$Q_{o(opt)} = 0.6\pi r_c^2 u_{o(opt.)} = \frac{0.6\pi r_c^2 2 D_m}{dp} \left[\frac{12\gamma\left((1+k')^2\right)}{1+6k'+11k'^2} \right]^{-0.5}$$

The factor, 0.6, takes into account that 40% of the cross-sectional area of a packed column is occupied by the solid packing and not available for flow of mobile phase. Thus the velocity of the mobile phase (u_t) flowing through a connecting tube of radius (r_t) will be given by

$$u_t = \frac{Q_{0(opt)}}{\pi r_t^2} = \frac{1.2 r_c^2 D_m}{r_t^2 dp} \left[\frac{12\gamma\left((1+k')^2\right)}{1+6k'+11k'^2} \right]^{-0.5}$$

Consequently, from equation (3.2) the variance per unit length, or height of the theoretical plate (H_t) of the uncoated connecting tube will be.

$$H_t = \frac{2 D_{m(o)}}{\dfrac{1.2 r_c^2 D_m}{r_t^2 dp}\left[\dfrac{12\gamma\left((1+k')^2\right)}{1+6k'+11k'^2}\right]^{-0.5}}$$

$$+ \frac{r_t^2}{24 D_{m(o)}} \frac{1.2 r_c^2 D_m}{r_t^2 dp} \left[\frac{12\gamma\left((1+k')^2\right)}{1+6k'+11k'^2}\right]^{-0.5}$$

$$H_t = \frac{1.66 r_t^2 dp}{r_c^2}\left(\frac{1+6k'+11k'^2}{12\gamma\left((1+k')^2\right)}\right)^{0.5} + \frac{0.05 r_c^2}{dp}\left[\frac{12\gamma\left((1+k')^2\right)}{1+6k'+11k'^2}\right]^{0.5} \quad (3.1)$$

Interface Conduits

Now, it is clear from the relative values of (r_c), (r_t), and (dp), that

$$\frac{0.05 r_c^2}{dp}\left[\frac{12\,\gamma\left((1+k')^2\right)}{1+6k'+11k'^2}\right]^{0.5} \gg \frac{1.66 r_t^2 dp}{r_c^2}\left(\frac{1+6k'+11k'^2}{12\,\gamma\left((1+k')^2\right)}\right)^{0.5}$$

Thus $\quad H_t = \dfrac{0.05 r_c^2}{dp}\left[\dfrac{12\gamma\left((1+k')^2\right)}{1+6k'+11k'^2}\right]^{0.5}$ \hfill (3.118)

Now, again from the Plate Theory [3], the peak variance ($\sigma_{t(v)}^2$) in volume units of a connecting tube of length (l_t) is given by

$$\sigma_{t(v)}^2 = \frac{(\text{Tube Volume})^2}{n_t} = \frac{\left(\pi r_t^2 l_t\right)^2}{n_t} = \frac{\pi^2 r_t^4 l_t^2}{n_t}$$

where (n_t) is the number of theoretical plates in the connecting tube. Now, by definition, $\left(\dfrac{l_t}{n_t}\right) = H_t$. Thus, replacing $\left(\dfrac{l_t}{n_t}\right)$ by (H_t) from equation (3.18),

$$\sigma_{t(v)}^2 = \frac{\pi^2 r_t^4 l_t 0.05 r_c^2}{dp}\left[\frac{12\,\gamma\left((1+k')^2\right)}{1+6k'+11k'^2}\right]^{0.5} \tag{3.19}$$

Employing equation (3.19), the standard deviation of the dispersion from connecting tubes, 0.005 in. I.D., can be calculated when used with columns operating at their optimum linear mobile phase velocity. Results for tubes of different length, connected to the small standard LC column 3 cm long, packed with particles 3 μm in diameter, are shown in Figure 3.3. The curves relate the standard deviation of the dispersion contributed by the tube to the tube length. The horizontal lines represent the maximum dispersion that can be tolerated by the column and is equivalent to one third of the column dispersion given in Table 3.2.

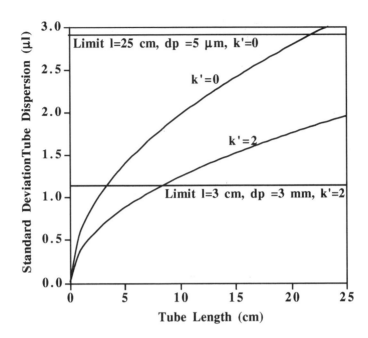

Figure 3.3 Graph of Standard Deviation of Dispersion from Connecting Tubes against Length (Standard Short Column)

It is seen that for solutes eluted at the dead volume, a tube length of about 2 cm is the maximum that can be tolerated. For solutes eluted at k'= 2, an extrapolated tube length of about 55 cm would be acceptable, but this would require that the chromatographic system was adjusted so that all the solutes of significance were eluted at (k') values greater than 2. Furthermore, the optimum column flow rate was also set for a solute eluted at k'=2. In fact, all eluted peaks should be given the same priority, and so it is clear that connecting tubes should be kept at an absolute minimum length, preferably less than 3 cm, and certainly less than 5 cm. Furthermore, the curves shown in Figure 3.3 assume that the column is operated at an optimum velocity appropriate for the solutes of interest. In fact, the optimum velocity is rarely used, because optimum columns are not usually designed for the specific analyses. It follows that the flow rate employed will be considerably in excess of the optimum flow rate to avoid extended elution times and long analyses. The effect of tube length on dispersion at flow rates more in line with those used in practice will be considered later.

The curves for tubes used with microbore columns operated at the their optimum flow rate are shown in Figure 3.4. It is seen that because the optimum flow rate for microbore columns is much smaller than that for standard columns, the tube dispersion is much less. In fact, even for the dead volume peak, a tube length of 35 cm can be employed without seriously denigrating the column efficiency. Nevertheless, the effect of the tube dispersion is only small for the microbore column when operated at the optimum flow rate. Unfortunately, the optimum flow rate is rarely used in practice, as the analysis time is too lengthy and at higher flow rates, the effect of tube dispersion becomes far more serious.

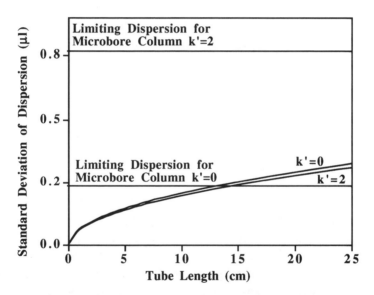

Figure 3.4 Graph of Standard Deviation of Dispersion from Connecting Tubes against Length (Microbore Column)

The variance per unit length of an open tube at velocities significantly higher than the optimum is given from equation (3.2) by

$$H = \frac{r_t^2}{24 D_m} u \qquad (3.20)$$

Now, $u = \dfrac{Q}{\pi r_t^2}$. Substituting for (u) in (3.20), $H = \dfrac{Q}{24\pi D_m}$

Now, $\sigma_{t(v)}^2 = \dfrac{(\text{Tube Volume})^2}{n} = \dfrac{\pi^2 r_t^4 l_t^2}{n_t}$

and $H = \dfrac{1}{n}$, thus, $\sigma_{t(v)}^2 = \dfrac{\pi^2 r_t^4 l_t^2}{n_t} = \pi^2 r_t^4 l_t H = \dfrac{\pi r_t^4 l_t Q}{24 D_m}$ (3.21)

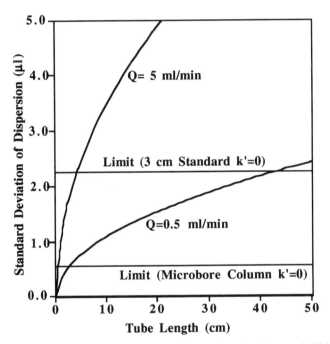

Figure 3.5 Graph of Standard Deviation of Dispersion from Connecting Tubes against Length for Different Flow Rates

Employing equation (3.21) the standard deviation of the dispersion from a tube 0.005 in I.D. (radius 0.00635 cm) was calculated for different tube lengths and flow rates. The results are shown in Figure 3.5, together with the limiting dispersion values for the microbore column and the short 3 cm standard column. Microbore columns are usually operated at about 0.5

Interface Conduits

ml per minute and standard columns at about 5 ml/min. It is seen that the acceptable length of tubing is considerably less than that for the same columns when operated at their optimum flow rates.

Equation (3.21) clearly indicates the procedure that must be followed to reduce the dispersion that arises from any open tube associated with a tandem interface. However, the magnitude of the flow rate (u_o) is determined by the chromatographic system and thus, is not a variable that can be used to control tube dispersion. In a similar manner the diffusivity of the solute, (D_m), is determined by the nature of the sample and the nature of the mobile phase that has been specifically selected for the particular separation. Consequently, (D_m), is also not a variable available for dispersion control. The remaining and in fact the major parameters that are free for controlling dispersion are the tube radius and, of course, the tube length. It is seen that the dispersion increases as the fourth power of the tube radius and thus, a reduction in the tube radius by a factor of two will reduce the dispersion by a factor of sixteen.

Unfortunately, there is a limit to the process of reducing (r) as, from Poiseuille's equation, the pressure drop (ΔP) across the tube is given by,

$$\Delta P = \frac{8\eta l u}{r^2}$$

where (η) is the viscosity of the mobile phase, and as $Q = \pi r^2 u$,

$$\Delta P = \frac{8\eta l Q}{\pi r^4} \quad (3.22)$$

It is seen from equation (3.22), that the pressure drop across an open tube increases inversely as the fourth power of the tube radius. Thus, as it is inadvisable to dissipate significant amounts of the available column inlet pressure across the interface, there will be a lower limit to which (r) can be reduced in order to minimize dispersion.

It is also interesting to note that changing the length of the tube has the same effect on both dispersion and pressure drop. Reducing (l) will linearly reduce variance of the dispersion and at the same time proportionally reduce the pressure drop across the tube. It follows that adjusting the tube length is by far the best method of controlling dispersion. Furthermore, by making (l) as small as possible, both the dispersion and the pressure drop can be minimized. However, depending on the overall geometry of the tandem system, the extent to which any tube component of the interface can be reduced in length may also be limited. In practice, the internal diameter of any tubular conduit contained in an interface, should not be made less than 0.012 cm (0.005 in. I.D.). This is not merely to limit the pressure drop that will occur across it, but for a more mundane, but very important reason. If tubes of less diameter are employed, they *will easily become blocked.*

An alternative approach to restricting the dispersion in tubular interface conduits is to utilize low dispersion tubing. Low dispersion tubing has only recently been used in chromatographic systems and is not, at this time, readily available commercially. Nevertheless, such conduits could be very useful in chromatography/spectrometer interfaces and consequently they will be briefly discussed.

Low Dispersion Connecting Tubes

As already stated, the dispersion in simple open tubes results from the parabolic velocity profile of the fluid flow. In order to reduce this dispersion, the velocity profile of the fluid must be disrupted to allow rapid radial mixing. The parabolic velocity profile can be disturbed, and secondary flow introduced into the tube, by deforming its regular geometry. The dispersion in geometrically deformed tubes (squeezed, twisted and coiled) has been extensively studied by Halasz [6-8], and the effect of radial convection (secondary flow) on the dispersion introduced in tightly coiled tubes has been examined both theoretically and experimentally by Tijssen [9]. The effect of secondary flow produced by employing serpentine shaped tubes has also been studied by Katz and Scott [10]. It was found that the dispersion characteristics of serpentine tubing

Interface Conduits

were far superior to those of coiled tubes. Furthermore, the dispersion that takes place in serpentine tubing is practically independent of the mobile phase linear velocity and consequently, such tubes can be used over a wide range of flow rates. The authors first examined the effect of secondary flow on band dispersion that took place in tubes of different radius coiled to different diameters. The theory of Tijssen [9] was successfully employed to qualitatively describe the relationship between the variance per unit length, (H), and the mobile phase velocity.

For the sake of simplicity, the equations that Tijssen derived for radial dispersion in coiled tubes are given in terms of conventional chromatographic terminology. At relatively low linear velocities (but not low relative to the optimum velocity for the tube) Tijssen derived the equation,

$$H = \frac{j r^2 u}{D_m} \quad (3.23)$$

where (j) is a constant over a given velocity range, and the other symbols have the meaning previously ascribed to them.

It is seen that the band variance is directly proportional to the square of the tube radius and the relationship is very similar to that derived by Golay [1] for a straight tube.

At high linear velocities, Tijssen deduced that,

$$H = \frac{b D_m^{0.14}}{\psi u} \quad (3.24)$$

where (b), is a constant for a given mobile phase
and (ψ), is the ratio of the tube radius to the coil radius, and was given the term the *coil aspect ratio*.

Consequently,
$$\Psi = \frac{r_{tube}}{r_{coil}}$$

It can be seen from equation (3.24) that, at the higher linear mobile phase velocities, the value of (H) depends on (D_m) taken to the power of 0.14 and inversely dependent on the coil aspect ratio and the linear velocity. According to equations (3.23) and (3.24), at low velocities the band dispersion *increases* with (u), whereas at high velocities the band dispersion *decreases* with (u). It follows that a plot of (H) against (u) should exhibit a maximum at a certain value of (H). By combining equations (3.23) and (3.24), an equation can be obtained that predicts the value of (u) at which (H) is a maximum, and is given by

$$u = \frac{c}{r\sqrt{\psi}}$$

where (c) is a constant for a given solute and given mobile phase.

The above equations were employed to investigate the effect of tube radius and coil aspect ratio on the onset of radial mixing in coiled tubes. The properties of the four different coils are shown in Table 3.3.

Table 3 Physical Dimensions of Coiled Tubes Examined

Tube	r (cm.)	L(cm)	r(coil cm)	(y)	L(coil)(cm.)
1	0.019	365.8	0.5	0.038	18.5
2	0.020	365.0	0.085	0.235	65.8
3	0.0127	998.0	0.0765	0.166	128.0
4	0.0127	337.5	0.0498	0.0255	73.7

The curves relating (H) and (u) are shown in Figure 3.5. It can be seen that at low linear velocities, where radial mixing is still poor, the values of (H) increases as (u) increases. Furthermore, the dispersion in coiled tubes (1) and (2) of larger radii is greater than that in tubes (3) and (4) which had smaller radii. At high linear velocities, where radial mixing commences, the values of (H) decrease as (u) increases. As the range of linear velocities is approached where radial mixing dominates, the solute dispersion becomes independent of the linear velocity (u). It is also seen

Interface Conduits

the maximum value of (H), for any particular coil, occurs at different values of (u) depending on the combined values of (r) and (ψ). In general, it would seem that a high coil aspect ratio reduces both the maximum value of (H) and the value of (u) at which it occurs.

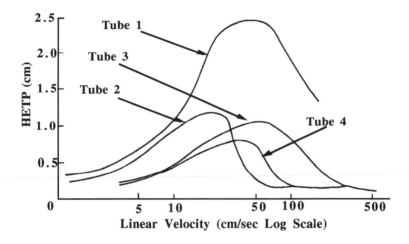

Figure 3.6 Curves Relating Plate Height against Mobile Phase Velocity for Different Coiled Tubes

It is interesting to note that although the straight tube theory of Golay is not applied to coiled tubes, his equation can be employed to qualitatively explain the shape of the curves given in Figure 3.5. At low values of the mobile phase velocity the effect of longitudinal diffusion dominates, but as the velocity tends to approach the optimum, the resistance to mass transfer term begins to increase and the value of (H) also rapidly increases. However, at higher velocities, the magnitude of the diffusivity of the solute in the solvent dramatically increases as a result of induced radial flow, eventually reducing the resistance to mass transfer factor to virtually zero. This results in a corresponding dramatic reduction in the value of (H). Finally, at very high velocities, the greatly reduced longitudinal diffusion effect again dominates. At this point, the value of (H) is very small and, in fact, decreases even further as the mobile phase velocity is increased.

Serpentine Tubes

The low dispersion serpentine tube, developed by Katz *et al* [10], was an alternative approach to the coiled tube, but was designed with the same intent, which was to induce and increase secondary flow across the tube. As opposed to the process of inducing radial flow in coiled tube, in a serpentine tube, the flow actually *reverses* in direction at each serpentine bend. This concept was evoked by the work of Halasz and his coworkers who introduced radial flow by crushing and crimping the tube. The purpose of the serpentine tube was to achieve the same goal, but by a simpler and more reproducible procedure. A diagram of a serpentine tube is shown in Figure 3. 7.

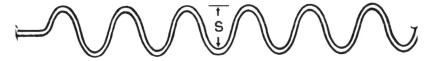

Tube Dimensions : (r), (internal), 0.0127 cm (0.010 in. I.D.), (r), (external), 0.025 cm (0.020 in. O.D.), (L), length (linear) 42.5 cm., (l), length (coil) 38.5 cm. and (S), (serpentine amplitude) 0.05 cm.

Figure 3.7 The Low Dispersion Serpentine Tube

A graph relating the variance per unit length (H),to flow rate, for a serpentine tube having the dimensions given in Figure 3.7, is shown in Figure 3. 7. The flow rate is employed as the independent variable, an alternative to the more usual linear velocity, as the flow rate is defined by the column with which the low dispersion tubing is to be used. It is seen that a similar curve is obtained for the serpentine tube, as that for the coiled tube, but the maximum value of (H) is reached at a much lower flow rate than that with the coiled tube. Furthermore, the variance remains more or less constant over a wide range of flow rates that encompass those usually employed in normal LC separations. It is now interesting to compare the dispersion characteristics of a straight tube with that of a serpentine tube. The variances of a straight tube and serpentine tube are plotted against flow rate in Figure 3. 8. The values of the variance for the straight tube were calculated from the Golay equation.

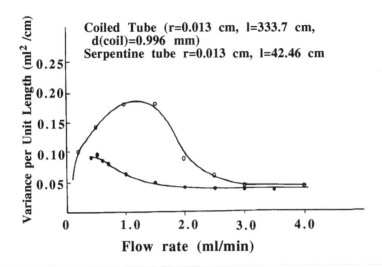

Figure 3.8 Graphs of Peak Variance against Flow Rate for Coiled and Serpentine Tubes

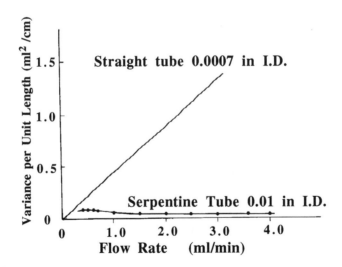

Figure 3.9 Graphs of Peak Variance against Flow Rate for a Straight and Serpentine Tube

It is clearly seen that the dispersion resulting from the serpentine tube is drastically reduced in comparison with the straight tube. According to the graph, the numerical value of the peak variance per unit length for the serpentine tube (0.010 in. I.D.) is 0.05 μL^2/cm and consequently, a tube 10 cm long would contribute a variance of 0.5 μl^2. In contrast, the dispersion of a straight tube of the same internal diameter and *only one centimeter in linear length* would be 5.5 μl^2, which is an order of magnitude larger.

It follows that in any conduit carrying eluent from a chromatographic system, great care must be taken to maintain the integrity of the peak dispersion and thus conserve the resolution. If straight open tubes must be employed, then they should be as short as possible and have the minimum radius that is practical. Unfortunately, it is often difficult in practice to have the column or detector exit close to the sample inlet of the mass spectrometer, and connecting tubes 30 cm or more long are often necessary. If long lengths of conduit are needed, then low dispersion tubing should be utilized, or at least the tubing should be tightly coiled to introduce radial mixing and reduce dispersion as much as possible.

It is also important to remember that all types of metal tubes can exhibit catalytic activity and cause pyrolysis or the molecular rearrangement of any labile compounds that might be eluted from the column. Consequently, fused silica or glass conduits should be used wherever possible. Unfortunately, it would appear that, to date, no low dispersion conduits have been fabricated from glass or fused quartz.

Synopsis

The integrity of the separation that is achieved in the column must be maintained during the passage of the eluent to the tandem instrument. The conduits used to connect the two instruments in tandem systems are usually simple tubes. The velocity profile in open tubes is parabolic in shape, the velocity at the center being a maximum, and that at the wall almost zero. As a result of the differential velocity across the tube, serious band dispersion can occur. The dispersion increases linearly with the tube length and as the fourth power of the tube radius. It follows that the tube should be as short as possible and have the minimum radius. In practice the minimum radius of the tubular conduit is constrained to about 0.0025 in. (0.0635 cm) to avoid the tube becoming blocked. The dispersion in the tube can be significantly reduced by bending or crimping the tube and low dispersion conduits can be constructed from coiled or serpentine shaped tubing. Wherever possible, connecting conduits should be made from low dispersion tubing, or the conduits must be kept short and have the minimum radius.

References

1. M. J. E. Golay, *Gas Chromatography. 1958*, (ed. D. H. Desty) Butterworths, London (1958)36.
2. K. Ogan and R. P. W. Scott, *J. High Res. Chromatogr.*, **7**(1984)382.
3. R. P. W. Scott, *Liquid Chromatography Column Theory*, John Wiley, Chichester (1992)130.
4. A. Klinkenberg, *Gas Chromatography 1960*, R.P.W.Scott (Ed.), Butterworths, London(1960)194.
5. J.J.Van Deemter,F.J.Zuiderwg and A.Klinkenberg,*Chem.Eng Sci.*,**5**(1956)271
6. I. Halasz, H. O. Gerlach, K. F. Gutlich and P. Walking, *U.S. Patent*, 3,820,660, (1974).
7. K. Hofmann and I. Halasz, *J. Chromatogr.*,**173**(1979)211.
8. K. Hofmann and I. Halasz, *J. Chromatogr.*, **199**(1980)3.
9. R. Tijssen, *Separ. Sci. Technol.*, **13**(1978)681.
10. E. D. Katz and R. P. W. Scott, *J. Chromatogr.*, **268**(1983)169.

Part 2

Gas Chromatography Tandem Systems

CHAPTER 4

GAS CHROMATOGRAPHY IR SPECTROSCOPY (GC/IR) TANDEM SYSTEMS

The gas chromatography/infrared spectrometer tandem system was the second type of tandem instrument to be developed, the first being the GC/MS combination. The GC/MS instrument, although very useful, was expensive, complex, and the early models were somewhat difficult to operate. In contrast, the IR spectrometer was less expensive, easy to operate and the IR spectrum was usually quite adequate to confirm the identity of an eluted component. In addition, the resolution obtainable from the mass spectrometers available in the early days of gas chromatography was inadequate for the elucidation of a completely unknown structure, so the need for supporting evidence from an IR spectrum was usually necessary.

Initially IR spectra were obtained off-line, by condensing the eluted solute in a cooled trap. The solute was then either dissolved in a suitable IR solvent, made into a 'mull' with 'Nujol' (a very pure form of a high boiling normal alkanes), or pressed into an alkali halide pellet and the spectrum obtained using standard techniques. The efficient collection of the solute by condensation, however, could be difficult as, due to the very low concentrations at which each solute was eluted, the partial pressure of the condensed material was often similar to its partial pressure as it left the GC column. The most efficient method of collecting the solute was to use argon as the carrier gas, and condense the argon and the solute simultaneously in a tube immersed in a liquid nitrogen bath. Subsequently, the argon was slowly evaporated leaving the solute as a residue ready for examination. Another procedure for improving the trapping efficiency was to retain the solute on an adsorbent contained in a short length of packed tube. After collection, the solute was regenerated by passing a stream of gas through the heated tube, and then it was either examined as a vapor sample, or it was bubbled through a suitable IR solvent and the spectrum obtained in the usual manner. An adsorbent that was frequently used for this purpose was a GC packing containing a heavy loading of a high boiling stationary phase, such as 'Apiezon Grease'. Apiezon Grease was a proprietary vacuum grease which was very popular in the early days of gas chromatography as a high-temperature stationary phase.

The Early GC/IR/MS Triplet System

The first automated on-line GC/IR system was described by Scott *et al.* [1] and involved the adsorption of each solute in a cold packed tube, followed by its thermal desorption into an infrared vapor cell. In fact, the instrument also provided mass spectra, the sample for the mass spectrometer being taken from the IR cell, after the IR spectrum had been obtained.

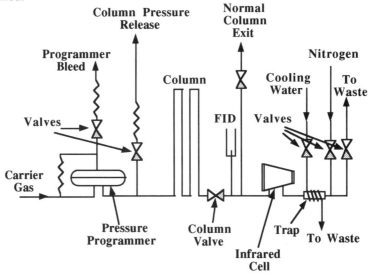

Figure 4.1 Diagram of an Automatic GC/IR Tandem System

Consequently, besides being the first tandem system to be described, the device was also the first triplet instrument to be reported (GC/IR/MS) and for this reason will be described in some detail. The layout of the pneumatic system of the Tandem Instrument is shown in Figure 4.1.

The procedure involves sensing the peak as it starts to elute and then diverting it through the IR cell to the packed trap. When the elution of the peak is complete, the carrier gas is arrested, the trap heated and the solute regenerated into the IR cell by a stream of nitrogen. The IR vapor spectrum is then obtained in the usual manner. After the IR spectrum is completed, a sample of the solute vapor is drawn from the IR cell into the mass spectrometer source, and the mass spectrum obtained. When the mass spectrum had been run, the carrier gas is turned on again and the chromatographic development continued. On the arrival of the next peak the procedure is repeated. Due to the nature of the stop/start technique, the process was given the term *interrupted elution development*.

In order to ensure that the quality of the separation was not impaired by the stop/start procedure, a carefully designed pneumatic system was used. When the elution of the peak was complete and the solute resided in the packed trap, a solenoid valve at the end of the column closed, and the pressure in the column was released by actuating a valve at the *front* of the column. This caused some slight back development of the solutes remaining in the column which was accompanied by some peak sharpening. The trap was then heated and a separate supply of nitrogen, controlled by another solenoid valve, was used to regenerate the solute into the heated IR cell. When the spectra had been obtained, the trap was cooled by a stream of cold water controlled by another solenoid valve and the column terminal valve opened. In order to prevent a sudden increase in the column pressure as the carrier gas was turned on again, a flow programmer was used that brought the column up to pressure over a period of about 10 seconds. A diagram of the IR cell is shown in Figure 4.2.

In the 1960s, there was a limited number of materials available that were transparent to infrared light and that could be used as cell windows for IR studies. The material used by Scott *et al.* was silver chloride (horn silver) and the cell was gold plated internally to reduce light absorption losses. This was a procedure introduced specifically for GC/IR systems and will be discussed in more detail later.

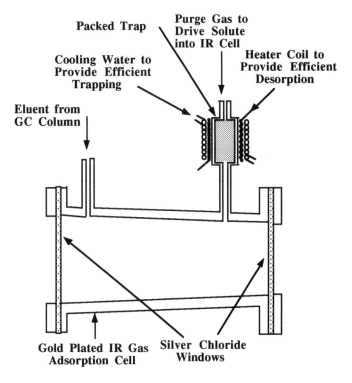

Figure 4.2 The Heated IR Flow-through Cell

The trap that was attached to the IR cell was electrically heated by appropriate heater coils so that the temperature could be maintained at 250°C. It was packed with a GC stationary phase that consisted of 15% w/w of Apiezon Grease on 60-80 BS. mesh brick dust (a high-temperature GC packing frequently used at that time). Around the outside of the trap were a series of cooling coils carrying cold water, the flow of which was also controlled by a programmed solenoid valve. The column was all glass 60 ft long, 4.0 mm I.D., and packed with 60-80 BS mesh brick dust

GC/IR Tandem Systems   137

carrying 15% w/w of Apiezon Grease, a photograph of which is shown in Figure 4.3.

Figure 4.3   The 60 ft Long Packed Glass Column

The column consisted of a series of U tubes and each U section was packed individually. After packing all the individual U sections, they were sealed together employing simple glass-blowing techniques [2]. The columns were made from thick-walled glass tubing which, if required, could tolerate pressures up to 200 psi. The column was mounted in a stainless steel rack for mechanical strength and rigidity. The chromatograph (a standard commercially available model from Phillips Chromatography Ltd.) was fitted with the normal temperature and flow programming facilities, but computer data acquisition and processing was not available at that time.

An example of the chromatogram and also some spectra obtained from the Triplet system are shown in Figure 4.4. The IR spectrometer used was the Unicam SP 200 dispersion instrument (now obsolete) and the MS10, a low resolution sector mass spectrometer (also now obsolete). It is seen that, although lacking in resolution compared with modern spectrometers, very useful spectra could be obtained and the collection was completely automatic. The device was used extensively for the separation and identification of essential oil constituents.

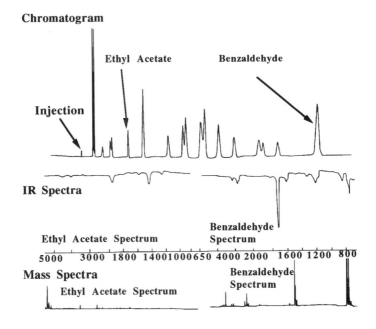

Figure 4.4 Chromatogram and Spectra Obtained from the First GC/IR/MS Triplet System

Modern GC/IR Systems

Light Pipe Interfaces

IR spectrometers are generally much less sensitive than mass spectrometers and, as sample sizes are often very limited when used in tandem with chromatographic instruments, special techniques must be employed to utilize the maximum sensitivity that is available. This is true even with the modern FTIR instruments although they can, indeed, be much more sensitive than the older dispersion type instruments. One device that is commonly used to increase the sensitivity of the IR instrument when used to examine vapor samples in a gas, (*e.g.* those eluted from a gas chromatograph) is the 'light pipe'.

The use of light pipes was introduced very early in the GC/IR development and one of the first references to the device, is that of Wilks and Brown in 1964 [3]. Light pipes are tubes of circular or rectangular cross-section with highly reflecting internal surfaces that are usually produced by gold plating. The light source is moved back, so that the focus, which coincides with the entrance of the light pipe, is transferred to the exit of the light pipe but the process is not completely efficient. Internal reflections at the walls of the light pipe increases the 'apparent path length' by about 33%. Many modern GC/IR systems employ light pipes to augment the IR signal, and an example of a contemporary GC/IR tandem instrument that utilizes light pipe amplification is that manufactured by the Perkin Elmer Corporation. A diagram of the optical system of the Perkin Elmer instrument, which is typical of many contemporary GC/IR tandem systems, is shown in Figure 4. 5.

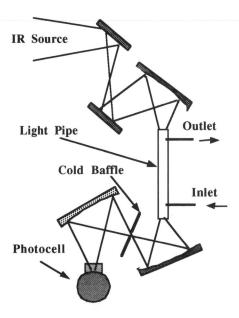

Courtesy of the Perkin Elmer Corporation

Figure 4.5 The Optical System of the Perkin Elmer GC/IR Tandem Instrument

Light from the IR source is focused on the front of the light pipe, and is subjected to continual reflections from the walls of the pipe as it passes through. The emitted light is focused by means of two curved mirrors onto the cooled IR sensor. Between the pipe exit and the sensor is situated a cold baffle, which ensures that none of the IR emitted by the hot oven that thermostats the tube falls on the sensor. This simple device reduces the noise generated by the oven and increases the signal-to-noise ratio and thus the sensitivity. Details of the light pipe are shown in Figure 4.6.

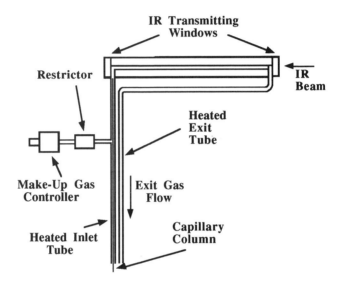

Courtesy of the Perkin Elmer Corporation

Figure 4.6 The Optical Arrangement of the Perkin Elmer Light Pipe Interface

The interface can be used with all types of GC columns including open tubular columns. The capillary column is led into the interface through a heated tube right up to the light pipe. Concentric to the column, and through the same heated tube, is fed a stream of scavenging gas that carries the solute through the IR light pipe. This maintains the integrity of the separation at the expense of some solute dilution and consequent slight loss of sensitivity. If the solute bands were not swept out by the scavenging gas, the solute peaks from the column would accumulate in the IR light pipe, and as a consequence, several solutes would be detected and measured simultaneously, and resolution would be lost. The light pipe itself is 120 mm long, 1 mm I.D and coated internally with gold. The oven surrounding the light pipe, can be operated up to 350°C, and is carefully designed to eliminate any cold spots on the tube, which might allow solute condensation. After the sample has passed though the light pipe, it can be returned to the FID in the chromatograph oven or passed to a second tandem instrument such as a mass spectrometer. It is claimed that satisfactory spectra can be obtained from a 1–5 ng of material in the light pipe which is supported by the spectrum shown in Figure 4.7. The spectrum from 10 ng of material, has more than sufficient IR absorption data to allow the identity of the solute to be confirmed by comparison with the library spectrum shown below the sample spectrum. Clearly, the system can be used, either to aid in structure elucidation, or to confirm compound identity.

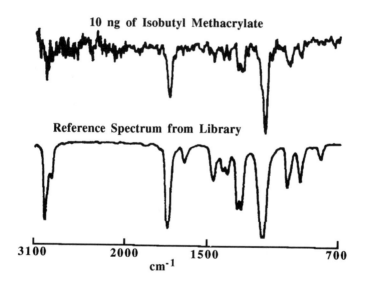

Figure 4.7 A GC/IR Spectrum of 10 ng of Isobutyl Methacrylate

Examples of the use of the device to identify some drugs of abuse are shown in Figure 4.8. The sample was separated on a GC column employing normal development techniques, and the spectra taken of the peaks as they were eluted through the light pipe. It is seen that the spectra that are obtained would be perfectly satisfactory for solute identification.

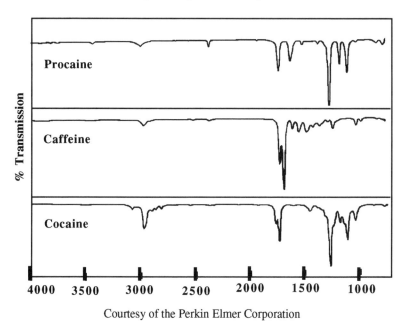

Courtesy of the Perkin Elmer Corporation

Figure 4.8 Spectra Taken from a GC/IR Analysis of Some Drugs of Abuse Employing the Perkin Elmer Light Pipe Interface.

Another typical example of the value of the instrument is for the analysis of the essential oils of *basil*. A sample of the oil was obtained by a supercritical fluid technique. Superfluid extraction is now a common method for obtaining samples of essential oils from botanical tissue. It has become popular due to it being chemically 'gentle' and does not readily cause the degradation of thermally labile materials. In the example given, the herb basil was extracted with liquid carbon dioxide at 60°C and 250 atmospheres pressure. The extract, a solution of the essential oil in liquid carbon dioxide, is decompressed through a length of silica capillary into

GC/IR Tandem Systems

an appropriate solvent. Alternatively, it can be trapped on a suitable adsorbent contained in a packed tube, and then thermally desorbed into the carrier gas of a gas chromatograph.

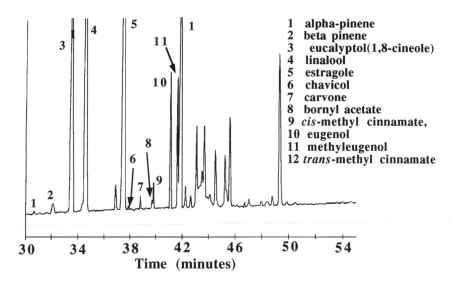

Courtesy of the Perkin Elmer Corporation

Figure 4.9 The Separation of a Sample of Basil Oil

The column that was used in this example, was a macro-bore open tubular column, 50 m long, 0.32 mm in diameter and carried a 5 μm film of a methyl silicone. The chromatogram was obtained by plotting the integrals of each adsorption curve against time. The chromatogram of oil of basil, obtained in this way, is shown in Figure 4.9. The light pipe was maintained at 250°C and the scavenger flow was set at 1 ml/min. Each point on the chromatogram resulted from the sum of two consecutive spectra. A spectrum, taken at the peak maximum of linalool (peak 4), is shown in Figure 4. 10. It is seen that a clean spectrum is obtained and the resolution from the *macro-bore capillary column* does not appear to have been denigrated. It has been claimed that spectra, with adequate resolution and signal-to-noise for solute identification, have been obtained from as little as 10 ng of material. However, the minimum sample size that can provide adequate data will depend strongly on the extinction coefficient of the solute at the critical adsorption wavelengths of the respective compounds.

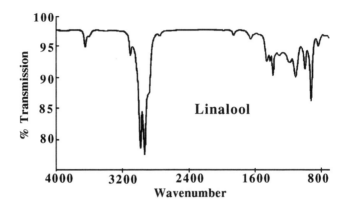

Courtesy of the Perkin Elmer Corporation

Figure 4.10 Spectrum of Linalool from a Light Pipe Interface

The use of the GC/IR tandem system for the analysis of gasoline is shown in figure 4.11. The chromatogram was obtained in the usual way by integrating the absorption curve and plotting the integral against time. The separation was carried out on a capillary column 250 μm I.D., and the low effective volume of the light pipe allowed scan rates of up to 20 scans per second to be used, so that even the very narrow peaks could be monitored without apparent distortion.

As the spectra were stored as they were scanned, they could easily be recovered and used for solute identification, or help in structure elucidation. They could be presented as shown previously as single spectra, or presented in the form of a stack plot as shown in figure 12. The stacked plot show each spectra taken throughout the elution of each peak and, as well as help confirm the identity of any given peak, can also indicate if a single elution curve contains more than one peak. It is clear that the combination of the two techniques, using a suitable interface, can provide extremely useful data for a wide variety of sample types.

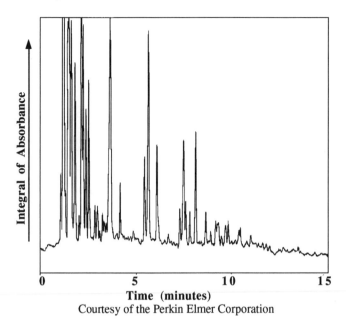

Figure 4.11 Chromatogram of Gasoline Monitored by IR Absorbance

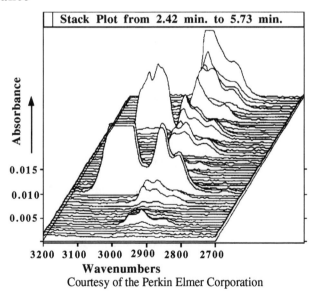

Figure 4.12 Stack Plot of Spectra from Gasoline Analysis

This particular type of application of the GC/FTIR tandem system was reported by Diehl et al [4], who used the system to determine the individual aromatic hydrocarbons in gasoline. By using selected wavelengths, the maximum sensitivity could be achieved for each aromatic or group of aromatic hydrocarbons. Diehl et al. used a Hewlett-Packard Model 5890 Series II GC/5965 IRD FTIR, with an open tubular column 60 m long and 0.53 mm I.D., carrying a 5.0 μm film of stationary phase on the internal surface. The column was programmed from 40°C to 190 °C at 2°C per min. and then to 300°C at 30°C per min. The light pipe and the transfer line were held at 300°C. Employing appropriate wavelengths, the elution of the aromatic hydrocarbons could be exclusively monitored. The separation was fairly lengthy, extending well over an hour but the aromatic hydrocarbons could be exclusively monitored, and the numerous aliphatic and unsaturated hydrocarbons, also present in the mixture, gave little or no response at the selected wavelength.

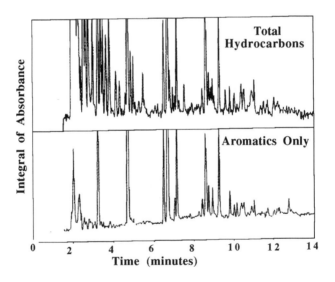

Courtesy of the Perkin Elmer Corporation

Figure 4.13. A Chromatograms Showing the selective monitoring of Aromatic Hydrocarbons in Car Exahust

GC/IR Tandem Systems 147

Additionally, a significant amount of spectral data on the double bond configuration of mono-unsaturated compounds have been reported by Attygalle *et al.* [5], which would also be useful for substance identification in similar types of sample. This procedure obviously greatly simplifies the quantitative estimation of the individual aromatic hydrocarbons present in a hydrocarbon mixture. It has been used in a very similar manner to determine the aromatics present in exhaust fumes. Samples of car exhaust were trapped onto a suitable absorbent, and then thermally desorbed onto a capillary column. The separation was monitored by IR absorption using the Perkin Elmer light pipe interface. The results obtained are shown in figure 4.13. The upper chromatogram depicts all the hydrocarbons as they are eluted, the lower chromatogram was obtained by plotting the absorption at the characteristic wavelength where aromatics exhibit maximum absorption. It is seen that in the lower chromatogram, the aromatic hydrocarbons are clearly and unambiguously depicted and the other hydrocarbons appear absent.

The Cryostatic Interface

The cryostatic interface was introduced as an alternative to the light pipe which, due to the catalytic nature of the gold surface, can cause some decomposition or molecular rearrangement with certain samples. In contrast, the total system in the cryostatic interface is sensibly inert to the sample, as it is contained in a solid argon matrix. A diagram of the Cryolect™ manufactured by Mattson Instruments Inc. is shown in Figure 4.14. Unfortunately, this instrument is no longer commercially available although a large number are in use at this time. The device is extremely sensitive and provide excellent spectra, for this reason, although not commercially available, the instrument will be discussed in some detail. The carrier gas is helium and contains a small amount of argon (*ca* 0.5 %), although nitrogen can also be used. On leaving the column, the carrier gas is allowed to impinge onto a rotating gold plated drum, situated in a evacuated box, thermostatted at about 12 °K by a liquid helium thermostat. The drum, as well as rotating, also moves very slowly in an axial direction, so the samples are deposited as a thin helical deposit on the outer walls of the drum. The frozen sample on the drum surface is trapped in a cage of solid argon and, as the argon insulates the sample from the gold surface, there can be no catalyzed decomposition, or molecular rearrangement. The carrier gas is removed by a low-pressure turbo-molecular pump that contains no oil, and thus eliminates the possibility of any long-term sample contamination.

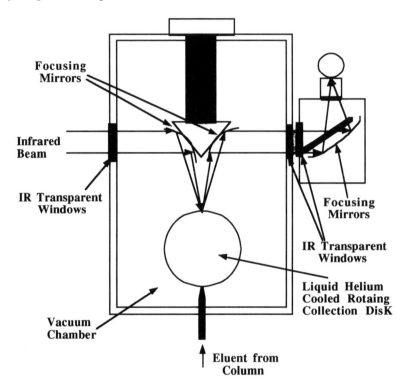

Courtesy of Mattson Instruments Inc.

Figure 4.14 The Mattson GC/IR Cryostatic™ Interface

Due to the extremely inert condition of the trapping system, the samples can be held for a very long time on the drum without loss or change. Consequently, the spectrum of any particular sample can be taken repeatedly, any number of times, to improve the signal-to-noise ratio and thus the overall sensitivity.

The interferometer's modulated IR beam focuses on the narrow argon 'stripe' that contains the sample. In practice, the IR sensitivity with this system of isolation is commensurate with that of the mass spectrometer. The improved IR sensitivity is partly due to the detector element being approximately the same size as the sample 'stripe' and partly due to the sample being in an inert argon matrix, which causes the absorption bands to be much sharper and thus much higher. An example of the relative absorption peak heights for a liquid sample, solid sample and a matrix isolated sample according to Mattson is shown in Figure 4.15.

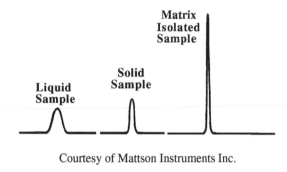

Courtesy of Mattson Instruments Inc.

Figure 4.15. Absorption Peaks for Liquid, Solid and Matrix Isolated Samples

It is seen that matrix isolation sharpens the peak considerably and will provide a very significant increase in sensitivity. The interferometer is contained in a vacuum chamber to keep the background free of spectral contaminants, which results in a remarkably stable output. The system is cleaned very rapidly by merely warming the drum and pumping out the argon and sample vapor, leaving the disk with absolutely no residue.

An interesting example of the use of a GC/IR tandem system, that can provide both high sensitivity and good spectroscopic resolution, is in studying the influence of molecular structure on physiological activity. Structural isomers exhibit widely different biological activity and traditionally such structures were identified by mass spectrometers employing GC/MS tandem instruments. However, the mass spectrometer does not differentiate well between certain important isomers, such as 1,4-

dimethyl naphthalene and 1,5-dimethyl naphthalene, whereas a GC/IR instrument with adequate sensitivity and resolution would do so unambiguously. The spectra of the two dimethyl naphthalenes collected from a gas chromatograph using the cryostat interface are shown in figure 4.16.

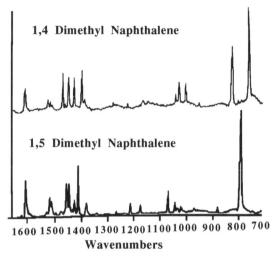

Courtesy of Mattson Instruments Inc.

Figure 4.16 Spectra of 1,4 and 1,5,-Dimethyl Naphthalenes Obtained Using the Cryostat Interface.

It is seen that the two spectra are distinctly different and contain sufficient detail to permit easy identification.

Another example, similar to the identification of the dimethylnaphthalene isomers, is the differentiation of the tetrachlorodibenzodioxin isomers. One particular isomer, the 2,3,7,8 tetrachlorodibenzodioxin, is the most potent carcinogen known today and, as a consequence, its identification in any environmental sample can be extremely important. The different dioxin isomers can not be reliably differentiated on the basis of their mass spectra alone, whereas their IR spectra exhibit clear and unambiguous differences.

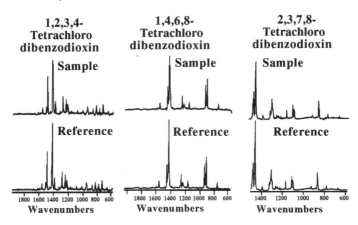

Figure 4.17 Spectra of Three Tetrachlorodibenzodioxins Isomers Obtained Using the Cryostat Interface.

The three spectra shown in figure 4.17 were taken of the components from a bulk sample of tetrachlorodibezodioxins that had been separated on a GC column. The standard reference spectra are included for each isomer so the different spectra can be compared. Each spectrum was derived from about 15 ng of the isomer and it would appear from the signal-to-noise ratio that a satisfactory spectrum could probably be obtained from as little as 100 pg of isomer.

Ragunathan *et al.* [6] evaluated a multidimensional system consisting of a GC, combined with the crysostatic interfaced FTIR spectrometer and a mass spectrometer, for use in essential oil analysis. The authors used cascarilla oil as their test material. Their apparatus took the form shown in Figure 4.18. The combination consisted of a modified Hewlett-Packard 5890 gas chromatograph, coupled to a Mattson Cryolect 4800 FTIR spectrometer and a HP 5970b mass selective detector. The preliminary separation was carried out on an open tubular column 30 m long, 0.032 mm I.D. carrying a film of polyethylene glycol 1 μm thick. Two alternative analytical columns were available, one 13 m long, 0.025 mm I.D. carrying a film 0.5 μm thick (Rtx-1701), and the other 8 m long, 0.025 mm I.D. carrying a film 0.25 μm thick (DB-5). Helium was used as the carrier gas and the raw sample was placed on the first column. The separation was monitored by the mass selective detector and when the solutes of interest began to elute, the flow was diverted to the cooled trap, and the solutes condensed and concentrated.

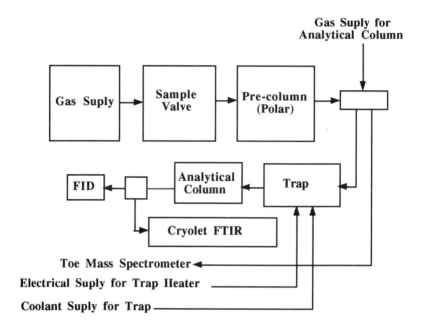

Figure 4.18 Multidimensional Gas Chromatography Coupled with a Mass Spectrometer and the FTIR Spectrometer

The trap was made of deactivated fused silica tubing, 10 cm long, 0.019 mm I.D contained inside a stainless steel tube 0.053 mm I.D. and was cooled to 50–100 K by a stream of cold nitrogen. When collection was complete, the carrier gas was returned to the mass selective detector, and the temperature of the trap raised to 200°C by passing a current through the heater. A second flow of carrier gas was passed through the trap expelling the collected solutes onto the analytical column. The carrier gas used in the second separation consisted of a mixture of 0.5% nitrogen in helium (in this case nitrogen was used as the IR matrix). The column eluent was split and passed into two capillary lines; one 100 μm I.D. that led to the FID, and the other 220 μm I.D. that led to the Cryolet FTIR spectrometer system. As a result, the separation was monitored by both the FID and the FTIR spectrometer simultaneously. The Cryolet transfer line was maintained at 250°C and the tip was positioned about 0.1 mm from the gold cryo-collector disk. The sample was normally diluted in n-hexane 1:10, and about 0. 2 µl was injected onto the pre-column using a split ratio of 1:10. The system dscribed by Ragunathan *et al.* is obviously complex and expensive but would provide reliable data from very small samples and relatively quickly.

Thermogravimetric Analysis (TGA) Coupled with GC/IR and GC/MS

The thermogravimetric analyzer, coupled with GC/IR and GC/MS, is another fairly complex combined instrument and might be considered as TGA coupled with two tandem systems, or even as a quintuplet system. The combination was assembled by McClennen *et al.* [7] who used it to study the thermal decomposition of a number of different polymers. A block diagram of their apparatus is shown in Figure 4.19.

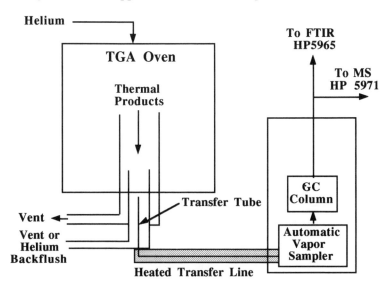

Figure 4.19 TG/GC/FTIR and TG/GC/MS Instrument Configuration

The configuration consists of a Perkin Elmer Model 7 thermogravimetric analyzer coupled to a standard Hewlett-Packard GC/IR/MS system. A heated transfer line directly couples the TGA to the gas chromatograph, by means of a special injection system that will be described below. The furnace was maintained at a pressure of 0.14 bar above atmospheric pressure, to drive the decomposition gases and vapors through the conduit to the sampling device. A specimen of the vapors, produced during the thermogravimetric analysis, is periodically taken during the heating program. The sample is separated on the capillary column and the exit gas is split into two streams. One stream goes directly to the FTIR spectrometer and the other to the mass spectrometer. In this way a total ion current chromatogram and/or a total IR absorption chromatogram can be generated in real time or, if preferred, can be reconstructed after the analysis is completed. In addition, IR and MS spectra from any selected peak can be generated from disc for identification purposes. The constituents are identified and quantitatively assayed from their spectra. Identification can be achieved by direct comparison with reference spectra, or by elucidating the structure from the data provided by the complementary IR and MS spectra.

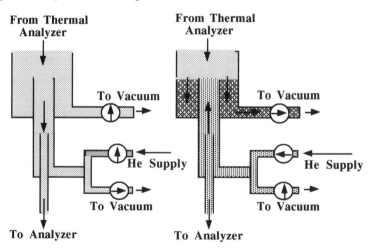

Figure 4.20 Diagram of the Automatic Vapor Sampler

GC/IR Tandem Systems

The sampling apparatus [8] is a simple but ingenious device, that allows short bursts of sample to be placed on the column at any pre-selected time by simple valve operations. A diagram showing the principle of the injection system is shown in Figure 4.20. The operational sequence of the valving system is extremely simple. The normal, non-sampling condition, is depicted on the right. The upper vacuum valve is open and the lower vacuum valve closed. Helium is fed into the center tube causing helium to flow through the column and forcing the sample flow to be diverted out through the top vacuum port. By closing the top vacuum valve and the helium supply valve, and simultaneously opening the lower vacuum valve, the sample contained in the helium flow from the thermal analyzer can be made to flow directly into the analyzer column (shown on the left). This flow is allowed to proceed for a prescribed time period that places a defined volume of sample on the column. The valves are then reversed to their original position, the sampling procedure arrested and the separation developed. The system ensures that the sample only passes through fused quartz tubing on its way to the column and, in doing so, does not come in contact with any valve material or pass through any valve. This eliminates the possibility of selective absorption or sample decomposition on the valve surfaces.

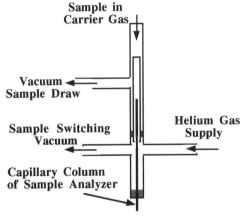

Figure 4.21 The Tube Arrangement of the Valveless Sample Injector

156   Tandem Techniques

The simple device is made solely of fused quartz tubing and the basic tube arrangement is shown in Figure 4.21. It is seen that the device is easy to construct, simple to operate and can be very compact. The authors demonstrated the use of the total TGA/GC/FTIR/MS apparatus by using it to examine the thermal degradation of poly-(α-methyl styrene), a styrene-isoprene block copolymer, and showed how it could be used to identify the decomposition products.

In anothe example of the TGA/GC/FTIR tendem combination, the Perkin Elmer Model 7 TG instrument was used and samples of 3-4 mg were employed for the analysis. The furnace was operated at a nominal heating rate of about 25°C per min. An HP-5 column 2.1 m long and 0.32 mm I.D was used for the GC/IR work and one 1.7 m long, 0.1 mm I.D column for the GC/MS analyses. These short columns provided very fast analyses and thus the thermal decomposition could be monitored extremely precisely.

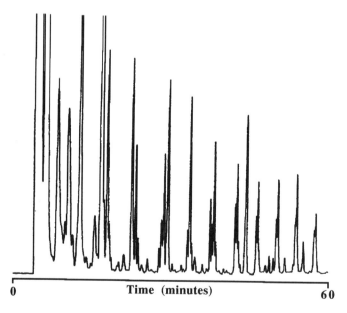

Courtesy of the Perkin Elmer Corporation

Figure 4.22 Chromatogram Obtained from the Pyrolysis of Polyethylene at 1000°C

GC/IR Tandem Systems

Thermal degradation products could also be examined using the light pipe interface which, although not nearly as sensitive or as fast, is much less expensive. Furthermore it can yield spectral data that is at least as accurate and precise as that obtained from the cryogenic interface. About 5 mg of a polyethylene film, 50 μm thick, dispersed on the surface of a platinum coil pyrolysis probe, was placed at the inlet of an open tubular column and was heated to a temperature of about 1000°C for a period of 10 seconds. The column was 25 m long, 0.53 mm I.D., and carried a 5 μm film of methyl phenyl silicone stationary phase It was operated at a flow rate of 4 ml/min, split 2:1 in a split/splitless injector. The separation monitored by a flame ionization detector is shown in Figure 4.22.

The column oven was initially held at 45°C for 3 min., and then programmed up to 280°C at 4°/min. 10% of the eluent passed to the detector and the remainder passed through the light pipe. The chromatogram obtained by reconstruction of the IR absorption spectra between 16 and 30 minute together with some of the associated spectra are shown in Figure 4.23. It is seen that a more than adequate sensitivity is achieved and well defined spectra are obtained from the four closely eluted peaks. The quality of the spectra are sufficient for solute identification, providing reference spectra were available. The spectra would also aid in the structure elucidation of a completely unknown substance providing complementary mass spectra were also available.

Thermal gravimetric analysis, in conjunction with a GC/FTIR tandem system utilizing a light pipe interface, has also been used to determine the pyrolysis and oxidation products of an alkylbenzimidazole. The system was operated both as a tandem TGA/FTIR combination and, by employing a suitable trapping procedure, as a triplet system TGA/GC/FTIR. The pyrolysis and oxidation process was carried out in a Perkin Elmer thermogravimetric analyzer and the FTIR spectra obtained from a PE 2000 FTIR instrument. Samples from the TGA instrument were either taken directly by a light pipe interface, to the FTIR spectrometer or, alternatively, the products were collected in a Tenax trap. The products were regenerated by rapidly heating the trap to 275°C in the capillary injector of the gas chromatograph, and the carrier gas stream transferred the solutes to the GC column.

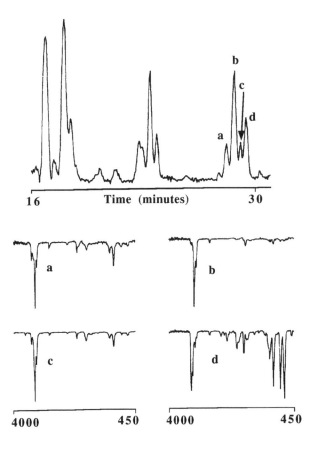

a; α, ω-diene component, b; an alkane component, c; 1-alkene component and d; an aromatic component.

Courtesy of the Perkin Elmer Corporation

Figure 4.23 the Reconstructed IR Chromatogram (16-30 min.) Chromatogram and Associated Spectra

The trap was then cooled back to room temperature after injection. The column was maintained at 60°C until the sample had been injected and then

GC/IR Tandem Systems 159

programmed from 60° to 200°C at 8°C per minute. The column used was 30 m long, 0.32 mm I.D., and carried a film of 5% phenyl silicone, 1 µm thick as the stationary phase. Samples of the pyrolysis or oxidation products were taken at regular intervals and their IR spectra obtained.

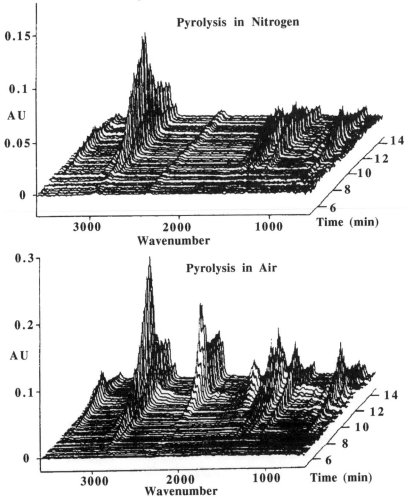

Courtesy of the Perkin Elmer Corporation

Figure 4.24 The Stacked Plots of the TGA/FTIR Tandem Analyses of Alkylbenzimidzole Carried Out in Nitrogen and Air

A stacked plot of the spectra, taken over a heating period between 8 and 14 minutes, is shown in the upper diagram of Figure 4. 24. The spectrum obtained from combining all the spectra taken during the TGA analysis between the temperatures of 290°C and 310°C is shown in Figure 4. 25. It is clearly seen that no absorption is present that might arise from any carbonyl groups, indicating a simple evaporation procedure was taking place

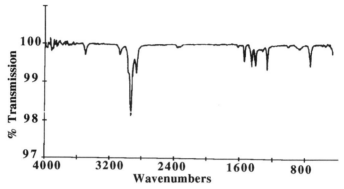

Courtesy of the Perkin Elmer Corporation

Figure 4.25 Co-added Spectra from Pyrolysis Products Sampled Between 290°C and 310°C.

In the lower part of Figure 4.24 is shown the stacked plot of the spectra taken during the TGA analysis carried out in air, over a heating period extending between 6 and 14 minutes.

It is seen that the spectra are quite different, one clear feature being the absorption peak for carbon dioxide arising from the oxidation of the sample. The spectrum obtained from combining all the spectra taken during the TGA oxidation analysis between 290°C and 310°C is shown in Figure 4.26. It is seen that the dominant absorption from the sample is carbon dioxide from the general oxidation of the material. Absorption in the carbonyl region (1800–1700 cm^{-1}) suggests incomplete oxidation has also taken place, generating some aldehydes, acids and possibly some ketones.

GC/IR Tandem Systems

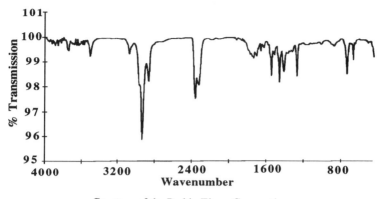

Courtesy of the Perkin Elmer Corporation

Figure 4.26 Co-added Spectra from Oxidation Products Sampled Between 290°C and 310°C

Integrated IR spectra of the eluent portions collected in the Tenax trap produced the upper chromatogram shown in Figure 4.27. The lower chromatogram is an expanded portion of the total separation from 14 to about 30 minutes. The oxidation products of greatest concentration are carbon dioxide, water vapor, hexanal, heptanal, benzaldehyde, some unresolved acids and aldehydes and some acid lactones. It is seen that tandem systems employing the light pipe interface for the FTIR spectrometer are very effective but significantly lower in sensitivity than the cryostatic system.

Without doubt, the most effective GC/FTIR tandem system employs the cryostatic interface and provides a sensitivity at least one if not two orders of magnitude greater than the contemporary light pipe interface. It is however, a mechanically complicated device and extremely expensive, relative to the light pipe alternative. So much so in fact, that the cryostat has been withdrawn from the market and at this time is no longer commercially available. This is most unfortunate and leaves the analyst with the light pipe as the only commercial alternative. Nevertheless, there is a need for a high-sensitivity interface, and this need will continue to grow as the demands of the environmental, forensic and biological sciences become greater. Hopefully, this increasing demand will evoke the development of another cryostatic GC/FTIR interface (perhaps at a reduced cost) and, sometime in the future, a new instrument will be manufactured and become commercially available. Without doubt, a device with similar performance specifications to that of the Cryostatic™ will eventually be required.

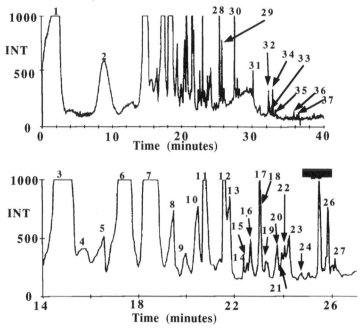

1:CO_2 and H_2O, 2:hexanal, 3:heptanal, 4:butyric acid, 5:pentanoic acid, 6: benzaldehyde, 7:acid and aldehyde, 8:acid, 9:aromatic aldehyde, 10:ketone, 11:unsaturated acid, 12: aldehyde and acid, 13: acid, 14: aldehyde, 15:siloxane, 16:aldehyde, 17: aldehyde, 18; unknown, 18: unknown, 19: acid, 20:acid, 21:acid, 22:aldehyde and acid, 23:acid, 24:acid, 25:acid lactone, 26:ketone, 27:unknown, 28:acid lactone, 29: unknown, 30: aliphatic hydrocarbon, 31:aliphatic hydrocarbon, 32:aromatic acid, 33:aliphatic hydrocarbon, 34;aliphatic hydrocarbon, 35:unknown, 36:aromatic hydrocarbon, 37:aromatic hydrocarbon.

Courtesy of the Perkin Elmer Corporation

Figure 4.27 Chromatograms Obtained From the Tenax Trap

Synopsis

The first GC/IR instrument was a triplet system that monitored both IR spectra and MS spectra simultaneously. Each peak was collected in a trap, the solute automatically regenerated into an IR cell, where the spectrum was taken and a separate sample passed to the mass spectrometer. The cell used for IR measurement was one of the early examples of the *light pipe*. The light pipe is a tube that has the internal walls made reflective, by coating with gold or some other appropriate material. Light pipe interfaces are the most commonly used interfaces for GC/IR today, and a number of different manufacturers produce efficient light pipe systems. The conduit to the light pipe, and the light pipe itself, must be maintained at a temperature that will prevent any solutes condensing on the surface. In general, the interface conduit and light pipe, should be maintained at least 20°C above the maximum temperature attained by the chromatography column. The light pipe interface has been successfully used for many commercial applications including pharmaceutical products, essential oils, and petroleum products. The *cryostat* interface, also popular, but considerably more expensive, consists of a drum cooled with liquid helium, on the surface of which the column eluent condenses. A mixture of argon and helium is often used as the mobile phase and so the sample is trapped in a solid argon matrix, which eliminates the possibility of catalytic decomposition or molecular rearrangement on the metal surface. The drum rotates, and a physical chromatogram is formed as a spiral of solutes, condensed on the drum surface. Reflective IR spectra are taken from the drum surface. The cryostat interface, which is by far the most sensitive GC/IR interface, has also been used with thermogravimetric instruments forming the TGA/GC/IR triplet system.

References

1. R. P. W. Scott, I. A. Fowliss, D. Welti and T. Wilkens, *Gas Chromatography 1966*, A. B. Littlrwood (Ed.) The Institute of Petroleum, London (1966)318.
2. R. P. W. Scott, in *Gas Chromatography 1964*, A. B. Littlewood(Ed.) The Institute of Petroleum, London (1966)25.
3. P. A. Wilks and R. A. Brown, *Anal. Chem.* **36**((1964)1896.
4. J. W, Diehl, J. W. Flinkbeiner and F. P. DiSanzo, *Anal. Chem.* **67(13)**(1995)2015.
5. A. B. Attygalle, A. Svatos and C. Wilcox, *Anal. Chem.* **66(10)**(1995)1696,
6. N. Ragunathan, A. T. Sasaki, K. A. Krock and C. L. Wilkins, *Anal. Chem.*, **66(21)**(1995)3751.

7. W. H. McClennen, R. M. Buchanan, N. S. Arnold, J. P. Dworzanski and H. L. C. Meuzelaar, *Anal. Chem.,* **65(29)**(1993)2819.
8. N. S. Arnold, W. H. McClennen and H. L. C. Meuzelaar, *Anal. Chem.* **63(3)**(1991)299.

CHAPTER 5

GAS CHROMATOGRAPHY/MASS SPECTROSCOPY (GC/MS) TANDEM SYSTEMS

Just four years after the first disclosure of GC as an effective separation technique by James and Martin in 1953, Holmes and Morrell [1], successfully combined the gas chromatograph with the mass spectrometer to produce the first tandem system. The authors connected the column outlet directly to the mass spectrometer employing a split system. The mass spectrometer was a natural choice for the first combination instrument, as it could easily accept samples presented as a vapor in a permanent gas. Prior to 1960, only *packed* GC columns were commercially available, and thus the major problem encountered, when associating a gas chromatograph with a mass spectrometer, was the elimination of the carrier gas. This was due to the relatively high flow rates that need to be used with packed columns (*ca* 25 ml/min or more). The contemporary vacuum pumps at that time had relatively low pumping rates (measured at atmospheric pressure), and thus only a small proportion of the eluate could be passed to the mass spectrometer, which resulted in a significant loss of sensitivity. This problem was solved by the use vapor concentrating devices.

Vapor Concentrators

In order to accommodate the relatively large flow rates that were used with packed GC columns, a number of concentrating devices were developed, *e.g.* the jet concentrator invented by Ryhage [2], and the helium diffuser developed by Bieman [3] later known as the Bieman concentrator.

The Ryhage Concentrator

A diagram of the Ryhage concentrator is shown in Figure 5.1. The concentrator consists of a succession of jets that are aligned in series but separated from each other by carefully adjusted gaps (usually a double jet combination is used but sometimes three or more can be employed). Each subsequent jet has a smaller aperture than the previous one. The helium passes through the center channel and diffuses away in the gaps between the jets. The helium that has left the main stream is removed by appropriate vacuum pumps (usually one for each jet). In contrast, the solute vapor, having greater momentum, continues into the next jet and finally into the mass spectrometer.

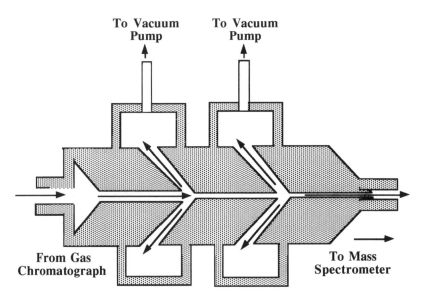

Figure 5.1   The Ryhage Concentrator

Despite first impressions to the contrary, the device proved to be very effective. The concentration factor is a little greater than an order of magnitude depending on the jet arrangement. Perhaps even more surprising, the sample recovery is usually well in excess of 25%. This device was produced commercially and used extensively for packed columns with GC/MS tandem systems, until they was eventually replaced by the open tubular column. Even today, the Ryhage concentrator is often still often used in GC/MS systems when packed columns are employed.

The Bieman Concentrator

The Bieman concentrator was developed at about the same time as the Ryhage concentrator and, due to its simplicity, was equally popular for use in GC/MS systems. A diagram of the Bieman concentrator is shown in Figure 5.2 and, as seen, functions on quite a different principle. The concentrator consists of a heated glass jacket surrounding a sintered glass tube. The eluent from the chromatograph passes directly through the sintered glass tube and the helium diffuses radially through the porous walls and is continuously pumped away. The helium stream enriched with solute vapor passes on to the mass spectrometer.

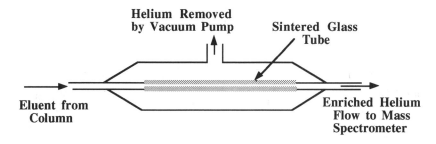

Figure 5.2 The Bieman Concentrator

Solute concentration and sample recovery is similar to the Ryhage device but the apparatus, though a little more bulky, is somewhat easier to operate. An alternative system was also devised, based on the same principle, that employed a length of porous polytetrafluorethylene (PTFE) tube as the concentrator, as opposed to a sintered glass tube; in other aspects the PTFE device functioned in the same manner.

The development of the open tubular columns eliminated much of the need for concentrating devices, as the mass spectrometer pumping system could easily dispense with the usual carrier gas flows that are employed with such columns and peak concentrations are usually adequately high. Consequently, the column flow can be passed directly into the mass spectrometer and the total sample will enter the ionization source.

Ion Sources Used in GC/MS

In order to discriminate between molecules of different masses by electrostatic or magnetic deflection systems, the molecules must be ionized and/or they must be fragmented into charged ions. There are two types of ion source in general use with GC/MS tandem systems and they are electron impact ionization (EI) and chemical ionization (CI). More recently, the inductively coupled plasma ionization process (ICPI), which is usually employed for liquid or solid samples, has also been introduced to GC/MS systems and has been used for certain types of inorganic analysis.

Electron Impact Ionization

The design and function of the electron impact ionization source has been discussed in Chapter 2. It is a relatively hard ionization procedure, and thus produces many fragments and often a relatively small parent ion. In fact, in some cases no parent ion may be produced at all. A diagram of the electron impact ion source similar to that given in Figure 2. 30 in Chapter 1 is shown in Figure 5.3.

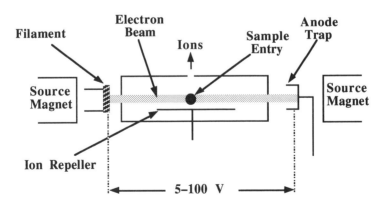

Figure 5.3 The Electron Impact Ion Source

GC/MS Tandem Systems

Electrons are formed by thermal emission from a heated tungsten or rhenium filament and accelerated by an appropriate potential to the anode trap. The magnitude of the potential may range from 5 to 100 V depending on the electrode geometry and the ionization potential of the substances being ionized. The filament current is sometimes automatically controlled by the magnitude of the trap anode current to maintain constant ionizing conditions. The sample is introduced into the gas stream at the center of the electron beam. The ions formed are repelled by a suitable potential, through a hole in the wall of the ion source enclosure and thus pass into the accelerating field of the mass spectrometer.

A magnetic field of a few hundred gauss is often maintained along the axis of the electron beam, to confine the electrons to a narrow helical path. In general only about 0.1% of the molecules entering the ion source are ionized. The optimum ionization energy of the electron varies with different compounds, but an average value appears to lie between 50 and 100 eV. The approximate relationship between ion current and electron energy takes the form shown in Figure 5.4.

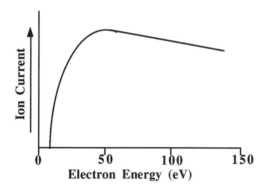

Figure 5.4 The Approximate Relationship Between Ion Current and Electron Energy

It is seen that the ionizing energy of the electrons can be easily controlled by adjusting the accelerating potential of the electrons in the ion source. This capability is important, because it allows conditions to be adjusted so that optimum fragmentation takes place that will provide the maximum information with respect to the structure of the compound being examined. An example of the effect of electron energy on the fragmentation pattern of a compound is shown in Figure 5.5. It is seen from the pattern obtained at low energies (*ca* 14 eV) that a large parent ion is produced but relatively few fragments. This means that the molecular weight of the material could be fairly easily identified but its structure would not be easy to determine as there were very few fragments to work with.

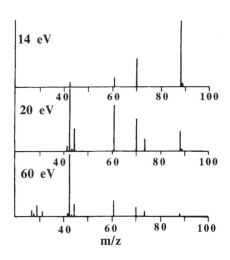

Figure 5.5 Fragmentation Patterns for a Molecule Ionized with Electrons Having Different Energies

In contrast, at high electron energies (*ca* 60 eV) there are a large number of fragments, particularly at low molecular weight, but the parent ion is hardly discernible. This means that although some of the secondary structure of the molecule may be revealed, the lack of a definite parent ion would again make the total molecular structure difficult to identify. However, at mid–electron energies (*ca* 20 eV) numerous fragments together with an unambiguous parent ion are produced, providing ample information for structural identification. It is seen that the electron energy in electron impact ionization is an important parameter on which to optimize, to ensure that the best possible data is generated for structure elucidation. If a more gentle form of ionization is required, however, chemical ionization should be used.

Chemical Ionization

Chemical ionization was first observed by Munson and Field [4], who introduced it as a technique for ionization in mass spectrometry in 1966. The procedure involves first the ionization of a reagent gas such as methane in a simple electron impact ion source. The partial pressure of the reagent gas is arranged to be about two orders of magnitude greater than that of the sample. The reagent ions collide with the sample molecules and produce ions. The process is considered to be a gentle form of ionization, because the energy of the reagent ions never exceeds 5 eV, including those reagent ions that are considered to have relatively high energies. As a consequence there is little fragmentation, and the major sample ion produced usually has a m/z value close to that of the singly charged molecular ion. The spectrum produced by chemical ionization depends strongly on the nature of the reagent ion and thus different structural information can be obtained by choosing different reagent gases. This adds another degree of freedom in the operation of the mass spectrometer. The reagent ion can take a number of forms. Employing methane as the reagent ion the following reagent ions can be produced

$$CH_4 \rightarrow CH_4^+, CH_3^+, CH_2^+$$

$$CH_4^+ + CH_4 \rightarrow CH_5^+ + CH_3$$

$$CH_3^+ + CH_4 \rightarrow C_2H_4^+ + H_2$$

Other reactions can also occur that are not useful for ionizing the solute molecules but, in general, these are in the minority. The interaction of positively charged ions with the uncharged sample molecules can also occur in a number of ways, and the four most common are as follows:

1. Proton transfer between the sample molecule and the reagent ion,

$$M + BH^+ \rightarrow MH^+ + B$$

2. There is an exchange of charge between the sample molecule and the reagent ion,

$$M + X^+ \rightarrow M^+ + X^+$$

3. There is simple addition of the sample molecule to the reagent ion,

$$M + X^+ \rightarrow MX^+$$

4. Finally there can be anion extraction,

$$AB + X^+ \rightarrow B^+ + AX$$

As an example (CH_5^+) ions, which are formed when methane is used as the reagent gas, will react with a sample molecule largely by proton transfer *e.g.,*

$$M + CH_5^+ \rightarrow MX^+ + CH_4$$

Some reagent gases produce more reactive ions than others. Consequently, some reagent gases will produce more fragmentation.

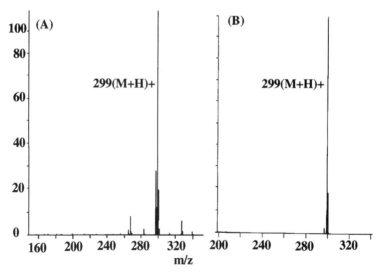

(A) Reagent Gas Methane; (B) Reagent Gas Isobutane.

Figure 5.6. The Mass Spectrum of Methyl Stearate Produced by Chemical Ionization

Methane produces more aggressive reagent ions than isobutane, and thus whereas methane ions produce a number of fragments by protonation, isobutane, by a similar protonation process, will produce almost exclusively the protonated molecular ion. This is clearly demonstrated by the mass spectrum of methyl stearate shown in Figure 5.6. Spectrum (A) was produced by chemical ionization using methane as the reagent gas and exhibits fragments other than the protonated parent ion. In contrast, spectrum (B) obtained with butane as the reagent gas exhibits the protonated molecular ion only.

The chemical ionization source is very similar to the simple ion impact source; in fact most mass spectrometer electron impact sources can perform the dual role, and in addition, act as a chemical ionization source. Dual-action sources do not perform quite as well as dedicated electron impact sources when used in the electron impact mode, but the loss of ionization efficiency is certainly no more than 50%. Continuous use of a chemical ionization source results in significant source contamination which impairs the performance of the spectrometer. This results from the build-up of residues from the ionization process and thus the source requires cleaning by baking-out fairly frequently. A diagram of a typical gas inlet system for a chemical ionization source is shown in Figure 5.7,

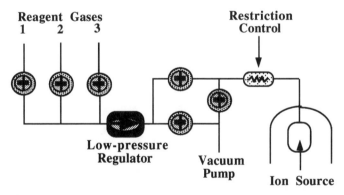

Figure 5.7 A Gas Inlet System for Chemical Ionization

The diagram illustrates the use of three different reagent gases but any number could be incorporated if desired. The source pressure is usually held at 0.1–0.5 torr and a low pressure regulator is incorporated to control the pressure to the required limits. The pressure regulator and valves are usually solenoid operated, so that they can be automatically actuated by the mass spectrometer control-computer. Consequently, it is easy to rapidly change from electron impact ionization to chemical ionization, if so desired. In GC/MS tandem systems both forms of sample ionization are commonly used. The sampling procedure is relatively simple as the sample enters the mass spectrometer as a vapor, in a gas stream, directly from the concentrator (if one is used) or from the column.

The Inductively Coupled Plasma Mass Spectrometer Ion Source

The inductively coupled plasma (ICP) mass spectrometer ion source evolved from the ICP atomic emission spectrometer, and is probably more commonly employed in LC/MS than GC/MS. However, it is occasionally employed in GC/MS, usually in the assays of organometallic materials and in metal speciation analyses, and so the ion source will be described in this chapter.

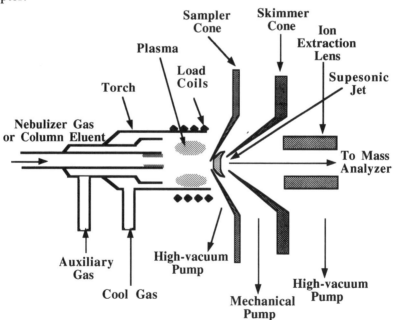

Figure 5.8 The ICP Mass Spectrometer Ion Source

GC/MS Tandem Systems

The ICP ion source is very similar to the volatalizing unit of the ICP atomic emission spectrometer, and a diagram of the device is shown in Figure 5.8. The argon plasma is an electrodeless discharge, often initiated by a Tesla coil spark, and maintained by rf energy, inductively coupled to the inside of the torch by an external coil, wrapped round the torch stem. The plasma is maintained at atmospheric pressure and at an average temperature of about 800 K.

The ICP torch consists of three concentric tubes made from fused silica. The center tube carries the nebulizing gas, or the column eluent from the gas chromatograph. Argon is used as the carrier gas, and the next tube carries an auxiliary supply of argon to help maintain the plasma, and also to prevent the hot plasma from reaching the tip of the sample inlet tube. The outer tube also carries another supply of argon at a very high flow rate that cools the two inner tubes, and prevents them from melting at the plasma temperature. The coupling coil consists of 2-4 turns of water cooled copper tubing, situated a few millimeters behind the mouth of the torch. The rf generator produces about 1300 watts of rf at 27 or 40 MHz which induces a fluctuating magnetic field along the axis of the torch. Temperature in the induction region of the torch can reach 10,000°K but in the ionizing region, close to the mouth of the sample tube, the temperature is 7000–9000 K.

The sample atoms account for less than 10^{-6} of the total number of atoms present in the plasma region, and thus there is little or no quenching effect due to the presence of the sample. At the plasma temperature, over 50% of most elements are ionized. The ions, once formed, pass through the apertures in the apex of two cones. The first has an aperture about 1 mm I.D., and ions pass through it to the second skimmer cone. The space in front of the first cone is evacuated by a high-vacuum pump. The region between the first cone and the second skimmer cone is evacuated by a mechanical pump to about 2 mbar and, as the sample expands into this region, a supersonic jet is formed. This jet of gas and ions, flows through a slightly smaller orifice into the apex of the second cone. The emerging ions are extracted by negatively charged electrodes (-100 to -600 V) into the focusing region of the spectrometer, and then into the mass analyzer.

The ICP ion source has several unique advantages; the samples are introduced at atmospheric pressure, the degree of ionization is relatively uniform for all elements, and singly charged ions are the principal ion product. Furthermore, sample dissociation is extremely efficient and few, if any, molecular fragments of the original sample remain to pass into the mass spectrometer. High ion populations of trace components in the sample are produced, making the system extremely sensitive.

Nevertheless, there are some drawbacks. The high gas temperature and pressure evoke an interface design that is not very efficient; only about 1% of the ions that pass the sample orifice pass through the skimmer orifice. Furthermore, some molecular ion formation does occur in the plasma, the most troublesome being molecular ions formed with oxygen. These can only be reduced by adjusting the position of the cones, so that only those portions of the plasma where the oxygen population is low, are sampled.

Although the detection limit of an ICP-MS is about 1 part in a trillion, as already stated, the device is rather inefficient in the transport of the ions from the plasma to the analyzer. Only about 1% pass through the sample and skimming cones and only about 10^6 ions will eventually reach the detector. One reason for ion loss is the diverging nature of the beam, but a second is due to space charge effects which, in simple terms, is the mutual repulsion of the positive ions away from each other. Mutual ion repulsion could also be responsible for some non–spectroscopic inter-element interference (*i.e.* matrix effects).

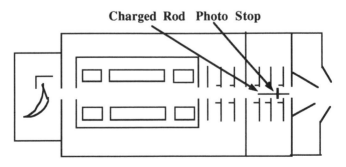

Figure 5.9 Modified Ion Lens for ICP-MS

The heavier ions having greater momentum suffer less dispersion than the lighter elements, thus causing a preferential loss of the lighter elements. Sandra *et al.* [5] introduced a simple modification to the ion lens system that increases the number of ions reaching the analyzer and this modification is shown in Figure 5.9.

The modification consists of the insertion of a short length of stainless steel rod 2.2 cm long and 2 mm in diameter along the axis of the existing lens system. The rod is mounted axially through the photon stop and isolated from it by a ceramic insulator. The optimum position of the photon stop was found to be about 1.5 cm from the front end of the rod and 0.7 cm from the rear end. It was found that a potential of 5 V applied to the rod improved the ion collection efficiency by about a factor of five (500%), with a corresponding reduction in the minimum detectable ion concentration.

Examples of Some Gas Chromatography/Mass Spectrometry Applications

One of the first gas chromatography/mass spectrometer systems to be developed was that described by Banner *et al.* (6) and, for historical interest, a diagram of their original apparatus is shown in Figure 5. 10. The mass spectrometer used by Banner *et al.* was a rapid scanning magnetic sector instrument that easily provided a resolution of one mass unit. Nowadays, mass spectrometers (giving vastly improved resolution) are mostly used with capillary columns, and operated in a very similar manner, with the column eluent passing directly into the ionization source of the spectrometer. Today the single, or triple quadrupole mass spectrometer are the most commonly used mass spectrometers in GC/MS tandem systems, and have been shown to give extremely impressive in-line sensitivity, extended mass range and a respectable resolution. GC/MS tandem systems have a number of attributes that make them particularly useful for certain types of applications. The chromatograph provides a high separating capability that can handle exceedingly complex mixtures, and the mass spectrometer can contribute both high resolution and very high sensitivity. As a consequence the components of a mixture can be identified with a high degree of confidence.

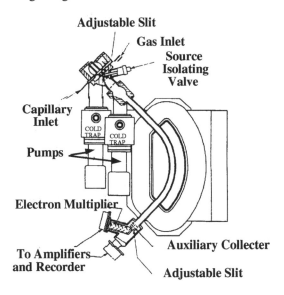

Figure 5.10 The First Tandem Capillary Column Mass Spectrometer System

It follows that the use of GC/MS has become a popular analytical procedure in forensic chemistry, toxicology and environmental studies. GC/MS has been employed for over thirty years in general analytical chemistry, and there is a plethora of applications in the literature that demonstrates the efficacy of the technique and its wide area or application. The following are a number of applications, taken from recent publications, that demonstrate some of the more recent uses of the technique.

Determination of Trace Concentrations of the 2,3,7,8-Chlorine-Substituted Dibenzo-p-Dioxins and Furans in Beef Fat

Recently a GC/MS system has been employed by Ferrario *et al.* [7] to determine the level of 2,3,7,8-chlorine substituted dibenzo-*p*-dioxins and furans in beef. The technique allowed the dioxins to be determined below the part per trillion level (ppt) in raw beef, and the individual solutes identified with a high degree of certainty. 200-300 g samples of back-fat were taken from freshly slaughtered cattle, and stored at -40°C. 100 g was homogenized and a 10 g sample taken for analysis. The samples were spiked with ^{13}C analogs of the various dioxins being determined. The crude extract was cleaned up, using simple column adsorption chromatography. Samples were separated on a Hewlett-Packard 5890 Series II, high-resolution gas chromatograph, employing a DM-5ms open tubular column, 60 m long, 0.32 mm I.D., carrying a 0.25 μm thick film of stationary phase. The initial oven temperature was 130°C, the injector temperature 270°C, and the interface temperature 300°C. The column was programmed at 5°C/min for one minute, then 15 min. at 6°C/min to 295°C. The Kratos high-resolution spectrometer was used as the tandem instrument. An amazingly high sensitivity was obtained, the actual limit of detection was found to be 0.05 ppt for the tetra-chloro-dibenzo-*p*-dioxins, and 0.1 ppt for the tetra-chloro-dibenzo-furans.

Analysis of Anabolic Steroids in Biological materials

Schoene *et al.* [8] employed a GC/MS tandem instrument to examine the metabolism of 17α-alkyl anabolic steroids in horses after oral administration. In an example, 100 mg of 17α-methyl testosterone together with the deuterated analog were fed to two horses *via* the usual food diet. Naturally voided urine was collected over 72 hours. 5 ml samples were extracted with ether and the solvent removed *in vacuo* to give the free urine fraction. The pH of the urine was adjusted to 6.8 and then heated with *E. Coli* (2000 U) at 50°C for three hours to hydrolyze the glucuronide conjugates. A solid phase extraction cartridge, Sep-Pak C_{18} was washed with water and *n*-hexane before applying the sample.

The hydrolyzed conjugates were eluted with ether, washed with 2 mol l^{-1} NaOH, dried over sodium sulfate and the solvents again removed *in vacuo*. The trimethyl silane derivatives were prepared by heating the fraction with 8% methoxyamine hydrochloride in pyridine at 80°C for 30 min, removing the pyridine *in vacuo* and heating the residue in N-methyl-N-(trimethylsilyl)-trufluoroacetimide at 80 °C for 1 hr. Excess reagent was also remove *in vacuo* and the residue dissolved in *n*-decane.

The mass spectrometer employed was the Finnigan MATTSQ 70 and the separation was carried out on a BPX5 column 25 m long, 250 μm I.D. The column was operated isothermally at 150°C for one minute, programmed at 5 °C per minute to 300°C and then maintained at 300°C for a further minute.

To follow the metabolic process, some knowledge of the basic nature of the fragmentation process that takes place in the mass spectrometer is helpful. In the electron ionization of the 17a-alkyl steroids the fragment pattern is dominated by the D-ring fragment ions. For the fragmentation of 17a-methyltestosterone, the process involves the cleavage at the C_{13} to C_{17} bonds and at the C_{14} to C_{15} bonds. This yields positively charged α, β–unsaturated ketone fragments at m/z 143. Consequently, observation of a pair of ions at m/z 143 and 146 in the EI mass spectra (the latter resulting from the labeled d_3 steroids) is indicative that the D-ring is unaltered in the metabolic process.

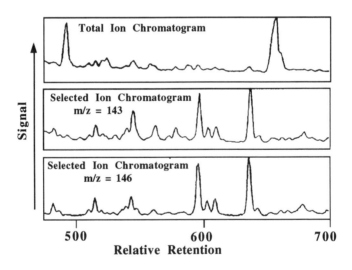

Figure 5.11 Separation of the Metabolites of 17α-methyl-testosterone [ref.8]

Examples of chromatograms from the equine urine samples are shown in figure 5.11. The upper chromatogram represents the total ion current and the two, lower chromatograms were monitored using the specific ions having m/z values of 143 and 146 respectively. It follows, that the peaks depicted on the single ion monitoring chromatograms represent metabolites in which the D-ring has been preserved. It is clear that the GC/MS technique can be a valuable tool for following the metabolic processes associated with the 17α-alkyl anabolic steroids.

Hooijerink *et al.* [9], used a similar technique to determine anabolic esters in oily formulations and in blood plasma. The work was carried out to develop a test that could be used for forensic purposes. Anabolic steroids are often illegally used as growth promoting agents in livestock breeding the use of which is prohibited in the European Community. The steroids are either injected intra-muscularly or applied topically and are slowly adsorbed into the blood stream. In either case the steroids can be detected in the blood plasma. The authors employed an automatic sample preparation apparatus, a Hewlett-Packard 5890 gas chromatograph and a Model 5970 mass detector. 5 ml of plasma was diluted with 5 ml of water and after centrifuging (1500g for 15 minutes) it was filtered through a 5 μm filter. The filtrate was then passed through an activated solid phase extraction column (SPE C_2), washed with 1.25 ml of 50% v/v methanol solution and the steroids were then eluted with 3 ml of pure methanol.

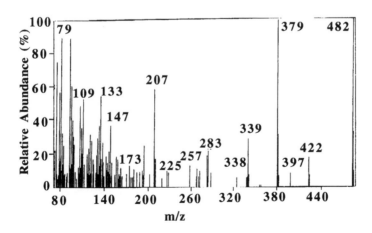

Figure 5.12 Mass Spectrum of Medroxyprogesterone Acetate Showing the Molecular Ion Peak at m/z 482 (ref.9)

The solvent was then removed and the steroids taken up in 10 µl of ethyl acetate and 100 µl of trifluoroacetic acid anhydride, to form the acetate derivatives. The mixture was kept at 70°C for about an hour, the reagent removed by evaporation and the residue dissolved in 15 ml of isooctane. This solution was use for analysis by GC/MS. The GC column was a DB-1 capillary column 30 m long, 250 µm I.D. carrying a 0.25 µm film of stationary phase. 2 µl of sample was injected directly onto the column with no split and the column was programmed from 130°C to 250°C, at 25°C per min, and from 250°C to 300 °C at 50° per min. The ion source of the MS was held at 200°C and the transfer line at 300°C. A spectrum obtained for medroxyprogesterone acetate is shown in figure 5.12 and the characteristic parent ion, mass 482 is clearly seen. The sensitivity of the technique can be very high and this is clearly demonstrated by the chromatogram shown in figure 5.13. The chromatogram was constructed by single ion monitoring using the m/z value of the molecular ion at 482.

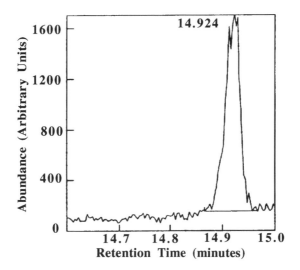

Figure 5.13 Single Ion Monitoring of Medroxyprogesterone Acetate Present at a Blood Plasma Level of About 6.5 ng/ml(ref.9)

The concentration of medroxyprogesterone acetate was about 6.5 ng per ml and the signal-to-noise ratio of the steroid ester peak seems to be about eight. Consequently, it would appear that the limiting sensitivity of the technique for measuring anabolic materials is likely to be well below 1 ng per ml.

Analysis of Waxes and Lipid Type Materials

The metabolism of fatty acids in cell cultures is important, particularly with regard to cancer cells. The possible conversion of the fatty acids to peroxides and the role of these compounds in carcinogenesis is still not completely understood. As a tool for this type of research Wallace and Coleman [10] developed a procedure employing a GC/MS tandem instrument for the fatty acid assay of cell tissue. Two human colon carcinoma lines, HT29/219 and HY115 together with a human breast cancer cell line ZR-75-1 were studied. The cells were grown in Dulbeccos's modified Eagles medium, augmented with 10% v/v foetal calf serum (DFC_{10})or horse serum (DH_{10}). They were seeded at a density of 1.9×10^4 cells per cm^2 and maintained at 37°C in a humidified atmosphere of 5% carbon dioxide and 95% air. The cells were harvested into a phosphate buffered saline at specific intervals up to 120 hours. Lipids were extracted with chloroform/methanol (2+1) containing 2, 6-di-*tert*-butyl-4-methylphenol as an antioxidant. The extract was separated on a thin layer plate using *n*-hexane-diethyl ether-acetic acid (70+30+1) as a solvent. The phospholipid, triglyceride and free fatty acid spots were scraped off the plates and the lipids removed by eluting with chloroform methanol (2+1). The phospholipids and triglycerides were trans-esterified [11] using sodium methoxide in methanol and the free fatty acids were methylated with diazomethane.

The authors found that a column that had a very low level of stationary phase bleed was essential for the GC/MS system to operate. Although the use of polymer coated capillary columns appear to be ideal, they were found to give inadequate separation and the optimum stationary phase was found to be a Carbowax polymer column (polyethylene glycol). It was found that the combination of the retention data with the mass spectrum gave virtually certain fatty acid identification. The authors used the technique to determine the fatty acid profiles of different cancer cell grown in the same media. An example of the separation of 15 different fatty acids using total ion current monitoring is shown in figure 5.14.

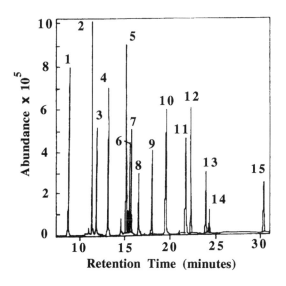

Figure 5.14 The Separation of a Series of Fatty Acids Using Total Ion Monitoring (ref.10)

Evershed *et al.* [12] developed a means of extracting and analyzing lipid type material from archaeological specimens using GC/MS techniques. Chemical analysis is now commonly used to investigate archaeological remains. Ceramics are of particular interest as they adsorb various substances into the pores of the material and, as a result, preserve them in an unchanged form for considerable periods of time. Hence the lipid content of pottery is a direct indication of the pots original contents and for what purpose it was used. Due to the pores becoming blocked with the very material that is being absorbed, little contamination from soil components takes place. The technique was developed using some freshly excavated ceramic potsherds, taken from the Raunds Area site in the Nene Valley Northhamptonshire UK.

Samples, about 2 g in weight, were scraped free from soil and ground to a fine powder in a carefully degreased pestle and mortar. A know mass of n-heptadecane was added as a standard, and the powder extracted twice with 10 ml of a mixture of chloroform and methanol (2+1) with supersonic agitation. After each extraction the mixture w centrifuged to remove the suspended solid. The extract was concentrated in a rotary evaporator and finally dried in a gentle stream of nitrogen. Trimethylester and ether derivatives were prepared by heating aliquots with excess N,O-bis(trimethylsilyl)trifluoroacetamide.

The GC/MS analysis was performed on a Pye 204 GC in conjunction with VG double focusing magnetic sector mass spectrometer and a Finnigan INCOS 2300 data system. The transfer line was maintained at 350°C, the ion source at 300°C and the electron impact ionization took place at 70 eV. The total ion current of a potsherd extract is shown in figure 5.15.

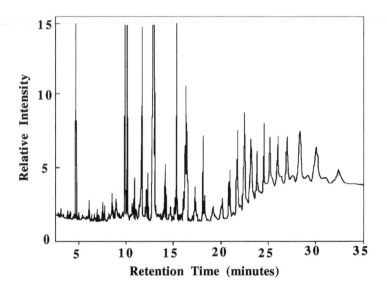

Figure 5.15. Total Ion Current Chromatogram of the Lipids from a Potsherd from a Late Saxon Ditch (ref 12)

It is seen that an interesting selection of lipid materials has been extracted from the potsherd. according to the authors, the technique is now a routine procedure for ceramic materials found at the various archaeological site under investigation at that time.

Laukó *et al.* [13] developed a GC/MS procedure for the analysis of crude cholesterol. Crude cholesterol is obtained from the spinal cord and brain of cattle and from wool grease. It is used in the pharmaceutical and cosmetic industries as an excipient in the production of ointments and creams, as an intermediate in hormone synthesis and in the manufacture of vitamin D_3 and liquid crystals. The gas chromatograph employed was the Hewlett-Packard 5890 using a capillary column 25 m long, 320 µm I.D, carrying a film of phenylmethylsilicone gum 0.52 µm thick and a 50–1 split injection system. The column was operated at 280°C and the detector and transit lines at 300°C. The internal standard was testosterone propionate. 1 µl samples of a 5% w/V solution of the crude cholesterol were injected. The mass spectrometer was the VG–TRIO–2 quadrupole used in the positive ion electron impact mode with an ionization energy of 70 eV, a total ion current chromatogram of the separation is shown in figure 5.16.

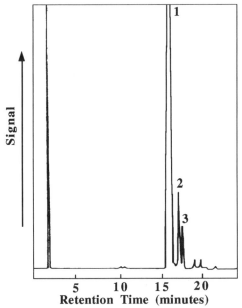

1.cholesterol; 2. desmosterol; 3. lathosterol

Figure 5.16. Total Ion Current Chromatogram of Crude Cholesterol (ref 13)

It is seen that a good separation is obtained from the individual impurities and so the actual cholesterol content can be easily assayed and with the accompanying mass spectra the identification of the individual impurities could be easily ascertained. The two major impurities were desmosterol (cholesta-5,24-dien-3β-ol) and lathosterol (5α-cholest-7-en-3β-ol). Other impurities that were also found present in certain batches were 5α-cholestan-3β-ol, cholesta-3,5-dien-7-one and cholest-4-en-4-one. A number of volatile lipids were also identified.

Some General Applications of GC/MS tandem Systems

Dugay et al. [14] employed a GC/MS tandem system to investigate the effect of hydrogen rearrangement on the determination of [^{15}N]leucine enrichment, when measuring the ratio of $^{15/14}$N-labeled leucine. Eleven different ester derivatives were examined, and it appeared that as result of the type of alcohol that was used in the derivatization process, hydrogen rearrangement could occur. The labeling ratio was found to increase with the length of the alkyl chain of the ester as well as with the number of hydrogen atoms at the β-site and, to a lesser extent, at the γ-site on this chain.

The leucine derivatives were made according to the method of Mackenzie and Tenaschuk [9,10]. The GC/MS system comprised the Fisons model MD 800 mass spectrometer used in conjunction with the 8000 series gas chromatograph. The capillary column employed was the DB-5, which was 30 m long and 0.25 mm I.D. The mass spectra of the 1-propyl N-(heptafluorobutyryl) and the 2-propyl N-(heptafluorobutyryl) esters of leucine were easily identified and the same major ion fragments at m/z, 240, 241, 282 and 283, which also appear in the mass spectra of other amino acid derivatives were also disclosed. Irrespective of whether hydrogen rearrangement takes place or not, it was shown that the ^{15}N abundance can be obtained from either the mass fragments 240 and 241 or 282 and 283, the ^{15}N fragments being shifted to 1 unit mass greater. For example, the $^{15/14}$N-labeled leucine ratio can be taken as proportional to the ratio of the 283 peak to the 282 peak. It is seen that GC/MS tandem systems are useful for solving a range of analytical problems, and the technique is not confined to simple substance-identity confirmation or structure elucidation.

C. de St. Etienne and J. Mettes used the GC/MS instrument to assay the purity of silane. Silane of high purity is required by the electronics industry for the fabrication of integrated circuits and as amorphous silica in photovoltaic applications. The author used a IGC gas chromatograph fitted with a katherometer detector and a packed stainless steel column, 13.2 m long, 3.17 mm I.D., containing Chromosorb PAW 45/60 mesh, coated with 28% of DC 200 stationary phase. Helium was used as the carrier gas at a flow rate of 20 ml/min. The column eluent was split through a low-dead-volume splitter directly to the mass spectrometer.

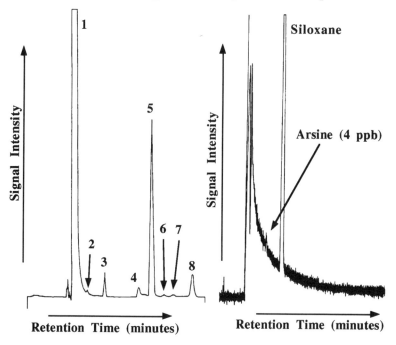

1. SiH_4; 2. C_2H_6; 3. SiH_3CH_3; 4. $(SiH_3)_2O$; 5. Si_2H_6; 6. $SiH_2CH_2CH_2$; 7. $Si_2(CH_3)_2$ and 8. $SiH_2CH_2CH_3$

Figure 5.17. Total Ion Current Chromatograms of Silane Samples (ref 15)

The mass spectrometer was a Riber AQX 156 quadrupole filter having a mass range of 0-300. Electron impact ionization was employed with energies of 70 eV, and the mass spectrometer ion source was maintained at about 100°C. An example of the total ion current chromatogram of a crude silane sample is shown in figure 5.17. It is seen that there are a number of low level contaminants in the crude material, and by the use of single ion monitoring, very small traces of contaminants can easily be identified. The chromatogram obtained by single ion monitoring at a m/z value of 76 discloses the presence of arsine at a level of 4 ppb in the original sample.

GC/MS has been used by Goldberg *et al.* [16] to assay *trans*-resveratol in wines. *Trans*–resveratol (3,5,4'-trihydroxystilbene) has potent antifungal properties and appears to be synthesized by vines, in response to fungal infection. Its incorporation in Japanese herbal medications has roused an interest in its use in fungal, inflammatory and lipid disorders. The *trans*–resveratol was extracted from the wine onto a bed of C18 reversed phase that had be preconditioned with ethyl acetate, followed by 96% ethanol, and then 10% ethanol in water. The solid phase extraction cartridge was then dried and the absorbed *trans*–resveratol extracted by ethyl acetate. The ethyl acetate extract was used directly for analysis.

The GC/MS comprised a Hewlett–Packard Model 5890 gas chromatograph and the Hewlett-Packard Model 5970 quadrupole MS detector. The column (DB-5) was 30 m long, 250 μm I.D. and carried a film of stationary phase, 0.25 μm thick. 1 μl aliquots of the ethyl acetate extract were injected; the column was held at 150°C for 6 min; then heated to 290°C at 20°C/min and held at 290°c for 2 min; then heated to 305°C at 25°C/min and finally held at 305°C for 5 min. The authors demonstrated that solid phase extraction, followed by GC/MS analysis using single ion monitoring, could be readily used for determining *trans*–resveratol in wines. Furthermore, although a little lengthy, the extraction procedure appeared to be generally useful for the analysis of other trace materials in aqueous solutions.

Urine Analysis by GC/MS.

The incidence of urine analysis has substantially increased in the contemporary analytical laboratory. It has been found very useful in toxicology studies, as a diagnostic tool, in pollution monitoring, and for forensic purposes. Some examples of the use of GC/MS in urine analysis will be given here, but it should be pointed out that new areas of application are being continuously reported in the literature.

4,4'-Methylenebis(2-chloroaniline) (MBOAC) is commercially important and is used in the polymer industry to cure urethane elastomers and epoxy-resins. It is a reported carcinogen and therefore exposure to the material is carefully controlled. One of the methods of monitoring exposure is to determine the level of MBOAC in urine. Jedrzejczak and Gaind [17] developed an extraction procedure followed by a GC/MS analysis to measure trace amounts of MBOAC in urine.

The procedure was based on the hydrolysis of MBOCA conjugates followed by solvent extraction after adding deuterium labeled benzidine-d_8 as an internal standard. To 5 ml of urine was added 0.2 ml of the standard and 2 ml of 1M phosphate buffer (pH=10). The mixture was heated at 80°C for 90 minutes to hydrolyze the conjugated MBOCA. 3 ml of ethyl acetate and 1 ml of 2.5% w/w aqueous ammonia were then added to the cooled mixture. The mixture was vigorously shaken with 50 µl of pentafluoro-propionic anhydride for 2 minutes and evaporated to dryness. The residue was then dissolved in 1 ml of ethyl acetate and a 1 µl taken for analysis. The separation was carried out on a 30 m capillary column 320 µm I.D. with a stationary phase film thickness of 1 µm. The column was directly coupled to the ion source of a Hewlett-Packard mass spectrometer which was operated in the negative ion chemical ionization mode using methane as the reagent gas. An example of the single ion monitoring of a sample is shown in figure 5.18.

The deuterated standard furnished an ion at m/z=464, whereas the MBOCA produced a major ion at m/z = 538. The peak at 538, shown in figure 5.18, is from a sample that contained 125 µg of MBOCA per liter.

It is seen that the signal-to-noise is still very high, so concentrations well below 125 µg per liter were detectable. Between 10 and 100 µg per liter the standard deviation appeared to range from about 0.6 µg per liter and 3.8 µg per liter respectively.

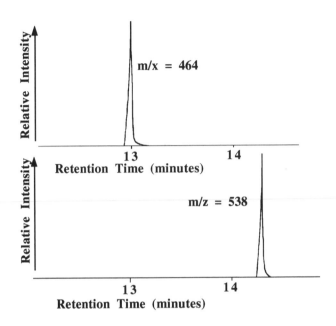

Figure 5.18. Selected Ion Current Chromatograms of 4,4'-Methylenebis(2-chloroaniline) (ref 17)

Several years later Brunmark et al. [18} developed a similar procedure to monitor absorption of 4,4'-methylene bisaniline from skin exposure by means of urine and blood plasma analysis. The urine samples were treated in a very similar manner to Jedrzejczak and Gaind [17]. 2 ml of urine, 50 µl of standard and 2 ml of 10 M NaOH were mixed and hydrolyzed at 80°C for 17 hours. The mixture was then treated with 3 ml of toluene shaken for 10 minutes and centrifuged. A 1 ml sample of the organic phase was treated with 20 µl of pentafluoro-propionic anhydride and shaken; the derivatization was complete in 5 minutes. The reaction mixture was then treated with 2 ml 1.M phosphate buffer, shaken, centrifuged and the organic layer was then used for GC/MS analysis.

In the plasma analysis 50 µl of standard was added to 1 ml of plasma and 1 ml of 10 M NaOH and the mixture hydrolyzed at 100°C for 16 hours. 1 ml of water and 3 ml of toluene were then added and the tube shaken for 10 minutes. A sample of the toluene layer was then worked up in the same way as that used for the urine sample. A Carlo Erba GC equipped with an auto sampler was employed with a capillary column 25 m long, 250 µm I.D. carrying a stationary phase film 0.25 µm thick. The column was heated at 100°C for 1 minute, then heated to 300°C at 15 ° per minute and then maintained at 300°C for two minutes. The mass spectrometer was the Trio 1000 quadrupole operated in the negative ion chemical ionization mode using ammonia as the reagent gas. The interface temperature was maintained at 300°C and the ion source at 200 °C. Chromatograms, obtained by selected ion monitoring, from both urine and blood serum samples from patients who had been exposed to 1.5 µmol of 4,4'-methylenebisaniline are shown in figure 5.19.

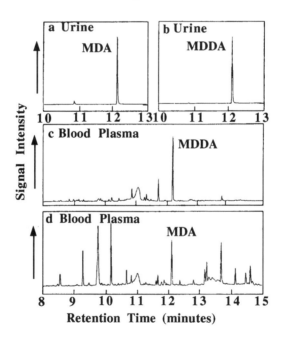

Figure 5.19 Selected Ion Monitoring of Urine and Blood Samples After Dermal Exposure to 1.5 µmol of 4,4'-Methylene Bisaniline [ref. 18]

The urine sample was taken between 6.5 and 8.5 hours after exposure and the blood serum samples 8.5 hours after exposure. The determined 4,4'-methylenebisaniline and the deuterated internal standard were monitored at m/z= 470 (MDA) and m/z=472 (MDDA) respectively, and each chromatogram is shown separately. The urine and plasma samples were found to contain 19 and 4 nmol of 4,4'-methylenebisaniline per liter respectively. It is seen that the technique is extremely sensitive with a high degree of specificity.

Wu *et al.*[19] developed yet another method for measuring 4,4'-methylenebis(2-chloroaniline) in urine using a thin layer chromatography stage in the preparation of the sample. A 20 ml sample of urine was heated to 80°C for 90 minutes and, after cooling, 10 ml of a mixture of diethyl ether–*n*-hexane (50+50) and 0.4 ml of 10 M caustic soda was added. The tube was shaken for 2 minutes and 0.5 ml of methanol was added dropwise across the surface to break up the foam. The mixture was then centrifuged for 10 minutes. The organic layer was removed and the extraction procedure repeated with fresh solvent. The combined extracts were evaporated to dryness in a stream of nitrogen. The residue was taken up in 2 ml of the same solvent mixture and 50 µl of pentafluoropropionic anhydride added. The sample was well mixed and heated to 60°C for 20 minutes. 2 ml of hexane and 4 ml of 0.1 M KH_2PO_4 buffer (pH=6) was added and the mixture well shaken. The mixture was separated by centrifugation and the organic layer collected. The solvent was removed and the residue taken up 0.5 ml of *n*-hexane. 0.4 ml of the solution was progressively loaded onto a thin layer plate and the spot dried. The plate was developed with an ether *n*-hexane mixture (25+75). The area associated with the derivatized 4,4'-methylene-bis(2-chloroaniline) was scraped off and placed in a glass column and washed with *n*-hexane and the washings discarded. The sample was then removed with a ether-*n* hexane mixture (50+50), evaporated to dryness taken up in 1 ml of ethyl acetate containing 0.3 µg of the standard.

The tandem instrument was similar to that used previously. It consisted of a Hewlett-Packard 5890 gas chromatograph fitted with a 25 m column 320 µm I.D. carrying a film of stationary phase 0.25 µm thick, The gas chromatograph was connected to a Hewlett-Packard 5989 mass spectrometer with a heated interface. The mass spectrometer was operated in the negative ion chemical ionization mode and the total ion chromatogram obtained from a sample of spiked urine is shown in figure 5.20

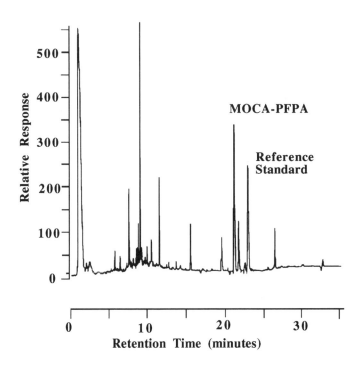

Figure 5.20 Total Ion Current from Urine Spiked with 4,4'-Methylene Bisaniline (ref.19)

It is seen that the procedure is protracted and complex but the peaks shown in figure 5.20 represent 0.095 and 0.24 ng of the 4,4'-methylene-bis(2-chloroaniline) derivative and the standard respectively. It is also seen that the signal to noise ratio is at least an order of magnitude which means that a mass of 10 pg would probably be detectable. The amount of effort put into developing this particular analysis indicates the range of different sample preparation procedures that can be used in GC/MS and, incidentally, the importance of determining the amount of 4,4'-methylene-bis(2-chloroaniline) in urine.

Aggarwal *et al.* [20] employed a GC/MS tandem system to determine the amount of tellurium in urine by isotopic dilution. AS-101 [ammonium trichloro(dioxoethylene-O,O')tellurate (IV)] was the first tellurium-containing compound to be shown to have immuno-modulating properties with little toxicity, consequently its assay in biological fluids was of considerable interest. The process was also quite complicated and can be briefly summarized as follows.

A sample of urine was spiked with a ^{120}Te enriched standard solution and digested with concentrated nitric acid. The dried residue was redissolved in deionized water, centrifuged and adjusted to pH 3.0 with ammonium hydroxide solution. The mixture was then buffered with acetate buffer, treated with a 20 mM solution of lithium bis(trifluoroethyl)dithiocarbamate, vortexed for a few minutes, and the chelate extracted with toluene. The extract was evaporated to dryness in a stream of argon and the dissolved residue, treated in ether solution with (4-fluorophenyl)magnesium bromide. The excess Grignard reagent was destroyed with isopropyl alcohol followed by nitric acid. The chelate was extracted with toluene, evaporated to dryness and taken up in methylene dichloride, a sample of which was placed on the GC column. The chromatograph used was a Varian 3700, with an SE-30 bonded phase fused silica capillary column, 10 m long, 320 µm I.D., carrying a stationary phase film 0.25 µm thick. On-column injection was employed and the column was programmed from 100°C to 300°C at 15 °C/min. The mass spectrometer was the Finnigan MAT Model 8230, and employed with electron impact ionization.

Molecular ion peaks from the sample were generated at m/z values of 312, 313, 314, 315, 316, 318 and 320 which corresponded respectively to derivatives from the isotopes, ^{122}Te, ^{123}Te, ^{124}Te, ^{125}Te, ^{126}Te, ^{128}Te and ^{130}Te. The position of the ^{120}Te(FC$_6$H$_4$)$_2$ peak from the ^{120}Te, used in the isotopic enrichment, was at 310. The method was found to be quite sensitive, and concentrations of the tellurium drug ranging from 100-500 ng/ml in urine could be easily determined.

Veillon and Patterson [21] used a similar technique to determine chromium in urine, employing a volatile chelate formed with trifluoroacetylacetone. Chromium is an essential element in human diet and appears to be associated with glucose metabolism and/or insulin response. The concentrations of interest are well below 1 ng/g and consequently there are a limited number of analytical techniques available for such assays. About 3 g of urine was spiked with an appropriate amount of a ^{50}Cr–enriched standard and 50 µl of magnesium nitrate. The mixture was evaporated to dryness at 90°C, 100 µl concentrated nitric acid was then added, and heating continued. When nearly dry, the residue was treated with 50 µl aliquots of hydrogen peroxide, and again heated until the dry residue was white. The residue was then dissolved in 50 µl of 6 M hydrochloric acid, and again evaporated to dryness. The residue was redissolved in 50 µl of 1 M hydrochloric acid and 1 ml of 1 M ammonium acetate buffer, and 100 µl of trifluoroacetylacetone added. The mixture was then heated in capped tubes at 70°C for 1.5 hours and the chelate then extracted with *n*-hexane. The excess reagent was removed by washing with 0.1 M NaOH, and the hexane layer washed with water. After evaporating to dryness, the residue was taken up in 25 µl of hexane and an appropriate aliquot of the resulting solution injected onto the column. The GC employed was the Finnigan Model 9610 and the mass spectrometer the Finnigan MAT Model 4000. The detection limit of the method was found to be 0.03 ng/g. By the nature of the analysis, the method is highly specific *i.e.* due to the combination of the chelating, gas chromatographic and mass spectrometric selectivity). The technique can also be used in tracer procedures that are employed to follow metabolic pathways in nutrition studies.

GC/MS Tandem Systems in Environmental Analysis

As many environmental contaminants have relatively low molecular weights and often involve organic chlorine or phosphorus compounds, they are often relatively volatile, and so they lend themselves to measurement by GC/MS tandem systems. The most important and frequently monitored environmental samples are those taken from the many and various sources of water. Water is not only an essential domestic requirement but is also the main means by which polluting materials are removed from the land. It follows, that a water analysis can provide the first indications of surface contamination and alert the authorities to a pollution problem in its early stages. Consequently, analytical techniques for measuring water contaminants at very low concentrations, have been the subject of much development research carried out over the last decade.

Buszka *et al.* [22] employed a GC/MS combination to determine trace amounts of organic chlorine compounds in ground water. They collected the samples by a purge-trap procedure and separated the extracted materials on a GC column. The column separation was followed by single ion monitoring. Single ion monitoring is a particularly valuable method of dealing with very complex mixtures, as the detection can be made highly specific and, as a consequence, less resolution is required from the separation technique.

The purge vessel had 100 ml capacity and was fitted with a sintered glass sparge inlet. The volatile materials sparged from the sample passed through a trap, 25 cm long, 2.7 mm I.D., packed with one third Tenax, one third silica gel, and one third charcoal, followed by a 1 cm length of OV-1 GC column packing. The gas chromatograph was the Hewlett-Packard 5996 Model, fitted with a DB-624 megabore capillary column, 30 m long and 530 μm I.D.; the stationary phase had a film thickness of 3 μm. The sample was sparged with helium for 15 min, at a flow rate of 40 ml/min, while the trap was held at room temperature. The trap was then heated to 180°C, and the contents desorbed through a transfer line maintained at 100°C, onto the column, which was held at 10°C. The column was held isothermally at 10°C for 15 minutes after sampling, and then programmed to 190°C at 6°C/min. The trap was baked at 190°C after sample regeneration to prepare it for the next sample. The GC/MS transfer line was held at 170°C and the source and analyzer at 200°C. The contaminant materials they were assaying were dichlorodifluoromethane, trichlorofluoromethane, *cis*-1,2-dichloroethane, trichloroethane, tetrachloroethylene and the isomers of dichlorobenzene. It was found that by employing selected ion monitoring, the GC/MS system provided detection levels that ranged from 1–4 ng/l (1 to 4 x 10^{-12} g/ml) in water.

Recoveries from organic free distilled water ranged from 70.5% for dichlorodifluoromethane to 107.8% for 1,4 dichlorobenzene.

Hiatt *et al.* [23] developed a vacuum distillation procedure for the isolation of volatile materials from natural sources, such as water, soil, oil, fish samples etc. The samples obtained were separated and identified by a GC/MS tandem system. The layout of their sample collection apparatus is shown in Figure 5.21. The samples were placed in a flask of appropriate size, which was connected in sequence to two valves. The first valve could connect the pump to the sample vessel either directly, or *via* a second valve and through a cryo-trap. The sampling procedure was as follows. The sample chamber valve was closed and the condenser, cryo-trap, and gas lines were evacuated. The sample flask was then cooled to the required temperature (usually -5°C). The pump valve was then rotated so that the vapors were drawn through the first valves to the cryo-trap, which was cooled to -196°C with liquid nitrogen. The volatile materials from the sample were condensed and collected in the trap. When collection was complete, the sample valve was again rotated, the cryo-trap warmed, and the contents eluted onto the GC column by a stream of helium.

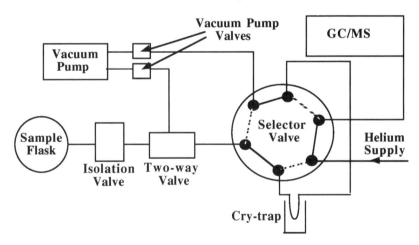

Figure 5.21 The Vapor Sampling Apparatus

Not surprisingly, the recovery of the individual volatile materials was found to be related to their boiling points. If the solubility of the substance was less than 1 g/l then it was shown that very low recoveries were likely. Nevertheless, for the more volatile contaminants, the technique gave results that were sufficiently accurate for many environmental analyses.

Boyd-Boland *et al.* [24] developed a solid phase micro extraction procedure for measuring 60 different pesticides in water, using a small-diameter optical fiber that had been coated with a polymeric stationary phase. The fiber is housed in a syringe assembly and the fiber exposed to the water sample inside the syringe for about 15 minutes. After the water contaminant have been absorbed into the fiber coating, the fiber is removed and placed in the injection system of the gas chromatograph. The fiber is then heated and the materials desorbed and passed onto the GC column.

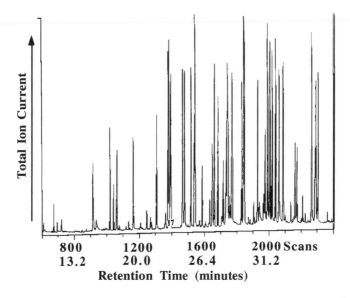

Figure 5.22 Total Ion Current Chromatogram Showing the Simultaneous Determination of 60 Pesticides (ref.24).

Two types of polymer coating were used, polydimethylsiloxane (100 μm thick) and polyacrylate (95 μm thick). The fibers were prepared in accordance with the manufacturers instructions which amounted to heating the fibers in the injection port at about 250°C until no peaks could be detected on a blank run.

The analysis was carried out on a Varian Model 3400 gas chromatograph which was coupled to a Varian Saturn ion-trap mass spectrometer. The capillary column was 30 m long, 250 µm I.D. and carried a film of SPB-5 stationary phase, 0.5 µm thick. The column was held at 40°C for 5 minutes, heated to 100°C at 50°/min., then to 275°C at 5°/min. and finally heated to 300°C at 30°/min. where it was held for 2 min. The injector and transfer lines were held at 250°C. the mass range scanned was 45-400 mu. A chromatogram showing the separation of all 60 pesticides, using the polyacrylate fiber for extracting the sample, is shown in figure 5.22. The concentration of each pesticide in the water was 100 ng/liter indicating a very high sensitivity and a very efficient extraction procedure. The authors claimed a linearity range of 0.1 to 1 µg per liter and demonstrated there was little difference in the extraction efficiency between the two types of polymer coated fibers.

A similar technique was developed by Vreuls *et al.* [25] using a macro extraction system involving a an extraction tube packed with polymeric material. A 1 ml sample was collected in an LC sample loop and the internal standard added. The sample was then displaced through a short column 1 cm long, 2 mm I.D. packed with 10 µm particles of a proprietary PLRP-S polymer (styrene-divinylbenzene copolymer) by a stream of pure water. The extraction column was then dried with nitrogen and the adsorbed materials displaced into the gas chromatograph with 180 µl of ethyl acetate. The sample passed to a short retention gap column and then to a retaining column. The GC oven was maintained at 70°C so that the ethyl acetate passed through the retaining column and was vented to waste. The solutes of interest were held in the retaining column at this temperature while the ethyl acetate was removed. When the ethyl acetate had been removed, the temperature was increased and the components in the residue separated on an analytical column using an appropriate temperature program. The eluents from the analytical column passed to a QMD 1000 mass spectrometer. An example of the chromatograms and spectra obtained are shown in figure 5.23

GC/MS Tandem Systems

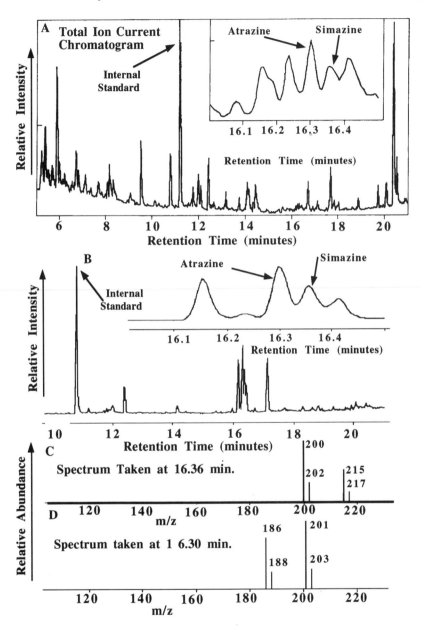

Figure 5.23 Chromatogram and Spectra from a Sample of River Water Containing 200 ppt of the Atrazine and Simazine (ref. 25)

Figure 5.23 A shows he total ion current chromatogram from a sample of Rhine river water containing 200 ppt of the herbicides Atrazine and Simazine. The peaks are shown enlarged in the insert. Figure 5.23 B shows a section of the same chromatogram presented in the selected ion mode. It is seen that the herbicide peaks are clearly and unambiguously revealed. In figures 5.23, C and D, the individual mass spectra of Atrazine (eluted at 16. 30 minutes) and Simazine (eluted at 16.36 min.) are shown. The spectra are clear and more than adequate to confirm the identity of the two herbicides.

Okumyura *et al.*[26] also developed an extraction procedure to determine carbamate pesticides in environmental samples of water. The extraction and analysis procedure, depicted in block diagram form, is shown in figure 5.24.

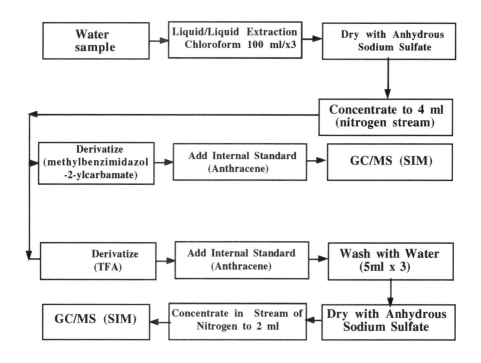

Figure 5.24. Extraction Procedure for Determining Carbamate Pesticides in Water

A sample of water 1 l in volume was extracted three times with 100 ml portions of dichloromethane for 5 minutes. The combined extracts were concentrated on a water bath at 70-80°C and then divided into two equal parts. Each part was evaporated to dryness. in a stream of nitrogen. One portion which was used to determine six types of carbamate, 3-(*sec*-butyl)phenyl *N*-methylcarbamate, 2-isopropoxyphenyl *N*-methyl-carbamate, 3,5-xylyl *N*-methyl-carbamate, 3-methylphenyl *N*-methyl-carbamate, 2-isopropyl-phenyl *N*-methyl-carbamate and 1-naphthyl *N*-methyl-carbamate. One part was dissolved in 0.2 ml ethyl acetate and reacted with 0.2 ml of trifluoroacetic anhydride for 2 hours at 50°C. 5 ml of a *n*-hexane-ether mixture (94 + 6), and 0.2 ml of an external standard solution of anthracene (1.0 ppm in n-hexane) was then added. The mixture was washed three times with 5 ml of water, and then concentrated to a volume of 2 ml. 3 µl of the solution was used for analysis by GC/MS employing the SIM mode. The other portion was dissolved in 0.5 ml of methanol and added to 0.5 ml of diazomethane in diethyl ether and allowed to stand for 1 hour at room temperature. The excess of diazomethane was decomposed with 0.1 ml of acetic acid and 0.3 ml of the anthracene standard added and shaken vigorously. 3 µl samples were used for analysis. The estimated detection limit was 14 to 180 pg per ml for a 1 liter water sample. The recover from the water was estimated to range from 83 to 127%, and the relative standard deviation for 11 analyses was between 2.6 and 22.6%

The identification of specific polychlorinated biphenyls is extremely difficult from mass spectra alone. In contrast vibrational spectroscopy is and extremely sensitive technique for identifying molecular structure and Hembree *et al.* [27] developed a triple system, involving a gas chromatograph, a mass spectrometer and an IR spectrometer. The use of a gold plated light pipe as an interface, can produce high sensitivities as previously discussed in chapter 4. However, the low temperature trapping techniques coupled with the high sensitive HgCdTe IR detector can permit excellent spectra to be obtained at the low picogram level. Furthermore, the condensed spectra that are produced are very similar to those produced by more conventional procedures at room temperature. The system was, in fact, operated as two tandem instruments in parallel.

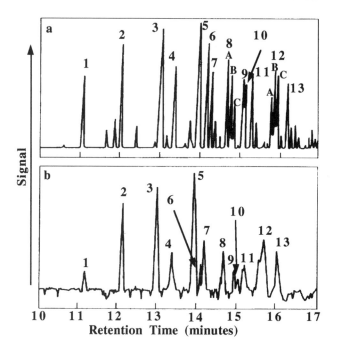

Peak Number	Retention Time(min)	IR identification
1	11.05	2.2'-Dichlorobiphenyl
2	12.01	2.4'-Dichlorobiphenyl
3	12.95	2,2'5-trichlorobiphenyl
4	13.31	2,2'3-trichlorobiphenyl + another trichloro isomer
5	13.88	2,4',5-trichlorobiphenyl
	13.91	2,4,4'-trichlorobiphenyl
6	14.10	2',3,4'-trichlorobiphenyl
7	14.21	2,3,4'-trichlorobiphenyl
8A	14.63	2,2',5,5'-tetrachlorobiphenyl
8B	14.74	2,2',4,5'-tetrachlorobiphenyl
8C	14.83	2,2',4,5-tetrachlorobiphenyl
9	15.04	2,2',3,5'-tetrachlorobiphenyl
10	15.06	3,4,4'-trichlorobiphenyly and a tetrachlorobiphenyl
11	—	a tetrachlorobiphenyl
12A	15.74	2,4,4',5'-tetrachlorobiphenyl
12B	15.81	2,3',4',5-tetrachlorobiphenyl
12C	15.85	2,3',4,4'-tetrachlorobiphenyl
13	—	a tetrachlorobiphenyl

Figure 5.25 Chromatogram of Arochlor:-(a) Total ion Current Chromatogram, (b) IR Functional Group Chromatogram (ref.27)

GC/MS Tandem Systems

A Hewlett-Packard gas chromatograph was used, equipped with two matched silicone gum columns and a single injection system. The matched columns gave a 50-50 split ratio of the injected sample. The FTIR instrument was a Digilab FTS-45 spectrometer, equipped with an interface that captured the eluted material onto a ZnSe plate held a low temperature ((Digilab Tracer). This plate acted as a transport medium to transfer the sample to the FTIR spectrometer. The mass spectrometer was the Hewlett-Packard 5970B mass selective detector. An example of chromatograms of the polychlorinated biphenyls obtained from the combined instrumentation are shown in figure 525. It is seen that excellent separations are displayed by both the ion current monitor and the integrated absorption spectra from the FTIR spectrometer. It is also seen that most of the isomers (all those for which reference spectra were available) were identifiable.

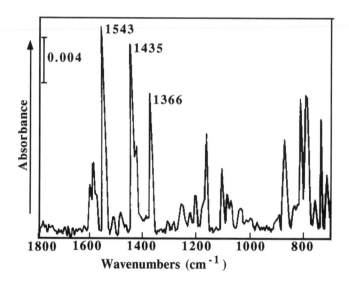

Figure 5.26 The IR Spectrum from 25 ng of 3,3',4,5-Tetrachlorobiphenyl obtained from the GC-FTRI Tracer Unit (ref.27)

The sensitivity of the cryostat interface and the FTIR was extremely high, the spectra shown in figure 5.26 was obtained from only 25 ng of sample

It is seen that the two spectroscopic techniques can be used in a complimentary fashion to identify the very small samples that frequently arise in pollution and environmental studies.

Another source of pollution involving aromatic material is the adsorption of polycyclic hydrocarbons from car exhaust and other sources onto street dust. Yang and Baumann [27] developed an analytical method, involving a GC/MS tandem combination for this purpose. The samples were collected form various sites through out Germany, and were subjected to liquid carbon dioxide extraction's followed by expansion and deposition in a series of short RP 18 columns [28]. The contents of each column could be individually selected and passed to the GC/MS instrument. A Hewlett-Packard 5890A gas chromatograph and a Hewlett-Packard 5970 mass spectrometer were employed for the analysis.

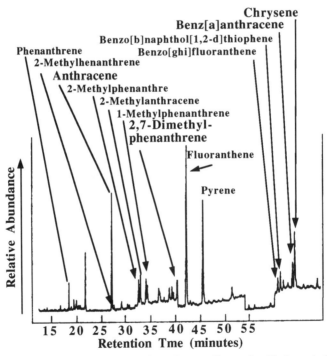

Figure 5.27 Chromatogram of a Dust Sample Extract (ref. 27)

The column was 50 m long, 250 μm I.D. and the separation was selective ion monitored. An example of a chromatogram obtained by selective ion monitoring, that displays the polycyclic hydrocarbon content of a dust sample, is shown in figure 5.27. This analysis would be extremely difficult to carry out with any degree of confidence without the tandem combination. The authors were able to monitor the change in polynuclear aromatic pollution with both the geographical area and the seasons.

De Beer *et al.* [29] developed a method for measuring herbicide contaminants in air using a GC/MS tandem system. They used a collection tube, (32 cm long, 6 mm I.D.) packed with 1 g of an absorbent resin XAD-2, and a 2 cm plug of silanized glass wool. The glass wool collected the dust particles, and the resin trapped any herbicide vapors in the air or were desorbed from the particles during sample collection. 5 to 8 m^3 were passed through the trap to collect the sample. 20 ml of n-hexane was pumped through the trap to elute absorbed material from the resin. The herbicide salts were eluted from the glass wool and from the particles by washing with four 5 ml volumes of alcoholic caustic soda. The eluates were concentrated to 10 ml on a rotary evaporator. The extract was then acidified with dilute sulfuric acid and left for 30 min. at room temperature to hydrolyze the salts to the respective acids.

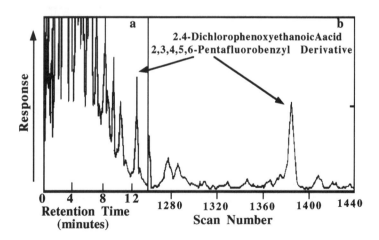

Figure 5.28 Chromatogram of Derivatized Air Contaminants (a) Monitored by Electron Capture Detector, (b) Displayed by Selected Ion Monitoring (ref. 29)

The sample was extracted with toluene and then derivatized in the manner described by Lee et al. [30]. A Varian 6000 gas chromatograph was employed in conjunction with a Finnigan ITD (model 800 cl) mass spectrometer. A DB-5 column was used, 30 m long, 312 μm I.D., with a film thickness of 1 μm. A example of the results of the analysis of an air sample is shown in figure 5.28. The peak for the herbicide is reported to be equivalent to 9 pg of derivatized material. The limit of detection was reported to be about 0.72 ng /m^3 of air.

Derivatization Techniques

Derivatization procedures are employed in GC/MS tandem analyses mainly for two reasons. First, to render the sample components volatile, so that they can be separated by development in a gaseous mobile phase, and second, so that they can be used with a simple direct-inlet MS interface. Involatility can arise from two causes. The sample may be strongly polar, *e.g.* aliphatic acids, carbohydrates or the higher molecular weight alcohols. Alternatively, the sample can have a very high molecular weight, *e.g.* waxes, synthetic polymers, biopolymers, etc. Substances that are involatile as a result of strong polarity can often be successfully derivatized and made relatively volatile. Unfortunately, substances that are involatile as a result of their high molecular weight can rarely be rendered volatile merely by derivatization. It follows that whereas it is easy to produce a volatile derivative of a low molecular weight aliphatic acid, it is virtually impossible to make a volatile derivative of a biopolymer. Such substances are usually analyzed on the complementary LC/MS tandem systems, using special interfaces which will be discussed later in this book.

GC derivatizing agents are classified on the basis of the chemical nature of the materials they render volatile. Different reagents are used to make volatile derivatives of diverse materials, for example, acids, alcohols, amino acids, etc.

Esterification

Esterification is a procedure commonly used to form volatile derivatives of organic acids. The most popular and useful method is esterification with an alcohol in the presence of hydrochloric acid. A few milligrams of the material is heated with 100-200 µl of the alcohol containing 3 M HCl, for about 30 minutes at 70°C. The alcohol is removed by evaporation in a stream of nitrogen (assuming the derivative is considerable less volatile than the alcohol). Boron trifluoride and boron trichloride are also useful esterifying reagents. BF_3 catalyzed reactions are fairly rapid, and after heating in a boiling water bath are complete in a few minutes. A few milligrams of the sample are added to 1 ml of the reagent, heated in boiling water for about 2 minutes and the esters extracted into *n*-heptane.

Diazomethane is another reagent used to produce methyl esters but, unfortunately, it is extremely carcinogenic and must be *handled with great care*. *N*-methyl-*N*-nitro-*N*-nitrosoguanidine is sometimes used as the starting material for generating diazomethane, but it too is exceedingly carcinogenic. The De Boer and Backer reagent (Diazald) (*N*-methyl-*N*-nitroso-*p*-toluenesulphonamide) [31] is to be preferred. Diazomethane is a yellow gas, sometimes employed as a solution in diethyl ether in the presence of methanol. It is usually prepared as required, employing the simple generating apparatus described by Schlenk and Gellerman [32].

A few milligrams of the acid are dissolved in 2 ml of ether containing 10% of methanol, and the gas generated from 2 mmol of Diazald is bubbled into the solution by a stream of nitrogen. When the yellow color in the sample solution persists, the ether/methanol is removed in a stream of nitrogen, providing a residue of the ester. The trimethyl silyl reagents, *e.g.* trimethylsilyl chloride, readily produce trimethyl silyl esters, although there are a number of other silicone compounds that can be used to produce similar esters. A few milligrams of the acid are dissolved in 600 µl of pyridine, and 200 µl of di(chloromethyl)tetramethyldisilazane are added, together with 100 µl of chloromethyldimethylsilyl chloride. The mixture is then allowed to stand for 30 minutes at room temperature to react. Another reagent that works well, and is frequently used for esterification using a similar procedure, is hexamethyldisilazane.

Acylation

Acylation can greatly reduce the polarity of amino, hydroxy, and thiol groups attached to small molecular weight substances. As a result, their chromatographic behavior can be significantly improved and the substances made more volatile. In particular, peak tailing can be reduced in GC and streaking in TLC. The two popular types of reagent for acetylation are the acid anhydrides and the acid chlorides. Between 1 and 5 milligrams of the sample is dissolved in chloroform (about 5 ml) and warmed with 0.5 ml of acetic anhydride and 1 ml of acetic acid for 2–16 hours at 50°C. Excess of reagent is removed by evaporation in vacuum (assuming the product is relatively involatile) and taken up in chloroform for subsequent GC analysis. Sodium acetate is also often used as a basic catalyst for acetylation, particularly for carbohydrates extracted from urine. The dried residue from the urine is oximated to derivatize the carbonyl groups and then the hydroxyl groups are acetylated with acetic anhydride in the presence of sodium acetate at 100°C. Acetyl chloride is not so widely used for acetylation, because hydrochloric acid is evolved in the reaction, so the presence of an appropriate base is considered necessary to scavenge the hydrochloric acid as it is produced. The N-acetyl methyl ester derivative of hydroxyanthranilic acid, which is a metabolic product of tryptophan, is often used for the GC analysis of the material. The sample is first esterified with diazomethane to give the methyl ester. The product is then evaporated to dryness in a stream of nitrogen and acetylated with a 50% benzene/acetyl chloride mixture [33]. The reaction mixture is dried in a stream of nitrogen and taken up in methanol for GC analysis. Only a fraction of the possible derivatizing reagents have been mentioned, for further details the reader is directed to Blau and Halket [34].

Chiu *et al.* [35] employed a derivatization technique to improve the detection limits of O^6-(hydroxyalkyl)guanines in a GC/MS tandem system, using electron capture as the GC detector. The O^6-(hydroxyalkyl)-guanines comprise a class of DNA products that appear to be strongly mutagenic as well as carcinogenic [36,37]. As a consequence these substances are of considerable interest even when present in trace amounts. In this case, a rather complicated derivatization procedure was employed. The O^6-alkylguanines adducts were treated with fluoroboric acid/nitrous acid to form the corresponding 2-fluoro-O^6-alkylhypoxanthines. The reaction products were then treated with pentafluorylbenzyl bromide, to form the fluorinated derivatives which are strongly electron capturing in character.

Derivatization can also be carried out *after* the separation has taken place, and this procedure is called post-column derivatization. However, post-column derivatization is not performed for chromatographic purposes but to either increase detector sensitivity or to augment the information provided by the mass spectra. It is clear that the derivatization must take place in the gas phase, and thus only relatively volatile derivatizing reagents can be used. The most common derivatizing procedures are bromination, deuterium exchange and acylation. A number of schemes for post-column derivatization have been reported. However, recently Ligon and Grade [38] described an instrument developed specifically for use with GC/MS systems. The principle of their derivatization apparatus is depicted in Figure 5.29.

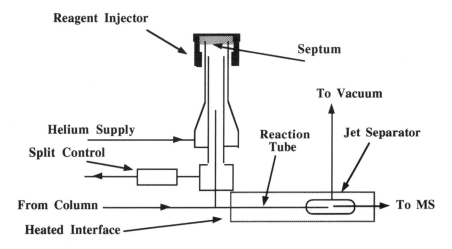

Figure 5.29 Post Column Derivatization Apparatus

In the original design the apparatus could take eluents from either of two columns, one for normal use, the other for thermal desorption analysis. The column outlet was connected to the mass spectrometer using a jet concentrator interface. The reagent was injected into a helium stream using a GC sample injector, and the helium stream, carrying the reagent vapor, joined the column eluent between the column and the jet interface. Reagent samples were injected though the system and the mass spectra taken, to ensure that no artifacts were produced. The reagent reached the mixing T in about 1 sec and the reaction time was also about 1 sec, so that peaks as little as 5 sec peak width at half height could be efficiently derivatized. The distinctive advantage of this system was that specific peaks could be chosen exclusively for derivatization. The sample was run normally, and the retention times noted. The separation was repeated and reagent injected at the appropriate times to derivatize the chosen peaks.

Some Special Applications of GC/MS

The presence of dispersive (nonpolar) toxicants in the many and various sources of aqueous effluents, is an ever present challenge for environmental chemists. Such materials are often in very low concentration and usually need extraction, and concentration, to bring them to a level that is convenient for assay. Some techniques that can be used for this purpose have already been discussed, but some further examples will be given that are of general use. Burkhard and Durham [33] developed a satisfactory method based on a GC/MS tandem system. In fact the sample was prepared for analysis by a preliminary liquid chromatography separation process, and so the method might be considered to be another triplet system. An appropriate volume of the sample, for example river water is passed through a preconditioned, C18 solid phase extraction tube, which selectively removes all the dispersive constituents of the water, on the front of the packing. The absorbed materials are then displaced from the packing using 25, 50, 75, 80, 85, 90, 95, and 100% methanol in water, providing 8 aliquots. The fractions are then concentrated, and each is then separated on an appropriate LC column, employing a suitable gradient. However, the gradient will need to be varied somewhat for each aliquot. LC fractions are collected at minute intervals and concentrated before injection onto the GC column. Due to preliminary LC separation, the fractions give very simple chromatograms to interpret, and, with the aid of the mass spectrometer, the sample can not only be quantitatively assayed, but the identity of the toxicants can usually be unambiguously confirmed.

Creaser et al.[40] developed an analytical procedure for assaying flavones in fruit juices. The flavenoids in citrus juices have been investigated extensively because of their pharmacological activity. The principal citrus flavanones, naringenin, hesperetin and eriodictyol do not occur in juices as free aglycones, but are combined through the C-7 hydroxy group with a sugar component such as β-neohesperidose or β-rutinose. The neohesperidosides are thought to contribute bitterness to the fruit, whereas not so the rutinosides.

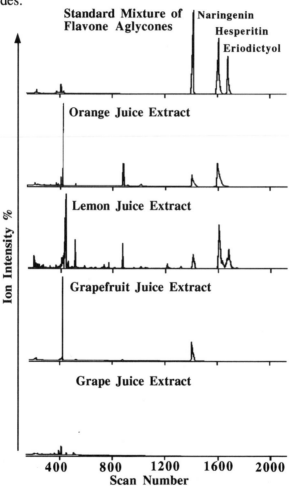

Figure 5.30. Selected Ion chromatograms of Fruit Extracts (ref. 40)

A 100 ml sample of hand squeezed, clarified fruit juice was extracted with two portions of 100 ml of ether and the extracts discarded. The aqueous phase was then extracted with 4 portions of 100 ml of ethyl acetate and the combined extracts evaporated to dryness in *vacuo* at 35-40°C. The residue was taken up in 10 ml of ethanol, and 5 ml of 5% HCl added to a 3 ml portion of the solution, and the mixture heated to 100 °C for two hours on an oil bath. The solution was extracted with three 10 ml portions of ethyl acetate, the extracts were combined, dried over anhydrous sodium sulfate and evaporated to dryness. The residue was dissolved in 0.2 ml of pyridine, 0.2 ml of hexamethyldisilazane and 0.1 ml of trimethylchlorosilane added, and the mixture heated overnight at 60°C. The prepared sample was separated on a Varian 3400 gas chromatograph which was directly interfaced with a Finnigan MAT ion trap mass spectrometer. The capillary column was 50 m long, 250 μm I.D. and carried a film of RSL 200 BP 0.2 μm thick. The oven was held at 130°C for 0.5 min, heated to 235 at 30°C/min and from 235 to 290°C at 1°/min. The transfer line was maintained at temperature of 280°C and the ion trap at 150°C. An example of the results obtained are shown in figure 5.30. In the chromatogram of the standards, each peak represented between 1 and 2 μg/ml of flavanone in the original juice. It is seen that the components are well separated and the technique is very effective for assaying flavanones in juices. It is interesting to note that the grape juice gave no indication of the presence of flavanones which occurred only in the citrus fruit juices.

McCalley and Torres-Grifol [41] investigated a head-space sampling technique for monitoring the condition of oranges by GC/MS analysis. The aim of the work was to provide a simple and rapid test to distinguish between good, damaged and diseased fruit. The technique was developed using Navel oranges of low and medium ripeness and of approximately 200 g in mass. The oranges were placed in a reaction vessel, 1.25 l in capacity, which incorporated a GC septum to allow the removal of a sample of the head space. The samples were taken at room temperature one hour after the fruit had been placed in the flask. Although trapping techniques would give greater sensitivity, the head space sampling offered a simpler procedure and a shorter overall analysis time. Retention of solutes on the syringe walls, while sampling the head space, was reduced to an acceptable level by heating the syringe to 60°C just before taking the sample. The gas chromatograph was the Hewlett-Packard 5890A fitted with 25 m capillary column, 320 µm I.D. containing a film of BP-20 that was 1.0 µm thick. The column was maintained at an initial temperature of 35°C for 5 min. and then programmed up to 200°C at 10°/min. A 500 ml sample of head space was introduced into the column through a split injection system with a split ratio of 3:1. The Hewlett-Packard 5995C mass spectrometer was employed with the chromatograph. The transfer line was maintained at 250°C, the source at 240°C and the separator at the same temperature 240°C. A chromatogram of the volatile materials contained in the head space over some oranges that had been incised with ten shallow cuts 20 cm long, is shown in figure 5.31.

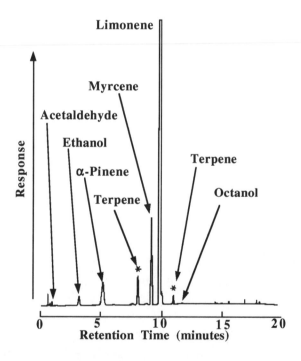

Figure 5.31 Chromatogram of the Volatiles Contained in the Head Space Over a Mechanically Damaged Orange (ref.41)

It is seen that an satisfactory chromatographic resolution is achieved and the solutes could be easily identified by spectra matching. The authors demonstrated that the head space pattern would be characteristic for bruised fruit, over-ripe fruit and fruit infected with *penicillium digitatum* Sacc. or *P. italicum* Wehmer.

Another interesting application of headspace analysis is in the measurement of certain types of contaminant found in pharmaceutical products that are volatile. These contaminants usually arise from the solvents used to purify the product by crystallization, or from flavor additives. It follows that as some solvents used in the crystallization process are toxic, their measurement is important. Schuberth [42] developed a procedure involving head space analysis followed by GC/MS evaluation, to assay drugs for volatile materials. The technique used has been simply described by Mulligan and McCauley [43]. A sample of the drug is carefully weighed and placed in a tube (often about 10 ml in volume) and usually filled with an inert gas such as helium, at a pressure of about 5 psi. The tube is then heated for a predetermined amount of time at an appropriate temperature, and an aliquot of the head space gas then allowed to flow into a gas sampling tube. The sample tube is then placed in line with the carrier gas to the GC column and the contaminants separated. Schuberth employed an ion trap mass spectrometer to detect and identify the volatile components recovered from the drug. Ethanol, acetone, 2-propanol, methyl acetate, toluene, eucalyptol and menthol were identified in various pharmaceutical products, at levels ranging from 1-10 μmol./kg.

Novel Interfaces for Specific Applications

Most of the interfaces described so far have been developed for general use and have been utilized for a wide range of different applications. However, a number of interfaces have been developed for use in specific types of analysis and, although all the different forms cannot be described in this book, a number of them will be discussed, particularly those that have been used in GC, LC and/or capillary electrophoresis tandem systems.

Affinity Membrane Mass Spectrometry Interfaces

The affinity membrane can be used as the separation system itself, or it can be used as the interface between a liquid chromatograph and a mass spectrometer. Nevertheless, as one form accepts the sample as a vapor into a gas stream that transports the sample to the mass spectrometer, it will be discussed in this section dealing with GC tandem systems.

The affinity membrane has been described by a number of workers, for trace analysis [44–46] and analysis of the chemical constituents of biological systems [47]. A system that describes a type of membrane that is very specific is that of Chen Xu *et al.* [48], who used it to selectively extract and detect aldehydes. A diagram of the selective membrane interface is shown in figure 5.32. The preparation of an affinity membrane to render it specific can be very complex, and the particular procedure used by the authors for forming an aldehyde-specific membrane will be described below. The membrane is supported diametrically across the center of a flat cylindrical chamber and held in place by two fluid-tight PTFE gaskets. The sample, which can be a gas or a liquid, and in this particular case is a liquid, passes across one side of the chamber that has been bisected by the membrane. The selected solute diffuses across the membrane and evaporates into a stream of helium gas that passes through the other half of the transfer chamber.

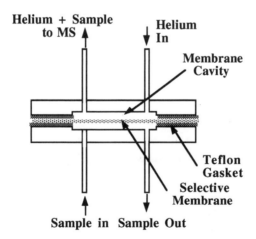

Figure 5.32 The Membrane Interface

The transfer film was prepared from a square sheet of cellulose dialysis-membrane (2 cm x 2 cm, 0.28 µm thick), having a molecular weight cut-off at 3500. The membrane was washed with water, methanol, and then water again, followed by washes with 30%, 50% and 70% (v/v) aqueous dioxane and finally pure dioxane. The activation and modification procedure was similar to that used for agarose beads [49]. 0.5 g of the membrane was suspended in 10 ml of dioxane containing 0.28 g of carbonyldiimidazole and the mixture stirred for 15 minutes at room temperature. The membrane was then washed with 50 ml of dioxane, followed by 50 ml of 0.3M sodium borate (pH 10). After washing, the membrane was suspended in 20 ml of borate buffer containing 3.25 g of hexanediamine, and the reaction allowed to proceed overnight at 4°C. After the reaction was complete, the membrane was washed exhaustively with water before use. It is seen that the preparation of the membrane is a complicated and tedious procedure, but for certain analyses may well be worth the effort expended. The publication does not indicate the life that could be expected from an average membrane. The interface is used with a jet separator (Ryharge separator), that concentrates the extracted materials in the helium purge gas. The output of the jet separator passes directly to the ion trap mass spectrometer. The layout of the total apparatus is depicted diagramatically in Figure 5.33.

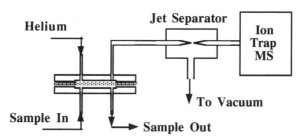

Figure 5.33 The Membrane Separator/Mass Spectrometer Tandem System

The tandem instrument was used to determine the concentration of a number of different aldehydes in aqueous solution, and it was found that the limit of detection or benzaldehyde was about 10 ppm. A detection limit of 10 ppm is not particularly high, but the system acts as a continuous analyzer and has some unique characteristics that might make it useful as a water pollution monitor.

A Combustion Interface for Isotope Ratio Monitoring with GC/MS

The use of a combustion process to determine the isotopic composition of gas chromatography eluents was reported in the 1970s [50,51]. However, more recently, Meritt *et al.* [52] have carried out an exhaustive examination of the method and from the results have suggested optimum conditions for a GC/MS tandem system. The basic system comprised a gas chromatograph, a high–temperature oxidizer, a water trap, a split sampling device and a mass spectrometer. The general layout of the apparatus used by Merritt *et al.* is shown in Figure 5.34.

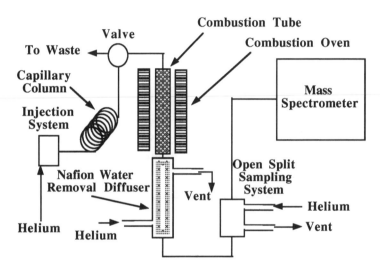

Figure 5.34 A Diagram of the Combustion Interface

The gas chromatographs employed were either the Hewlett–Packard Series 5890A, or the Varian Model 3400 and samples were injected either on-column or by a 1 µl sample valve. After passing through the column, the carrier gas could either be diverted to waste or passed directly to the combustion unit. The micro-volume combustion reactor consisted of a non-porous alumina tube, 30 cm long, 0.5 mm I.D., packed with a metal oxide. In one embodiment, the tube was packed with three lengths of copper wire and one length of platinum wire centered inside the tube. In an alternative combustion tube the packing consisted of nickel wires. The tubes were placed in a resistively heated quartz tube furnace and all transfer capillaries were constructed from deactivated fused silica tubes, 0.32 mm I.D. The water formed by combustion was removed by a tubular Nafion membrane, which was mounted coaxially in a stainless steel tube. The annular volume round the transfer tube was purged free of water vapor by means of a stream of helium gas. The flow of eluent from the water separator to the mass spectrometer was controlled by an open split valve, similar to that described on page 154, in Chapter 4. The mass spectrometers used were the Delta S and the Finnigan MAT 252.

It was found that the organic carbon contained in peaks, 5 sec or more in width, containing as much as 40 nmol of carbon, could be quantitatively converted to CO_2, in tubular combustion reactors 200 x 0.5 mm, packed with either copper oxide or nickel oxide. No auxiliary supply of oxygen was required, the necessary oxygen for oxidation being provided by the oxides. Due to the relatively high partial pressure of oxygen over copper oxide at high temperatures, this material could not be used above 850°C. Combustion was complete using nickel oxide at 1150°C and, if some free oxygen was present, the reaction temperature could be reduced to 1050°C. It was found that the standard deviation of the data for the measurement of $^{13}C/^{12}C$ was less than twice the shot noise, and so the combustion system did not affect the analytical precision. It was also found essential to remove all the water produced during combustion, otherwise protonation of the CO_2 would result in significant inaccuracies.

Pyrolysis GC/MS of Biological Particulates Collected During Space Shuttle Missions

The absence of significant gravitational forces when in orbit allows particles of human food, skin flakes, paint chips, lint, dust etc., to remain suspended in the shuttle atmosphere, until removed by air filters. These free-floating particles, and those arising from in-space experiments, are potential sources of eye, skin and other forms of irritation, and could also transport and spread infectious diseases and possibly produce other types of hazard. Concern over the possible release of particles from animal and other experiments, and the likely and consequent effects on health has instituted the collection of dust particles during orbit for analysis on return to earth. Microscopic examination and elemental analysis has proved to be of limited value, and so means of identifying particulate sources, by tests that would provide 'fingerprint' identification of source materials, were consequently investigated.

Pyrolysis/GC/MS has been successfully applied to the study of a variety of biological materials [53], including polysaccharides [54], lignins [55], and soils [56]. With these publications in mind, a pyrolysis/GC/MS system was developed by Matney and Limero [57], that could provide chromatographic and spectrographic patterns that might allow particulate source identification. The triplet combination comprised a pyrolysis unit, a gas chromatograph and a mass spectrometer. A diagram of their apparatus is shown in Figure 5.35.

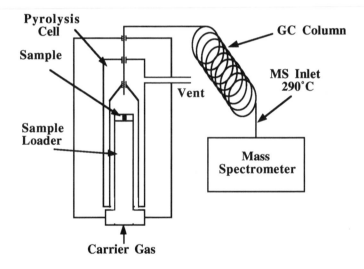

Figure 5.35 Pyrolysis GC/MS System for Particulate Analysis

The quartz pyrolysis cell is connected directly to a capillary column which, in turn, is coupled directly to a Hewlett–Packard 5971 mass selective detector. Samples were loaded automatically into the pyrolysis chamber and heated at 30°C for 7 minutes to establish thermal equilibrium. The cell was then heated to 600°C at a rate of 30°/min. A helium flow of 10 ml/min was passed through the cell, and the pyrolysis products were condensed on the front of the capillary column, which was held at -55°C. A DB-5 column was employed, 30 m long, 320 µm I.D., carrying a film of stationary phase 0.25 µm thick.

After pyrolysis was complete, the column was programmed from -55°C to 200°C at 5°/min. The mass spectrometer scanned the eluent at 1.3 scans/sec over the mass range 25-450 amu. It was found that particles 0.5-1.0 µm in diameter and weighing as little as 40 µg could be identified using this technique. The triplet combination was able to identify rat food and soilless plant-growth media, as two sources of particles collected from the shuttle atmosphere taken during flight.

Table 5.1 Mass Spectrometer Suppliers

Mass Spectrometer	Supplier
Automass 150	ATI Instruments, 1001 Fourier Drive, Madison, WI 53717, 608-831-5515.
Magnum	Finnigan MAT, 355 River Oaks Pkwy, San Jose, CA 95134, 408-433-4800.
MD 800 EI/CI+/CL-	Fissons Instruments, 55 Cherry Hill Dr., Beverly, MA 01915, 508-524-1000.
5972A	Hewlett Packard, 2850 Centreville Road, Wilmington, DE 19808, 302-633-8000.
3DQ[C]	Hitachi Instruments, 3100 North First St., San Jose, CA 95134, 408-432-9520
Q-Mass 910	Perkin Elmer, 761 Main Avenue, Norwalk, CT 06859, 203-762-1000.
QP-5000	Shimadzu, 7102 Riverwood Drive, Columbia, MD 21046, 410-381-1227.
3DQ Discovery[C]	Teledyne, 1274 Terra Bella Avenue, Mountain View, CA 94043, 415--691-9800
Saturn 3	Varian, 2700 Mitchell Drive, Walnut Creek, CA 94598, 510-939-2400.

GC/MS is an extremely powerful analytical procedure and has a wide range of applications. In general it cannot unambiguously elucidate the structure of a completely unknown substance, although this might be possible in certain cases, if a high resolution mass spectrometer is employed. An IR spectrum simultaneously produced by the GC/MS/IR triplet system would help significantly in the structure elucidation. The GC/MS system, however, can confirm the identity of a given solute with a high degree of confidence and for this purpose the GC/MS tandem system is most commonly used. Some supplier of mass spectrometers that have been employed with tandem instruments are given in Table 5.1.

Synopsis

The GC/MS combined instrument was introduced soon after the invention of GC itself, and initially, the mass spectroscopists looked upon GC as a novel sampling system for MS. In the early days, packed columns were used which required high flow rates, and thus sample concentrators were developed. The two most popular were the Ryharge concentrator, that operated on the principle of differential gas diffusion between concentric jets, and the Bieman concentrator, that operated on the principle of differential gas diffusion through a porous membrane. There are two popular types of ionizing processes used in GC/MS systems, electron impact ionization, and chemical ionization. Electrons, having energies between 5 and 100 V, are usually employed in electron impact ionization, and the ions formed are magnetically and/or electrostatically impelled into the analyzer region of the mass spectrometer. The accelerating voltage is adjusted to provide the optimum energy that would give the desired ion fragment pattern. Chemical ionization is achieved by producing ions from a reagent gas, such as methane by electron impact, and allowing the reagent ions to collide with sample atoms, producing sample ions. The ionization is gentle, (ca 5 eV), so very few fragment ions are produced, and those that are produced have m/z values close to that of the parent ion. Different reagent gases give different ions, and can provide different structural information. Most chemical ionization sources have a number of reagent gases available, and many ion sources can perform the dual role of electron impact and chemical ionization. The inductively coupled plasma ionization source, produces ions in an argon plasma maintained by inductively coupled rf energy, and about 50% all the elements present in the sample produce ions in the plasma region. The sample atoms account for less than 10^{-6} of the total number of atoms present in the plasma region, and thus there is little or no quenching effect due to the presence of the sample. The ions pass through a series of skimmers to remove the gas, and are eventually electrostatically deflected into the analyzer of the mass spectrometer. This type of ionization has the advantages that ionization proceeds at atmospheric pressure, all the different elements are ionized, and only singly charged ions are produced. The detection limit is about 1 part in a trillion. Today, capillary columns are mostly used in GC/MS instruments, in conjunction with single or triple quadrupole mass spectrometers. Employing the above interfaces, the GC/MS tandem system has been successfully applied to wide range of analytical problems, but has been particularly useful in environmental and pharmaceutical analysis and in forensic chemistry. A number of other less common interfaces have also been developed. Some of these include the membrane interface that is fabricated to specifically transport selected materials from one gas or liquid stream to a stream of helium; a combustion interface for isotope ratio monitoring; and a pyrolysis interface developed for identifying particulate matter from its chromatographic and mass spectrometric fingerprints. Derivatization techniques are sometimes used to increase the partial pressure of certain samples, and thus make them amenable to separation by GC, and allow them to be introduced into the MS by a direct inlet probe. Samples that have low volatility due to their high molecular weight cannot be rendered volatile by derivatization. Conversely, strongly polar, low molecular weight materials can usually be derivatized to provide volatile products. Acids can often be esterified with appropriate alcohols, catalyzed with hydrochloric acid or boron trifluoride and also reacted with diazomethane. Hydroxyl compounds can be acetylated with acetic anhydride or acetyl chloride. Derivatization can take place before the separation (pre-column) or after the separation, (post-column) depending on the nature of the sample.

References

1. J. C. Holmes and F. A. Morrell, *Appl. Spec.*, **11**(1957)86.
2. R. Ryhage and E. von Sydow, *Acta Chem. Scand.*, **17**(1963)2025.
3. J. T. Watson and K. Bieman, *Anal. Chem.* **37**(1965)844.
4. M. S. B. Muson and F. H. Field, *J. Am. Chem. Soc.*, **88**(1966)2621.

5. S.L Bonchin-Cleland, T. J. Cleland, L. K. Olson and J. A. Caruso, *American Lab.,* **Feb**(1996)34P.
6. A. E. Banner, R. M. Elliott and W. Kelly, *Gas Chromatography 1964,* A. Goldup (Ed.), J. Inst Pet., London, (1964)180.
7. J. Ferrario, C. Byme, D. McDaniel and A. Dupu, Jr., *Anal. Chem.,* **68(4)**(1996)647.
8. C. Schoene, A. N. R. Nedderman and E. Houghton, *Analyst,* **119**(1994)2537.
9. D.Hooijerink, R.Schilt, E van Bennekom and B. Brouwer, *Analyst,* **119**(1994)2617.
10. H. M.Wallace and C. S. Coleman, *Analyst,* **115**(1990)517.
11. W. W. Christie, *J. Lipid Research,* **23**(1982)1072.
12. R. P. Evershed, C. Heron and L. J Goad, *Analyst,* **115**(1990)1339.
13. A. Laukó, E. Csizér and S. Görög, Analyst, **118**(1993)609.
14. A.Dugay,B.Dang-Vu,J.C. Moreau and F. Guyon, *Anal. Chem.,* **67(21)**(1995)4000.
15. C. de St. Etienne and J. Mettes, *Analyst,* **114**(1989)1649.
16. D. M. Goldberg, J. Yan, E. Ng, E. P. Diamandis, A. Karumanchiri, G. Soleas and A. L. Waterhouse, *Anal. Chem.,* **66(22)**(1995)3959.
17. K. Jedrzejczak and V. S. Gaind, Analyst, **117**(1992)1417.
18. P. Brunmark, M. Dalene and G. Skarping, *Analyst,* **120**(1995)41.
19. Weh S. Wu , R. S. Szklar and R. Smith, *Analyst,* **121**(1996)321.
20. S. K. Aggarwal, M. Kinter, J. Nicholson and D. A. Herold, *Anal. Chem.,* **66(8)**(1995)1316.
21. C. Veillon and K. Y. Patterson, *Anal. Chem.,* **66(6)**(1994)856.
22. P. M. Buszka, D. L. Rose, G. B. Ozuna and G. E. Groschen, *Anal. Chem.,* **67(20)**(1995)3659.
23. M. H. Hiatt, D. R. Youngman and J. R. Donnelly, *Anal. Chem.,* **66(6)**(1994)905.
24. A. A. Boyd-Boland, S. Magdie and J. B. Pawliszyn, *Analyst,* **121**(1996)929.
25. J. J. Vreuls, A-J. Bulterman, R. T. Ghijsen and U. Th. Brinkman, *Analyst,* **117**(1992)1701.
26. T. Okumura, K. Imamura and Nishikawa, *Analyst,* **120**(1995)2675.
27. D. M. Hembree, N. R. Smyrl, W. E. Davis and D. M. Williams, *Analyst,* **118**(1993)249.
28. K. Schäand and W. Baumann, *Fresenius'. Z. Anal. Chem..,* **332**(1989)884.
29. P. R. de Beer, E. R. I. C. Sanddmann and L. P. van Dyk, *Analyst,* **114**(1989)1641.
30. H. B. Lee, Y. D. Stokker and A. S. Y. J. Chau, *J. Assoc. Off. Anal. Chem.* **69**(1986)557.
31. T. J. de Boer and H. J. Backer, *Recl. Trav. Pays. Bas.* **73**(1954)229.
32. H. Schlenk and J. L. Gellerman, *Anal. Chem.,* **32**(1991)8.
33. D. P. Rose and P. A. Toseland, *Clin. Chim. Acta.,* **17**(1967)235.
34. *Handbook of Derivatives for Chromatography* (Ed.Karl Blau and John Halket), John Wiley, New York (1993).
5. C. S. Chiu, M. Saha, A. Abushama and R. W. Giese, *Anal. Chem,* **65(21)**(1993)3071
36. G. Isowa, K. Ishizaki, T. Sadamanto, K. Tanaka, Y. Yamaoka, K. Ozawa and M. Ikenaga, *Carcinogenisis,* **12**(1991)1313.
37. A. T. Natarajan, S. Vermeulen, F. Darroudi, M. B., Brent, T. P. Brent, S. Mitre and K. Tano, *Mutagenisis,* **7**(1992)83.
38. W. V. Ligon and H. Grade, *Anal. Chem.,* **63(3)**(1991)255.
39. L. P. Burkhard and E. J. Durham, *Anal. Chem.,* **63(3)**(1991)277.
40. C. S. Creaser, M. R. Koupai-Abyzania and G. R. Stephenson, *Analyst,* **117**(1992)1105.

41. D. V. McCalley and J. F. Torres-Grifol, *Analyst,* **117**(1992)721.
42. J. Schuberth, *Anal. Chem.*, **68(8)**(1996)1317.
43. K. J. Mulligan and H. McCauley, *J. Chromatogr. Sci.*, **33**(1995)49.
44. P. J. Savickas, M. A. LaPack and J. C. Tou, *Anal. Chem.*, **55**(1983)813.
45. G. J. Kallos and N. H. Mahle, *Anal. Chem*, **61**(1989)2332.
46. B. J. Harland, P. J. Nicholson and E. Gillings, *E. Water Res.,*, **21**(1987)107.
47. M. Gazda, L. E. Dejarme, T/. K. Choudury, R. G. Cooks and D. W. Margerum, *W. Environ. Sci., Technol.*, **27**(1993)557.
48. Chen Xu, J. S. Patrick and R. G. Cooks, *Anal. Chem.*, **67**(1995)724.
49. T. K. Choudhury, T, Kotiaho and R. G. Cooks, *Talanta*, **39**(1992)573.
50. M. Sano, Y. Yotsui, H. Abe, and S. Sasaki, *Biomed. Mass Spectrum,.* **3**(1976)1.
51. D. E. Mathews and J. M. Hayes, *Anal. Chem.,* **50**(1978)1465.
52. D. A. Merritt, K. H. Freeman, M. P. Ricci, S. A. Studley and J. M. Hayes, *Anal. Chem.*, **67(14)**(1995)2461.
53. W. J. Irwin, *Analytical Pyrolisis; A Comprehensive Guide*, Marcel Dekker, New York, (19820.
54. A. D. Pouwels, A. D. Eijkel and G. B. Boon, *J. Anal. Appl. Pyrolysis*, **14**(1989)237.
55. O. Faix, D. Meier and I. Fortman, *Holz Roh-Werkst*, **48**(1990) 281.
56. J. J. Boon and J. W. De Leeuw, *J. Annal. Appl. Pyrolysis*, **11**(1987)313.
57. M. L. Matney and T. F. Limero, *Anal. Chem.*, **66(18)**(1995)2820.

CHAPTER 6

GAS CHROMATOGRAPHY/ATOMIC SPECTROSCOPY (GC/AS) TANDEM SYSTEMS

The combination of a gas chromatograph with an atomic spectrometer is not an arrangement that is as popular as many other tandem instruments. However, this particular form, does have certain areas of application where it can be extremely useful. There are three types of analyses in which the GC/AS system is most often employed. The first, and probably the most important, is to provide elemental analysis of the eluted components, and consequently their empirical formulae, to aid in identifying unknown substances. The second is its use as a specific detector to reduce the need for high-resolution columns when analyzing highly complex mixtures. Third the GC/AS tandem system is often used for trace analysis and, in particular, for following the speciation of selected elements in a mixture. Both the absorption and emission spectrometers have been used in conjunction with the gas chromatograph. However, whereas the absorption spectrometer has been frequently used in LC/AS tandem systems, the combination of the gas chromatograph with the atomic emission spectrometer is probably the more common.

Atomic emission spectrometers can basically take two forms the spark discharge spectrometer and the plasma spectrometer. The plasma emission spectrometers are the most popular types of atomic emission spectrometer and the plasma can be generated in three ways. Plasma can be formed by direct current discharge, inductively coupling and by microwave induction. The inductively coupled and the microwave-induced plasma instruments are by far the most common, and both instruments will be described.

The Microwave-induced Plasma (MIP) Atomic Emission Spectrometer in GC/ES Systems

The microwave-induced atomic emission spectrometer has been briefly described in Chapter 2, but the actual interface, and the resonance cavity of the spectrometer, will be discussed further, in order to fully understand its operating characteristics. The design chosen will be that developed relatively recently by Quimby and Sullivan [1], which is similar to the system originally described by Beenakker [2,3]. A diagram of the re-entrant cavity of their emission spectrometer is shown in Figure 6.1 and, as its construction is fairly critical, it will be described in some detail.

The cavity differs slightly from that of Beenakker, in that the pedestal is situated in the center of the cavity, which is much smaller, and the coupling loop is much thicker. In addition, a quartz cooling jacket is placed in the center of the cavity surrounding the discharge tube. The main body of the cavity consists of a stainless steel block, with the cylindrical section and pedestal machined out of it, and a cover plate. The two main cavity plates are fitted together with a silver-filled silicone rubber gasket to ensure a good seal. The quartz water jacket (carrying water at 60°C) is situated between the two halves of the cavity, and held in place by O-rings. The inlet and outlet water connections are radially drilled through the stainless steel. Stainless steel plates are attached to the outside of both halves of the main cavity, and sealed with two more O-rings. As a consequence, the water passes through an annular space between the discharge tube and the water jacket and, in addition, the discharge tube is held concentrically inside the water jacket.

The discharge tube is 4.2 cm long, 1.25 mm O.D., and 1.0 mm I.D., made of fused quartz coated with polyimide. However, the polymer coating does not extend over that portion of the tube in which the discharge is taking place. It is secured by a polyimide ferrule on the GC side of the tube and by a Viton O-ring on the optical side of the tube, which is compressed by the exit chamber. The exit chamber is made from a machinable ceramic, Macor, and screws into the water-seal plate. A sparker wire, that is used to initiate the discharge, contained inside an insulating PTFE tube, also passes into the exit chamber.

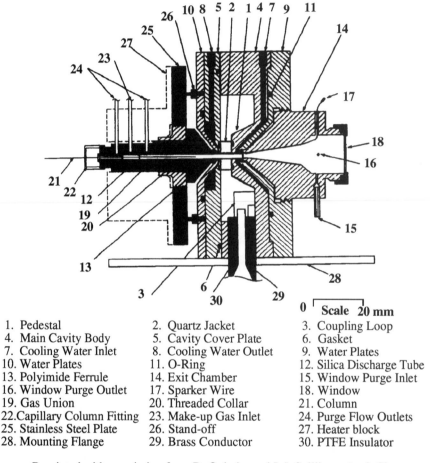

1. Pedestal	2. Quartz Jacket	3. Coupling Loop
4. Main Cavity Body	5. Cavity Cover Plate	6. Gasket
7. Cooling Water Inlet	8. Cooling Water Outlet	9. Water Plates
10. Water Plates	11. O-Ring	12. Silica Discharge Tube
13. Polyimide Ferrule	14. Exit Chamber	15. Window Purge Inlet
16. Window Purge Outlet	17. Sparker Wire	18. Window
19. Gas Union	20. Threaded Collar	21. Column
22. Capillary Column Fitting	23. Make-up Gas Inlet	24. Purge Flow Outlets
25. Stainless Steel Plate	26. Stand-off	27. Heater block
28. Mounting Flange	29. Brass Conductor	30. PTFE Insulator

Reprinted with permission from D. Quimby and J. J. Sulllivan, *Anal. Chem.*, **62**(10)(1990)1027, Copyright 1990 American Chemical Society

Figure 6.1 The Re-entrant Cavity of the Atomic Emission Spectrometer

The window of the exit chamber is made from UV transparent fused silica and is about 2.2 cm in diameter, and about 1.6 mm thick. A heated assembly on the GC side of the cavity provides connections to column and gas supplies. The central assembly is built round a union which is held by a threaded collar that seals it against the polyimide ferrule on the discharge tube. The ferrule is tapered at both ends, and seals the gas union to the discharge tube and excludes water from the heated area. The column is fed through the end of the union, and extends into the discharge tube terminating at a point, 5 mm from the ferrule. The column is sealed with a standard capillary column fitting and a graphite ferrule. The make-up gas and reagent gas enter the center connection to the union, and the other two connections are used for purge flow. The threaded collar of the union fits into a stainless steel plate, which is mounted to the cavity with stand-offs that prevent heat from the gas-union passing to the cavity. Heat is provided by two stainless steel blocks, containing cartridge heaters, that fit over the union. Temperature monitoring sensors are appropriately situated in the block. The sampling procedure utilizes a valveless venting system which is shown in Figure 6.2.

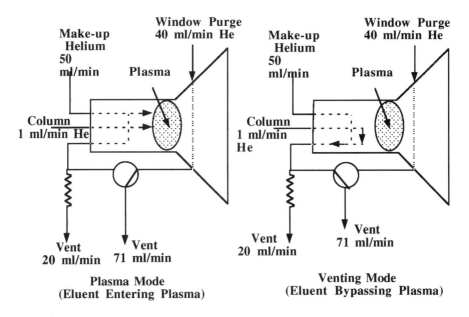

Reprinted with permission from D. Quimby and J. J. Sulllivan, *Anal. Chem.*, **62(10)**(1990)1027, Copyright 1990 American Chemical Society

Figure 6.2 The Venting System

The gas flows and pressures from the various connections must be carefully adjusted, and in the sampling mode (the left-hand diagram) the sample, helium make-up gas and reagent pass directly through the plasma. On rotation of the valve (the right-hand diagram) the flow is diverted and the sample, helium make-up gas and reagent now pass back out of the left-hand vent, and thus are prevented from passing through the plasma. In this way, flow switching is achieved without the sample coming in contact with any valve surfaces. The total arrangement of the GC/AE tandem system, using a microwave-induced plasma torch, is shown in Figure 6.3.

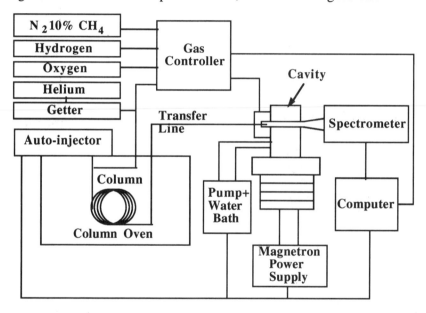

Figure 6.3 A Block Diagram of the GC/AE Tandem Combination

Microwave-induced plasma is not nearly so energetic as inductively induced plasma, and so reagent gases are employed to help energy transfer between the plasma atoms and molecules, and the sample atoms. For this reason, a number of reagent gases are made available with their respective flow controllers, and these are shown in the top left of Figure 6.3. Quimby and Sullivan used their microwave torch in conjunction with a Hewlett-Packard 5890A gas chromatograph. A capillary column was employed, 25 m long, 0.32 mm I.D., carrying a film of methyl silicone as the stationary phase, 0.17 μm thick. The column eluent was transferred to the torch by means of a heated interface tube. Microwave energy was generated by a magnetron power supply which was passed to the torch by means of appropriate wave guides. The tandem system was computer controlled, and a diode array sensor system was used, so that reconstructed chromatograms could be generated at wavelengths pertinent to specific elements.

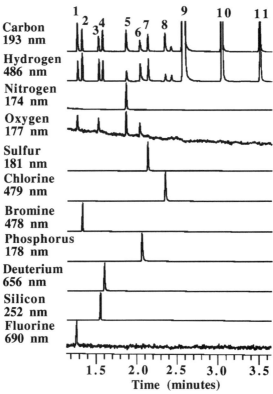

1. 4-fluoroanisole, 2.5 ng
2. 1-bromohexane, 2.6 ng
3. tetraethylorthosilicate, 2.1 ng
4. *n*-perdeuterodecane, 1.9 ng
5. nitrobenzene, 2.7 ng
6. triethyl phosphate, 2.4 ng
7. *tert*-butyl disulfide, 2.1 ng
8. 1,2,4-trichlorobenzene
9. *n*-decane, 17 ng
10. *n*-tridecane, 5.1 ng
11. *n*-tetradecane, 5.1 ng

Reprinted with permission from D. Quimby and J. J. Sulllivan, *Anal. Chem.*, **62(10)**(1990)1027, Copyright 1990 American Chemical Society

Figure 6.4 Chromatograms of an Eleven-component Mixture by Element Monitoring

The chromatograms of an eleven-component synthetic mixture, monitored for elements, carbon, hydrogen, deuterium, nitrogen, oxygen, sulfur, fluorine chlorine, bromine, silicon and phosphorus, are shown in Figure 6.4. The advantages of the technique are clearly seen. The chromatograms monitored by measuring the carbon and hydrogen elements are very similar as would be expected. The other chromatograms show single peaks, indicating the unique nature of their element content.

The GC/ES combination, employing microwave plasma emission, has been used by a number of analysts in various environmental studies. In particular, Lobinski *et al.* [4] have used it for determining tin compounds in soils and ground water. There has been a growing interest in the determination of organotin, due to the increasing number of anthropogenic sources. Tin is used largely in the plastics industry, the paint industry and in agriculture. The toxicity of tin compounds depends on their structure and in the $R_nSnX_{(4-n)}$ series of tin compounds, where (R) and (X) are alkyl groups, the maximum biological activity appears to occur when $n = 3$.

The use of GC as the chosen separation process appears to arise from the very high efficiencies that are readily available and the associated short analysis times. Nevertheless, most tin compounds are highly polar, consequently they must be derivatized before separation on a GC column. Popular preparation procedures involve extraction of the ionic tin compound after chelation, followed by reaction with an ethyl, pentyl or hexyl Grignard reagent [5-7]. The alkyl tin compounds so formed are then injected onto the column, and the separation monitored with an atomic emission spectrometer. As the extracts are usually loaded with a wide variety of interfering organic compounds, monitoring the chromatogram with a conventional FID detector would provide a bemusing mass of peaks. By selecting a particular emission wavelength associated with elemental tin, a chromatogram showing only tin compounds can be reconstructed. An example of the procedure to separate and identify three organotin compounds is shown, as a two-dimensional emission wavelength/retention time chromatogram, in Figure 6.5.

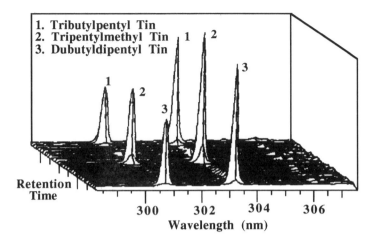

Reprinted with permission from R. Lobinski, W. M. R. Dirkx, M. Ceulemans and F. C. Adams, *Anal. Chem.*, **64(2)**(1992)159., Copyright 1992 American Chemical Society

Figure 6.5 A Three-dimensional Representation of the Separation of Three Organotin Compounds

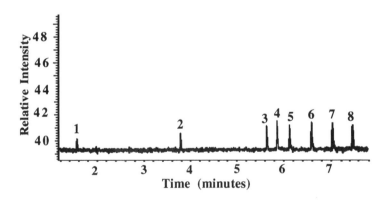

1. Me_3SnPe, 0.18 pg   2. Me_2SnPe_2, 0.23 pg   3. Bu_4Sn, 0.26 pg   4. $MeSnPe_3$, 0.31 pg
5. Bu_3SnPe, 0.23 pg   6. Bu_2SnPe_2, 0.25 pg   7. $BuSnPe_3$, 0.22 pg   8. Pe_4Sn, 0.19 pg

Reprinted with permission from R. Lobinski, W. M. R. Dirkx, M. Ceulemans and F. C. Adams, *Anal. Chem.*, **64(2)**(1992)159., Copyright 1992 American Chemical Society

Figure 6.6 Chromatogram Demonstrating the Sensitivity of the GC/ES Tandem System for Monitoring Traces of Organic Tin Compounds

GC/AS Tandem Systems   235

The combination can be extremely sensitive and lends itself ideally to the assay of tin compounds in environmental samples. The results obtained from the GC/ES tandem system, when operated close to the overall noise level of the total instrumentation, is shown in Figure 6.6.

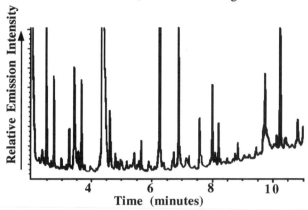

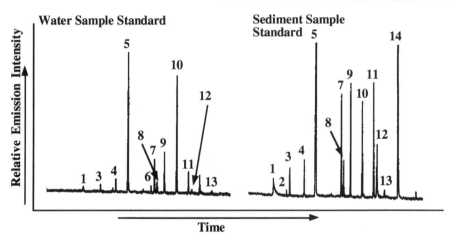

1. Unidentified
2. unidentified
3. Dimethyldipentyl Tin
4. Triethylpropyl Tin
5. Tripropyl Tin
6. Tetrabutyl Tin
7. Dipropyldipentyl Tin
8. Methyltripentyl Tin
9. Trubutylpentyl Tin
10. Dibutyldipentyl Tin
11. Tripentylbutyl Tin
12. Unidentified
13. Tetrapentyl Tin

Reprinted with permission from R. Lobinski, W. M. R. Dirkx, M. Ceulemans and F. C. Adams, *Anal. Chem.*, **64(2)**(1992)159., Copyright 1992 American Chemical Society

Figure 6.7 Speciation of Tin in Environmental Samples

It is clear from Figure 6.6, that the minimum detectable mass of the overall apparatus can be as low as 2×10^{-13} g of organotin present in the injected sample. This level of sensitivity for assaying an inorganic environmental contaminant can be extremely useful. Lobinski *et al.*[4] examined the speciation of tin in water and sediments. The sample preparation process involved the extraction of the methyl and butyl tin compounds into pentane, as diethyl dithiocarbamate complexes at a pH of 5. The solvent was then removed by evaporation under reduced pressure, after which the residue was derivatized with an *n*-pentyl Grignard reagent in *n*-octane. This produced derivatives of the form $R_nSnPe_{(4-n)}$. Test samples showed recoveries of 95-100%. The separation of the extracted and derivatized components monitored at the emission wavelength of carbon is shown in the upper chromatogram of Figure 6.7.

It is seen that the extract is exceedingly complex, and it would be extremely difficult, if not impossible, to locate and measure the specific peaks for the tin compounds. In contrast, by monitoring the separation on a wavelength specific for tin, the tin compounds are clearly and unambiguously selected for detection. The sensitivity of the analysis can be assessed, for example, from peak 10 (dibutyldipentyl tin), which is present in the sediment sample at about 1.8 ng/g, while that in the water sample was about 0.6 ng/L. The GC/AS tandem instrument is clearly very useful for inorganic element identification, and can easily accommodate the very low concentrations of contaminants that can occur in environmental samples.

Liu *et al.* [8] also used a GC/AS tandem system for determining the speciation of organotin compounds in soils and sediments, but in addition used a complexing/supercritical fluid extraction procedure to isolate the tin derivatives. The soil or sediment sample was amended with a complexing agent (diethylammonium diethyldithiocarbamate) and was then extracted with supercritical carbon dioxide, carrying 5% of methanol. The extraction was carried out at 450 atm. and at 60°C. The extracted material, after the carbon dioxide and methanol had been removed, was then treated with pentylmagnesium bromide, which converted the ionic organotin compounds to their neutral derivatives.

Samples of the derivatives were then injected into the gas chromatograph, and the eluents monitored with the tandem atomic emission spectrometer. The extraction efficiency for the tin compounds was generally about 95%, with a number of exceptions. For example, the monoalkyl tin compounds were particularly recalcitrant and could not be recovered, even with a complexing reagent. An example of the separations that were obtained are shown in Figure 6.8, which includes a chromatogram monitored by the emission at 271 nm, the characteristic emission wavelength of atomic tin.

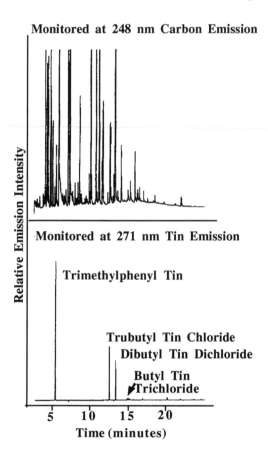

Reprinted with permission from Y. Liu, V. Lopez-Avilla and M. Alcaraz, *Anal. Chem.*, **66(21)**(1995)3788, Copyright 1995 American Chemical Society

Figure 6.8 Chromatograms of Derivatized Supercritical Fluid Extracts of Sediments Spiked with Organo Tin Compounds

It was also shown that the carbon dioxide containing 5% methanol was more efficient for extracting the tin chelates than carbon dioxide alone. Furthermore, the extraction efficiency was also shown to increase with the supercritical fluid pressure. Increasing the pressure also reduced the amount of water extracted and increased the necessary extraction time. Increasing the temperature from 60°C to 100°C did not appear to change the extraction efficiency significantly one way or the other. The use of a complexing reagent was essential to achieve high recoveries.

Emteborg *et al.* [9] also employed a supercritical extraction procedure followed by a GC/AS analysis to develop a method for assaying mercury in sediments. 0.5–1 g of sediment was placed in the extraction thimble and diluted with clean sand and extracted with liquid carbon dioxide.

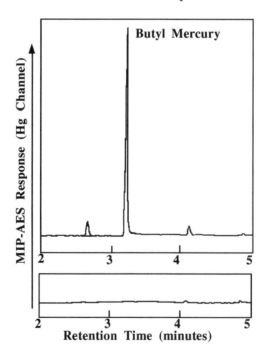

Figure 6.9 GC–MIP–AES Chromatograms from Standard Sediment Samples Monitored on the Mercury Channel (ref. 9)

The extract was depressurized in a ODS trap and the methylmercury eluted from the trap with toluene. The toluene extract was then placed in a

10 ml centrifuge tube, cooled to 0°C and then treated with 200 µl of the Gringard reagent (2 M solution of butylmagnesium chloride in tetrahydrofuran). The reaction was allowed to proceed for 10 minutes, and then the excess Gringard reagent was decomposed by the addition of 300 µl of 1 M HCl. The mixture was then centrifuged and the organic layer separated and 2–20 µl portions were used for GC/MS analysis. An example of the results obtained are shown in Figure 6.9. The concentration of methylmercury was about 5.4 ng/g of sediment and it is seen that the signal to noise ratio was still fairly high. The authors claimed that the detection limit for methylmercury in sediments, using the procedure described, was about 0.1 ng/g. This Figure was based on a sample size of 0.5 g, an injection volume of 20 µl and a signal to noise ratio over that of the blank of 3.

Snell et al. [10} developed an alternative method for determining traces of mercury in natural gas concentrates analysis using GC/AS. The interface comprised a combustion tube and an amalgamation unit to trap the mercury compounds situated between the gas chromatograph and the MIP atomic emission spectrometer. A diagram of the interface is shown in Figure 6.10

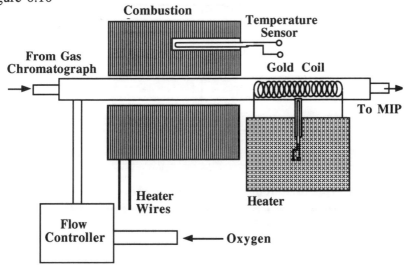

Figure 6.10 The Amalgamation Trap for Measuring Mercury in Natural Gas Condensates (ref. 10)

The trap comprised a quartz tube, 23 cm long, 3.7 mm O.D., 1.9 mm I.D. which was fitted with two heaters. The first heater maintained the temperature at 850-900°C Mercury free oxygen was fed into the gas stream from the capillary column of the gas chromatograph, and the solutes were burnt in this first heated section. A valving system between the capillary column and the interface permitted selected peaks to be diverted to the trap as required. Subsequent to the combustion furnace was a section of tube containing gold-platinum wire (85% gold–15% platinum) that trapped all the mercury compounds by amalgamation. The section of the tube was also surrounded by a heater. When trapping was complete the mercury was desorbed from the gold-platinum wire by heating, and the mercury vapor was swept into the MIP atomic emission spectrometer. The results from this interface together with those obtained by the solid phase extraction procedure previously described are shown in Figure 6.11.

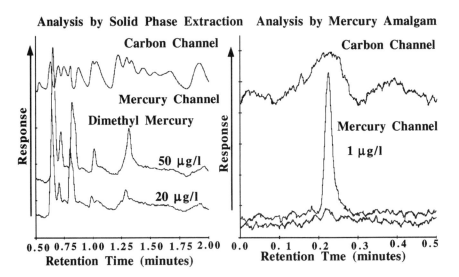

Figure 6.11 Measurement of Traces of Mercury in Natural Gas Condensate by Super Critical Fluid Extraction and by Selective Amalgamation (ref. 10)

It is clear that the amalgam interface produces the highest sensitivity and it is seen that he back ground noise is extremely low. The trace from the blank run indicates a signal to noise ratio of at least 8 and consequently, the lower sensitivity limit would appear to be about 200 pg/l.

Microwave induced plasmas suffer from certain drawbacks, one being the easy manner in which the plasma can become extinguished in the presence of excess organic material. This plasma quenching often occurs in tandem systems as a result of the elution of the solvent peak from the gas chromatograph.

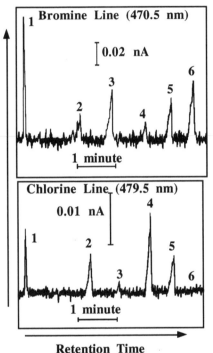

1	Pentane Solvent	4	o-Dichlorobenzene (0.27 µg)
2	Chlorobenzene (0.23 µg)	5	o-Bromochlorobenzene (0.34 µg)
3	Bromobenzene (0.31 µg)	6	o-Dibromobenzene (0.41 µg)

Figure 6.12 The Analysis of Five Halogenated Aromatic Compounds by Gas Chromatography/Capacity-Coupled Microwave Plasma Atomic Emission Spectrometer (ref. 12)

To obviate this problem Winefordner et al [12] developed the capacity coupled microwave plasma employing a tubular electrode, and obtained a very stable discharge using argon and helium as plasma gases. Uchida *et al.* [12] utilized the capacity coupled microwave plasma system to provide multi-element detection for a capillary gas chromatograph. It was employed very successfully for monitoring the elution of different halogenated aromatic compounds and the separation of a group of these is shown in Figure 6.12. By monitoring on both the chlorine and bromine lines, the compounds containing exclusively bromine or chlorine can be clearly discerned form those compounds that contain both halogens. More importantly, it is seen that the plasma persisted in a stable form throughout, and after, the elution of the pentane solvent which was present in the plasma at a very high concentration.

Inductively Coupled Plasma (ICP) GC/ES Systems

The basic difference between the ICP torch and the MIC torch is the method of energy transfer, the physical arrangement of both torches being fundamentally similar. In addition, the ICP torch may not provide good spectra for the common elements, but is very sensitive to elements of higher atomic weight and in particular the metallic elements. A diagram of the ICP torch is shown in Figure 6.13.

The ICP torch is often employed with a capillary column system, but can also be used with packed columns providing argon is used as the carrier gas. The torch is made of three concentric tubes of silica, through the center of which passes the column flow. If a capillary column is used, a scavenger flow of argon must also be mixed with the carrier gas, to ensure the sample bands are swept cleanly from the system, and do not accumulate in the torch. The next outer tube also furnishes make-up argon to provide sufficient for forming a plasma, and also helps cool the inlet tube. The outer tube carries a large flow of argon or nitrogen that keeps the temperature of the front part of the torch from reaching the melting point of silica.

The high-frequency transducer consists of a few turns of a water cooled coupling coil, the rf power being supplied by an appropriate frequency generator and power amplifier. Although energy transfer between the plasma and the elements being examined is more efficient with inductively coupled plasma sources, the power requirement is, nevertheless, considerably greater than that needed by the micro-wave torch, which accounts to some extent for the latter's popularity.

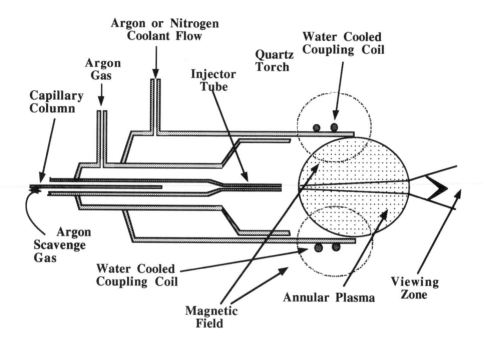

Figure 6.13 A Diagram of the ICP Torch for the Atomic Emission Spectrometer

The spectrometer arrangement is very similar to that used with a microwave induced spectrometer and is shown in Figure 6.14. Figure 6.14 depicts the emission spectrometer fitted with the older type of diffraction grating optical system. Most modern ICP atomic emission spectrometers utilize a diode array sensor that is far more compact but, depending on the number of diodes, may have somewhat less optical resolution. Virtually any type of well-designed gas chromatograph can be fitted to the spectrometer, using an appropriately heated transfer line

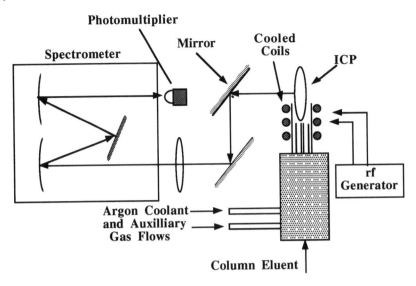

Figure 6.14 The Inductively Coupled Plasma/Atomic Emission Spectrometer

Precautions must be taken to ensure the transfer line does not contribute significantly to band dispersion as discussed in Chapter 3.

The ICP atomic emission spectrometer was coupled to the gas chromatograph as a tandem combination by Duebelbeis *et al.* [13], who used it to develop a method for surveying the distribution of different elements in a variety of organic matrixes. To test the method, a sample of coal was 'spiked' with a number of different metal elements, and carbonized in a laboratory facsimile of a coal gasification plant. The apparatus consisted of bed of the spiked coal, through which was passed a mixture of oxygen and steam. The coal was ignited by heating a section of the tube externally with a Nichrom wire heater coil. The effluent from the tube was passed through a series of traps. The first trap contained quartz wool maintained at 90°C to retain the heavy tars. The gases were then bubbled through a solvent held in a cold ice methanol bath. A third trap was used to check the efficiency of the first two traps. The samples from the solvent trap were injected onto the GC, and the column eluent monitored with the ICP atomic emission spectrometer, using the interface shown in Figure 6.15.

The interface was very simple, and consisted of a heated stainless steel tube that conducted the column eluent directly into the ICP torch. The chromatograph was a Trecor Model 560 fitted with a capillary, split/splitless injector. A fused capillary column (DB-5) was employed, 30 m long and 0.25 mm I.D., and argon gas was used as the mobile phase at a flow rate of 30 ml/min. The ICP emission spectrometer was the Baird Model PS-1. To produce the plasma, 1.5 kW of power was supplied to the cooled coupling coils from an rf generator operating at 27.1

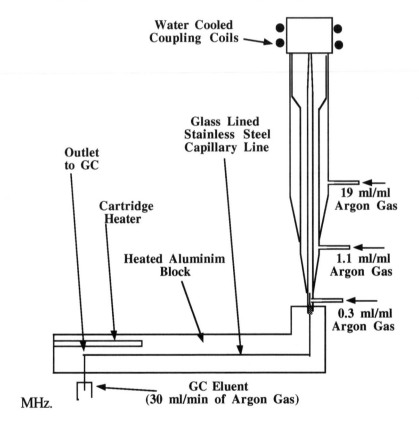

Figure 6.15 The Gas Chromatograph/ICP Torch Interface Used by Duebelbeis *et al.*

The elements used for testing the equipment were tin (tetraethyl tin), iron (ferrocene), lead (tetraethyl lead) and chromium (chromium trifluoroacetylacetonate) and the results obtained are shown in the four chromatograms depicted in Figure 6.16.

It is seen that the tandem system clearly, and unambiguously, identifies the different organometalic compounds as they are eluted. It is also seen that the system is generally very sensitive, the minimum detectable amounts being about 10–100 pg. The lowest sensitivity was shown for the organchromium compound. However, this lower sensitivity is partly due to chromium only representing a relatively small proportion of the total organochromium complex. Some coal was spiked with selenium sulfide and burned in the laboratory coal gasification apparatus and the coal oxidized in a wet oxygen stream.

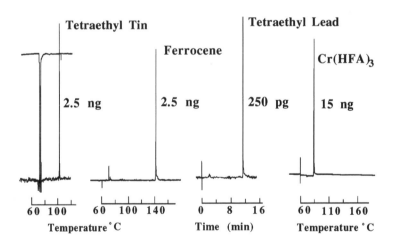

Figure 6.16 Chromatograms Showing the Response of the GC/ICPAE Tandem Instrument to Tin, Iron, Lead and Chromium from a Test Run

The coal bed was burned in two directions, one a forward burn where the burn proceeded in the direction of the gas flow, and thus the gaseous products were swept away from the burning zone as soon as they were formed. The second was a reverse burn, that allowed the oxidation products to pass back through the burning zone after they had been formed and were thus subject to further reaction or oxidation. Comparing the retention times of the peaks in the forward and reverse burn chromatograms (Figure 6.17) it is clear that very different compounds are formed in the two methods of burning. The peaks that are starred in the reverse burn chromatogram have the same retention times as the two large peaks, F1 and F2, in the forward burn chromatogram. By treating samples of the condensate with different reagents, that would remove compounds of specific chemical types, and re-running the treated sample, the peaks resulting from particular selenium compounds could be identified. This type of application gives some idea of the diverse ways in which tandem instruments can be utilized.

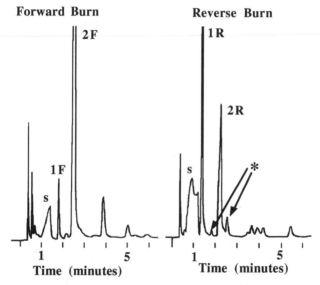

Figure 6.17 Separation of the Volatile Selenium Products from Coal Gasification

Forbes *et al.* [14] coupled the atomic emission spectrometer that utilized an ICP torch to a supercritical fluid chromatograph and demonstrated its efficacy for monitoring the separation of some silicon compounds. Their chromatograph/emission spectrometer tandem combination was very similar to those used with the microwave torch. A diagram of the apparatus is shown in Figure 6.18.

The SFC separation apparatus consisted of a modified gas chromatograph, fitted with a syringe pump and a liquid carbon dioxide supply. The fused silica capillary column was coated internally with a film, 0.05 µm thick, of poly(dimethylsiloxane) which comprised the stationary phase. The column was 20 m long and 200 µm I.D. and the sample valve consisted of an electronically actuated loop valve, having a 0.1 µl capacity, manufactured by Valco Inc. A split injection system was used in conjunction with a retention-gap injection-tube, which was placed between the valve and the column to reduce band distortion on injection. The outlet of the column passed through a heated transfer line (the last part of the column) and was butt-joined to a capillary restrictor. The capillary restrictor was made of a length of uncoated fused silica tubing that penetrated the torch and terminated close to the argon plasma.

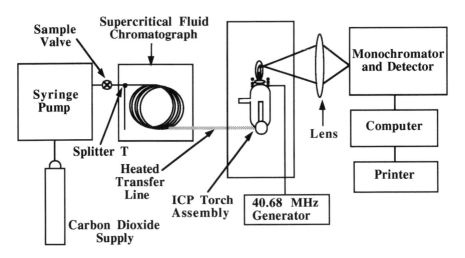

Reprinted with permission from K. A. Forbes, J. F. Vecchiarelli and P. C. Uden, *Anal. Chem.*, **62**(18)(1990)2033, Copyright 1990 American Chemical Society

Figure 6.18 Diagram of a Supercritical Fluid Chromatograph Combined with an Inductively Coupled Plasma Atomic Emission Spectrometer.

The emitted light was measured at a number of different wavelengths and chromatograms reconstructed from carbon and silicon emission. The optimum emission wavelength for measuring silicon was found to be 251.6 nm. An example of the separation of a pair of silicon compounds, monitored by atomic emission, using the ICP torch is shown in Figure 6.19. It is seen that the combination provides good resolution, a high sensitivity and an extremely selective response. The use of the emission spectrometer in a tandem combination with the gas chromatograph, is becoming more and more common for the detection and measurement of trace metals. Nevertheless, the atomic spectrometer appears to be equally popular for use in tandem with the liquid chromatograph, where it is important in environmental and toxicology studies.

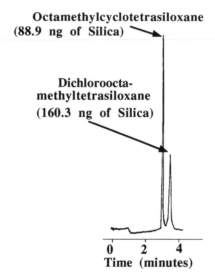

Reprinted with permission from K. A. Forbes, J. F. Vecchiarelli and P. C. Uden, *Anal. Chem.*, **62(18)**(1990)2033, Copyright 1990 American Chemical Society

Figure 6.19 The Separation of Two Tetrasiloxanes Monitored by Atomic Emission Employing an ICP Torch

Synopsis

There are three types of analyses in which the GC/AS system is most often employed. The most important is to provide elemental analysis of the eluted components, and consequently their empirical formulae to aid in substance identification. The system is also used as a specific detector, to reduce the need for high-resolution columns when analyzing highly complex mixtures. Finally, the GC/AS tandem system is also used for trace analysis and, in particular, for following the speciation of selected elements in a mixture. The inductively coupled and the microwave induced plasma emission spectrometers are those most often associated with the gas chromatograph. Microwave induced plasma is not nearly as energetic as inductively induced plasma, and so reagent gases are employed to help energy transfer between the plasma atoms and molecules, and the sample atoms. For this reason, a number of reagent gases are usually made available for the plasma torch. One problems with the microwave induced plasma is that it is easily quenched by excessive loads of high carbon content material and, in particular, the solvent front from a gas chromatograph. An alternative, energy coupling technique, called capacitively coupled microwave plasma has largely overcome the problem of plasma extinguishing by solvent overload. Employing a GC/AS tandem system chromatograms can be reproduced based on the presence of a specific element. The high sensitivity coupled with the extreme element selectivity has made the GC/AS instrument extremely valuable for environmental studies. In particular the tandem system has been used effectively in the examination of tin contamination in water supplies, soils and sediments. The basic difference between the ICP torch and the MIC torch in GC/MS is the method of energy transfer, the physical arrangement of both torches being fundamentally similar. The ICP torch may not always provide good spectra for the common elements, but is very sensitive to elements of higher atomic weight, and in particular the metallic elements. In tandem systems it is commonly used in conjunction with capillary columns. A GC/MS employing the ICP torch has been used to monitor chemical pathways of heavy metals in coal during coal gasification processes.

References

1. D. Quimby and J. J. Sulllivan, *Anal. Chem.,* **62(10)**(1990)1027.
2. C. I. M Beenakker, *Spectrochim. Acta,* **32B**(1977)173.
3. C. I. M Beenakker and P. W. J. M. Boumans, *Spectrochim. Acta,* **33B**(1978)53.

4. R. Lobinski, W. M. R. Dirkx, M. Ceulemans and F. C. Adams, *Anal. Chem.*, **64(2)**(1992)159.
5. J. Asjby, S. Clark and P.J. Craig, *J. Anal. At. Spectrom.*, **3**(1988)735.
6. M. D. Müller, *Anal. Chem.*, **59**(1987)617.
7. M. A. Unger, W. G. MacIntyre, J. Greaves and R. J. Huggett, *Chemosphere*, **15**(1986)461.
8. Y. Liu, V. Lopez-Avilla and M. Alcaraz, *Anal. Chem.*, **66(21)**(1995)3788.
9. H. Emteborg, E. Björklund, F. Ödman, L. Karlsson, L. Mathiasson, W. Frech and D. C. Baxter, *Analyst*, **121**(1996)19.
10. J. P. Snell, W. Frexh and Y. Thomasson, *Analyst*, **121**(1996)1055.
11. S. Hanamura, B. W. Smith and J. D. Winefordner, *Can. J. Spectrosc.* **29**(1984)13.
12. H. Uchida, A. Berthod and J. D. Winefordner, *Analyst*, **115**(1990)933.
13. P. O. Duebelbeis, S. Kapila, D. E. Yates and S. E. Manahan, *J. Chromatogr.*, **351**(1986)465.
14 K. A. Forbes, J. F. Vecchiarelli and P. C. Uden, *Anal. Chem.*, **62(18)**(1990)2033.

PART 3

LIQUID CHROMATOGRAPHY TANDEM SYSTEMS

CHAPTER 7

LIQUID CHROMATOGRAPHY/UV SPECTROSCOPY FLUORESCENCE SPECTROSCOPY (LC/UV/FS)TANDEM SYSTEMS

LC/UV Tandem Systems

Today, the liquid chromatograph/UV spectrometer combination is hardly recognized as a tandem instrument, as it is in such common use. The UV diode array spectrometer is one of the most commonly used LC/UV instruments and consequently, the system is regarded more as a multi-wavelength detector than as a tandem instrument. In fact, its use as a device to produce UV spectra is relatively limited compared with its use to select the optimum wavelength for maximum light absorption, and consequently allow the separation to be monitored at the maximum sensitivity. The limitations of UV spectra for confirming solute identity have already been discussed in Chapter 2. Certain compounds, such as those containing aromatic rings, give spectra with sufficient fine structure to aid in solute identification. Unfortunately, the spectra of the majority of substances are very similar in shape, have little or no fine structure, and are therefore virtually useless for this purpose.

In addition, the resolution obtainable from the diode array spectrometer is limited by the number of diodes in the array, and consequently unless a large number of diodes are employed, the limited resolution could impair the certainty of identification. This limitation will be particularly pertinent when fine structure is present in the spectrum.

An example of the use of the diode array detector to select a specific wavelength, to give a high response to specific compound types and consequently allow the separation to be monitored at the maximum sensitivity, is shown in Figure 7.1. The separation is that of a series of common fatty acids and was carried out on a reversed phase column, using water buffered with phosphoric acid as the mobile phase. In order to achieve adequate sensitivity, the total separation was monitored by the diode array detector, the UV absorption at 210 nm recalled, and the chromatogram obtained by plotting the absorption at this wavelength against time.

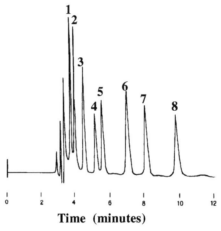

Courtesy of Supelco Inc.

Column: Spherisorb® Octyl, 25 cm x 4.6 mm I.D., 5 μm particles. Mobile phase: 0.2 M phosphoric acid. Flow rate 0.8 ml/min. monitored at 210 nm. 1. tartaric acid, 2. lactic acid, 3. malic acid, 4. formic acid, 5. acetic acid, 6. citric acid, 7. succinic acid, 8. fumaric acid.

Figure 7.1 The Separation of Some Carboxylic Acids Monitored by UV Absorption at 210 nm

Due to the relatively low absorption of carboxylic acids at higher wavelengths, (*e.g.* at 254 nm, which is the wavelength of most fixed wavelength detectors), the acids are usually separated by ion exchange chromatography, and the separation monitored by the electrical conductivity detector. Using the diode array detector, and choosing the optimum wavelength to reconstruct the chromatogram, the high chromatographic selectivity of the reversed phase can be exploited. It should be said that, in the past, the multi-wavelength dispersive detector has also proved extremely useful for this purpose. It has provided adequate sensitivity, good versatility, a linear response and a resolution that is not limited by diode content. It was found, however, to be somewhat bulky (due to the need for a relatively large internal 'optical bench'). Furthermore, it has a *mechanically operated* wavelength selector and requires a stop/flow procedure to obtain spectra 'on-the-fly'. More important, the dispersive instrument can only provide absorption data at one specific wavelength in the continuous monitoring mode. Consequently, if the separation is to be monitored at another wavelength the separation must be repeated. The diode array detector has the same advantages as the dispersive instrument, but none of the disadvantages and provides multi-wavelength detection in a single chromatographic separation.

The manner in which the resolution of the diode array detector is restricted can be examined as follows. The resolution of the diode array detector ($\Delta\lambda$), will depend on the number of diodes (n) in the array, and also on the range of wavelengths covered ($\lambda_2 - \lambda_1$).

Thus
$$\Delta\lambda = \frac{\lambda_2 - \lambda_1}{n}$$

It is seen that the ultimate resolving power of the diode array detector will depend on the semiconductor manufacturer, and on how narrow the individual photo cells can be commercially fabricated. Most instruments have a data processing package that will permits the separation to be monitored in real time by at least one diode, so that the chromatogram can be followed as the separation develops. By noting the elution time of a particular peak, a spectrum of the solute can be obtained by recalling the output of all the diodes from memory, at that particular time. This gives directly the spectrum of the solute, *i.e.* a curve relating adsorption and wavelength. The liquid chromatograph/UV spectrometer tandem system (the diode array detector or the dispersive UV spectrometer) can be used in a number of unique ways to aid in chromatographic analysis. The most common uses of the tandem system are as follows.

1. To monitor the separation at an optimum wavelength and thus provide the maximum chromatographic sensitivity.

2. To test the purity of a solute peak by calculating the absorption ratios across the peak.

3. To extract the spectrum of a peak and compare it to a library standard to confirm the identity of a substance.

An example of the use of the tandem instrument to enhance the sensitivity of detection has already been given, and other examples will be provided later. The method for testing the purity of a peak using the tandem combination is particularly interesting. If the composition of the peak is chemically homogeneous throughout the whole elution curve, then the ratio of the absorption at any particular wavelength to that taken at another significantly different wavelength will be constant throughout the entire peak. This constant ratio will be independent of the nature of the spectra and whether there is any fine structure present or not. It follows that if this ratio is plotted against time, across the breadth of a *pure* peak, a perfectly rectangular trace will be obtained. In contrast, if another substance is co-eluted, under the peak, then the rectangular shape will be significantly distorted. An example of this method of purity-testing is given in Figure 7.2.

The chromatogram, which was monitored at 274 nm, is shown in the lower part of Figure 7.2. As a diode array detector was employed, it was possible to ratio the output from the detector at different wavelengths and plot the ratio simultaneously with the chromatogram monitored at 274 nm. Now, as already explained, if the peak was pure and homogeneous, the ratio of the adsorption at the two wavelengths (those selected being 225 and 245 nm) would remain constant throughout the elution of the entire peak. The upper diagram in Figure 7.2, shows this ratio plotted on the same time scale, and it is seen that a clean rectangular peak is observed. This rectangular shaped curve unambiguously confirms the purity of the chlorthalidone peak. It should be pointed out that, to improve the confidence limits of purity assurance, the wavelengths chosen to provide the confirming ratio should be based on the UV adsorption characteristics of the substance concerned, relative to those of the most likely impurities to be present,

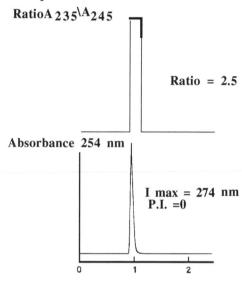

Courtesy of the Perkin Elmer Corporation

The chlorthalidone was isolated from a sample of tablets and separated by a (C18) reverse phase on a column, 4.6 mm I.D., 3.3 cm long, using a solvent mixture consisting of 35% methanol, 65% aqueous acetic acid solution (water containing 1% of acetic acid). The flow rate was 2 ml/min.

Figure 7.2 Dual Channel Plot from a Diode Array Detector Confirming Peak Purity

Perhaps a more practical example of the use of the diode array detector to confirm the integrity of an eluted peak, is afforded by the separation of the series of hydrocarbons shown in Figure 7.3. This separation illustrates the results that are obtained when an impure peak is identified. It is seen that the separation appears to be satisfactory and all the peaks represent

260   Tandem Systems

individual solutes. As a result, and without further evidence, it would be reasonable to assume that all the peaks were pure.

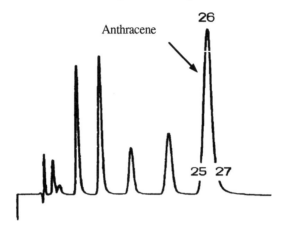

The separation was carried out on a column, 3 cm long, 4.6 mm in diameter, and packed with a C18 reversed phase on particles 3 μ in diameter.
Courtesy of the Perkin Elmer Corporation

Figure 7.3 The Separation of Some Aromatic Hydrocarbons

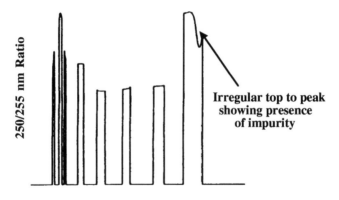

Courtesy of the Perkin Elmer Corporation

Figure 7.4 Curves of Adsorption Ratio, at $\dfrac{250 \text{ nm}}{255 \text{ nm}}$ to Time

However, by adopting the same procedure, and plotting the adsorption ratio, (250/255), for the anthracene peak, it will become clear that the peak tail contains an impurity, as the clean rectangular shape of the other peaks is not realized. The absorption ratio peaks are shown in Figure 7.4. The ratio peaks in Figure 7.4, confirm the presence of an impurity in the anthracene peak by its sloping top. The existence of an impurity, is confirmed unambiguously by the difference in the spectra for the leading and trailing edges of the peak, shown superimposed in Figure 7.5. It is obvious that the substances eluted at the beginning of the peak is quite different from that eluted at the tail of the peak. This type of information is extremely valuable in forensic chemistry, where confidence in solute identity is at a premium. In point of fact, further work identified the impurity as t–butyl benzene present at a level of about 5%.

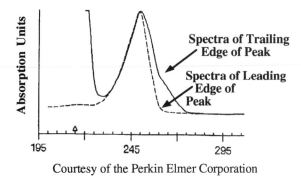

Courtesy of the Perkin Elmer Corporation

Figure 7.5 Superimposed Spectra Taken at the Leading and Trailing Edges of the Anthracene Peak

Another example of the use of the diode array detector to determine peak purity is shown in Figure 7.6. The spectra and chromatogram depicted display the facility that the tandem system provides to extract specific spectra from the different parts of a peak, or peaks in the elution curve. The chromatogram shows the separation of five solutes. In order to confirm peak purity, spectra have been taken of peak (a) and peak (b), at points halfway up the rising side of each peak, at the top of each peak, and halfway down the trailing edge of each peak. It is clearly seen, that although the spectra are obtained at differing concentrations of solute, all the spectra for peak (b) are similar. It can be concluded therefore that peak (b) is homogeneous and represents the elution curve of a single solute. In contrast, the spectra for peak (a) all differ extensively, and thus confirms that the peak is *not* pure, and is a mixture of at least two unresolved substances.

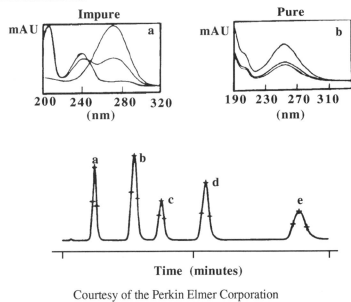

Courtesy of the Perkin Elmer Corporation

Figure 7.6 Diode Array Spectra Demonstrating Peak Purity

Monitoring the column eluent, by selecting a specific wavelength from which to reconstruct the chromatogram, is similar in principle to single ion monitoring in mass spectrometry, but not nearly so discriminating. Whereas single ion monitoring can be used to detect a specific molecular ion or molecular fragment only, single wavelength monitoring can only broadly enhance the sensitivity of the system to certain chemical types. Nevertheless, this feature can be extremely useful, particularly in environmental analysis where maximum sensitivity is often essential. However, the wavelength selected for high sensitivity detection may also restrict the choice of solvents that can be used, and thus directly effects the nature of the chromatographic process. For example, in Figure 7.7 it is seen that light of relatively short wavelengths (210 and 220 nm) is used for detection, because both the alkaloids and the cardiovascular drugs absorb well at these wavelengths. In order to achieve the separation so that the absorption at 210 nm can be used for chromatogram reconstruction, however, it was necessary to employ a mobile phase that is transparent to light of this wavelength.

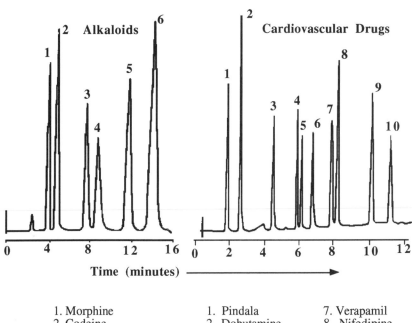

1. Morphine	1. Pindala	7. Verapamil
2. Codeine	2. Dobutamine	8. Nifedipine
3. Cryptopine	3. Ovprenaolo	9. Lidoflazine
4. Thebane	4. Digoxin	10. Flunanzine
5. Narcotine	5. Dipyradinamol	
6. Papaverine	6. Diltiazem	

The Alkaloids
The stationary phase was Nucleosil CN, packed in a column 25 cm long, 4.6 mm I.D. having particles 5 μm in diameter. The mobile phase consisted of a mixture of an aqueous ammonium acetate buffer (pH 5.8) 80%, acetonitrile 10% and dioxane 10% and the flow rate was 1.5 ml/min.

The Cardiovascular Drugs
The stationary phase was Supersil LC-18, packed in a column 3.3 cm long, 4.6 mm I.D. having particles 3 μm in diameter. The mobile phase was changed from 25 mM potassium dihydrogen phosphate with 0.02% of triethylamine phosphate (pH 3.0) and 10% acetonitrile, to a mixture containing 50% acetonitrile over 10 min. the flow rate was 2 ml/min.

Courtesy of Supelco Inc.

Figure 7.7 The Separation of a Selection of Drugs Monitored at 210 and 220 nm

Hence the mobile phase largely consisted of an aqueous acetate buffer (80%) and 10% each of acetonitrile and dioxane respectively, all three solvents being very polar in nature. In consequence, the stationary phase had to be chosen that would retain the solutes in the presence of these very polar solvents, and in this case the polar cyano bonded phase was employed. In fact, only a limited number of solvents are transparent to UV at 210 nm, and if it is deemed necessary to employ light at even lower wavelengths to monitor the separation, then this places severe restrictions on the solvents that can be chosen for the chromatographic process.

The separation of the cardiovascular drugs was monitored with UV light at 220 nm and it is seen that a C18 reversed phase could now be employed (a highly dispersive stationary phase) and the acetonitrile content of the mobile phase is raised to a level of 50% at the end of the separation. At this wavelength, the amount of solvent in the mobile phase can be higher and other solvents that are transparent at 220 nm can be used. This gives more freedom to the analyst in the choice of solvents that can be employed to obtain the best separation. The cut-off wavelengths of a range of solvents commonly used in liquid chromatography are shown in Table 7.1.

Table 7.1

Some Physical Properties of Solvents in Common Use in Liquid Chromatography

Solvent	Cut-off (nm)	Solvent	Cut Off (nm)
n-pentane	205	nitromethane	380
n-heptane	197	n-propyl ether	200
cyclohexane	200	ether	215
carbon tetrachloride	265	ethyl acetate	260
n-butyl chloride	220	methyl acetate	260
chloroform	295	acetone	330
benzene	280	tetrahydrofuran	225
toluene	285	acetonitrile	190
dichloroethane	232	n-propanol	205
tetrachloroethylene	280	ethanol	205
1,2–dichloroethane	225	methanol	205
2–nitropropane	380	water	180
		acetic acid	210

It should be noted that the wavelength of the light used cannot even be *close* to the cut-off wavelength as a very poor response will be obtained. It is seen that, as with many other analytical techniques, the choice of operating conditions must always be a compromise between the different demands of the tandem instruments.

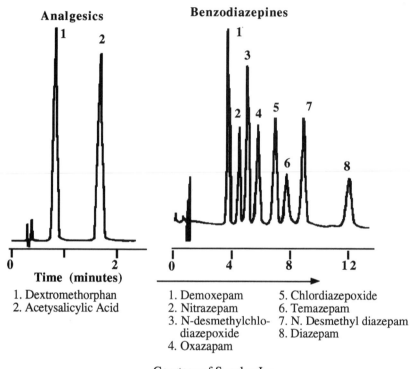

Courtesy of Supelco Inc.

The Analgesics
The stationary phase was Supersil LC-ABZ, packed in a column 5 cm long, 4.6 mm I.D. having particles 5 μm in diameter. The mobile phase consisted of a mixture of an aqueous 25 mM potassium dihydrogen phosphate (pH 2.3) 80% and acetonitrile 20% and the flow rate was 2 ml/min.

The Benzodiazepines
The stationary phase was Supersil LC-8, packed in a column 15 cm long, 4.6 mm I.D. having particles 5 μm in diameter. The mobile phase consisted of methanol, 26.5%, acetonitrile, 16.5% and 0.1M ammonium acetate (pH to 6.0 with acetic acid) 57% the flow rate was 2 ml/min.

Figure 7.8 The Separation of a Selection of Drugs Monitored at 230 and 245 nm

Figure 7.8, depicts the separation of some analgesics and benzodiazepines, that the wavelengths used for chromatogram reconstruction are 230 and 245 receptively. This means, that at these wavelengths, the choice of solvents has been extended and, for example, small quantities of tetrahydrofuran could have been used in the mobile phase if so desired

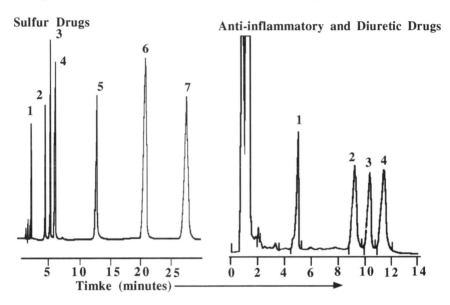

Courtesy of Supelco Inc.

Some Sulfur Compounds
The stationary phase was Supersil LC-ABZ, packed in a column 25 cm long, 4.6 mm I.D. having particles 5 μm in diameter. The mobile phase consisted of 10% acetonitrile and 90% 25 mM potassium hydrogen phosphate buffered to pH 2 with phosphoric acid and the flow rate was 2 ml/min.

Anti-inflammatory Drugs
A Hisep was used, column 15 cm long, 4.6 mm I.D. having particles 5 μm in diameter. The mobile phase consisted of a gradient from 20% acetonitrile and 180 mM ammonium acetate (pH 5) to 30% acetonitrile and the flow rate was 2 ml/min.

Figure 7.9 The Separation of a Selection of Drugs Monitored at 254 and 267 nm

LC/UV/FS Tandem Systems

It is seen that solvents having the extremes of polarity, for example the most polar solvents, such as water and the lower alcohols, and the least polar solvents (strongly dispersive), the n-alkanes, all have relatively low cut-off wavelengths. This means that the solvents of intermediate polarity are those that exclude the use of light of lower wavelengths in LC/UV tandem systems. It follows that the restrictions on solvent selection are greatest when separating substances that require solvents of intermediate polarity for satisfactory elution.

The use of the specific monitoring wavelengths 254 nm and 267 nm in the separations shown in Figure 7.9, does not restrict the choice of solvent, or the concentration of the solvents, as their cut-off wavelengths are well below those used for monitoring. The choice of wavelengths for the separations in Figure 7.9, is based solely on the response of the solutes, and is not influenced by the nature of the separation.

The versatility and advantages of the liquid chromatograph/UV spectrometer combination are obvious. Largely due to the relatively high cost of the first models, they were originally used mainly as research instruments or for method development. However, the cost of the liquid chromatograph/UV spectrometer combination has now been reduced significantly. Consequently, the LC/UV tandem instrument is now considered simply as a liquid chromatograph fitted with a multi-wavelength UV detector, for general use in LC analysis.

The performance of both types of multi-wavelength detectors are very similar, both have a sensitivity of about 1×10^{-7} g/ml and a linear dynamic range of about 1×10^{-7} to 5×10^{-4} g/ml.

LC/Fluorescence Spectroscopy Tandem Systems

The fluorescence spectrum is not often used for identifying compounds, although it can be equally if not more useful for this purpose than the UV spectrum. The fluorescence spectrometer, at present, is available in two forms: the first has a monochromator that allows the excitation light to be selected and the total fluorescent light is measured (this model is merely a fluorescence detector, that can permit the excitation light to be selected, but cannot provide a fluorescence spectrum); the second has two monochromators that allow the excitation light to be selected and also the fluorescent spectrum to be obtained. It is fairly obvious that the second model is far more complex and, consequently, costs significantly more than the first. If both the wavelength of the excitation light and the wavelength of the fluorescent light can be selected, the tandem system can be arranged to provide exceedingly high sensitivities for chosen compounds. In addition monitoring a chromatogram at both the optimum excitation wavelength and the optimum emission wavelength makes the system very selective and, indeed, begins to approach the selectivity of the single ion monitoring technique used in mass spectrometry.

Two of the companies that have designed and produced fluorescence spectrometers for tandem use, with the liquid chromatograph, are the Hewlett-Packard (HP) Corporation, and the Perkin Elmer Corporation (PE). The spectrometer developed by the Perkin Elmer corporation has already been discussed in chapter 2, and so the Hewlett-Packard instrument will now be described.

The Hewlett-Packard Fluorescence Spectrometer Designed for Use as a LC/FS Tandem System

Fluorescence instruments usually employ either the deuterium lamp, the low pressure-mercury lamp, or the xenon discharge lamp as the excitation source. The mercury lamp emits light at essentially only a few discrete wavelengths at high intensity, and the deuterium lamp, although a broad wavelength light source, has maximum emission between 200 nm and 300 nm which is rather limited. The deuterium lamp, however, can be operated continuously. The xenon lamp emits light at high intensity over a broader range of wavelengths, *i.e.*, 150-600 nm. The xenon lamp, however, can only be operated intermittently, and thus the discharge must be pulsed. The pulsing, however, has the advantage that, as considerably less energy is dissipated in the lamp, base line drift from the thermal changes is virtually eliminated. Nevertheless, due to its wider emission

LC/UV/FS Tandem Systems 269

range the HP fluorescence spectrometer utilizes the xenon lamp. A diagram of the HP fluorescence spectrometer is shown in Figure 7.10.

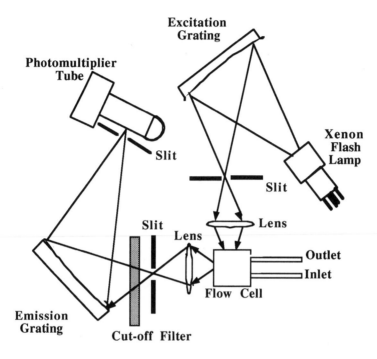

Courtesy of the Hewlett Packard Corporation

Figure 7.10 The Hewlett-Packard Fluorescence Spectrometer

Light from the xenon source is focused onto an excitation grating which is movable to enable the wavelength of the excitation light to be selected. Light of the selected wavelength passes through a slit and is focused on the liquid chromatography flow cell. Light emitted at right angles to the excitation light is focused through a slit onto the emission grating and then through a third slit onto the photomultiplier. The emission grating is also movable, to enable the wavelength of the emission light to be selected or the fluorescence spectrum to be scanned. Between the second slit and the emission grating, there is a cut-off filter that prevents interference from stray light.

The flow cell, for maximum fluorescence light sensitivity, should be as large as possible. However, all post column volumes must be minimized to prevent peak dispersion and to preserve the peak shape. In practice, the cell volume was chosen to be 5 µl, as compromise between these conflicting demands. A diagram of the HP flow cell is shown in Figure 7.11.

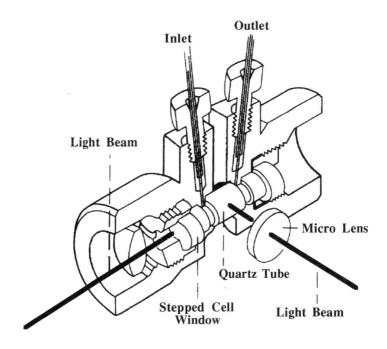

Courtesy of the Hewlett Packard Corporation

Figure 7.11 The HP Fluorescence Spectrometer Flow Cell

The cell is made from a quartz tube with stepped quartz windows. The optics are arranged such that the whole of the cell volume is illuminated, and thus all the sample can be excited and will fluoresce. Temperature stability is achieved by passing the eluent through a heat exchanger, thus helping to establish base line stability. The cell windows are spring loaded against their sealing faces by a pressure of about 1200 p.s.i., to protect the cell from rising column back-pressure

In general, fluorescence detectors are about 100 times more sensitive than the UV absorbance systems, and thus can detect substances present in the sub pico-gram range. The sensitivity of the HP tandem instrument is clearly demonstrated by the results shown in Figure 7.12.

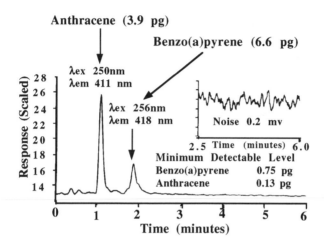

An ODS-Hypersil column 10 cm long, 2.1 mm I.D., packed with 5 µm particles, was employed. The mobile phase was a mixture containing 20% water and 80% acetonitrile operated at a flow rate of 0.5 ml/min.

Courtesy of the Hewlett Packard Corporation

Figure 7.12 Demonstration of the Sensitivity of the HP LC/FL Tandem Instrument to Anthracene

The sample mixture contained a mixture of benzo(a)pyrene and anthracene. An amplified section of the base line (included in Figure 7.12) shows the noise to be about 0.2 mv. Consequently, the minimum detectable mass is quoted as 0.75 pg for benzo(a)pyrene, and 0.13 pg for anthracene. Unfortunately, the more useful specification, the minimum detectable *concentration*, was not reported and is difficult to estimate with any degree of accuracy from the chromatogram.

The greater sensitivity provided by fluorescence, over that provided by UV absorption, is not the only advantage of monitoring a separation by fluorescence. As many compounds do not fluoresce, the chromatogram provided by fluorescence monitoring is simpler than that by UV absorption, as only a fraction of the solutes are detected. This allows the peaks of interest to be highlighted against a relatively flat base line, and renders quantitative estimations far more accurate. This is clear from the separation depicted in Figure 7.13.

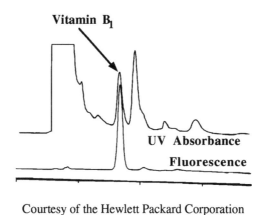

Courtesy of the Hewlett Packard Corporation

Figure 7.13 Comparison of the Chromatograms of a Milk Analysis Monitored by UV Absorbance and Fluorescence

It is clearly seen that the area or height of the vitamin peak can be measured with far greater accuracy with the fluorescence chromatogram, as it is virtually devoid of interfering substances.

The fluorescence spectrometer, used in conjunction with the liquid chromatograph, can often be programmed with respect to time. Thus, once the separation has been developed, the associated spectrometer can be programmed to provide the optimum excitation and fluorescence wavelengths for each peak as it emerges. This procedure provides the ultimate in sensitivity when using fluorescence detection. The principle of optimizing both the excitation and emission light wavelengths to obtain maximum sensitivity, however, can become quite complex as shown by the separation of some priority pollutants carried out on the PE LC/FL tandem instrument and depicted in Figure 7.14. The separation was carried out on a column 25 cm long, 4.6 mm in diameter, and packed with a C18 reversed phase. The mobile phase was programmed from a 93% acetonitrile, 7% water mixture to 99% acetonitrile, 1% water mixture over a period of 30 minutes.

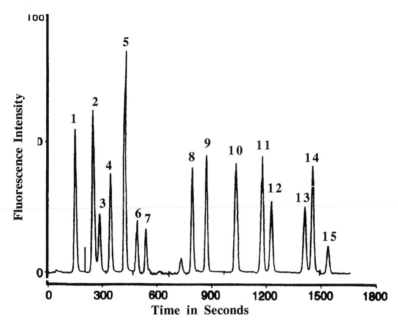

Fifteen Priority Pollutants

1 Naphthalene
2 Acenaphthene
3 Fluorene
4 Phenanthrene
5 Anthracene
6 Fluoranthene
7 Pyrene
8 Benz(a)anthracene
9 Chrysene
10 Benzo(b)fluoranthene
11 Benzo(k)fluoranthene
12 Benzo(a)pyrene
13 Dibenzo(a,h)anthracene
14 Benzo(ghi)perylene
15 Indeno(123-cd)pyrene

Courtesy of the Perkin Elmer Corporation

Figure 7.14 Separation of a Series of Priority Pollutants with Programmed Fluorescence Detection

The gradient was linear and the flow rate was 1.3 ml/min. All the solutes are separated and the compounds, numbering from the left, are given in the Figure 7.14. The separation illustrates the clever use of wavelength programming to obtain the maximum sensitivity. The program is shown in Table 7.2.

Table 7.2 The Fluorescence Detector Program

Time (seconds)	Wavelength of Excitation Light	Wavelength of Emitted Light
0	280 nm	340 nm
220	290 nm	320 nm
340	250 nm	385 nm
510	260 nm	420 nm
720	265 nm	380 nm
1050	290 nm	430 nm
1620	300 nm	500 nm

It is seen that during the development of the separation both the wavelength of the excitation light and that of the emission light was changed to an optimum for each particular solute. This ensured that each solute, as it was eluted, was excited at the most effective wavelength and then monitored at the strongest fluorescent wavelength.

Monitoring the separation in this way is a somewhat elaborate procedure and must be carried out with a complex and relatively expensive instrument. Nevertheless, if the analysis is sufficiently important, this type of instrumentation may be essential. The tandem system can also provide fluorescence or excitation spectra, by arresting the flow of mobile phase, allowing the solute to reside in the detecting cell, and scanning the excitation and/or fluorescent light. This is the same technique as that used to provide UV spectra with the variable wavelength dispersion UV spectrometer. Thus it is possible to take excitation spectra at any chosen *fluorescent* wavelength, or fluorescent spectra at any chosen *excitation* wavelength. Consequently, many hundreds of spectra can be produced, any or all of which (despite many spectra being very similar) can be used to confirm the identity of a compound.

Merely using a single optimum excitation wavelength and monitoring on the complementary optimum emission wavelength can provide extremely valuable signal enhancement for specific analyses. An example of this type of selective fluorescence detection is shown in the assay of aflatoxins using reversed phase chromatography and depicted in Figure 7.15.

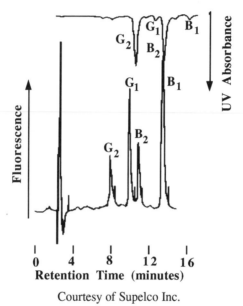

Courtesy of Supelco Inc.
(Supplied to Supelco Inc. by Dr. J. Hurst, Hershey Foods Corp. Hershey, PA USA.)

Figure 7.15 The Separation of Some Aflatoxins by UV Absorbance and Fluorescent Emission.

An LC 18 reversed phase column was employed with a mobile phase consisting of 20% acetonitrile, 20% methanol and 60% water and a flow rate of 1.1 ml/min. The UV absorbance was monitored at 365 nm and the fluorescence excited at 365 nm and the emission measured at 455 nm. It is seen that fluorescence response is three to five times greater than that provided by he UV absorbance. However, as the magnitude of the noise level in each case is almost impossible to discern the actual difference in sensitivity can not be determined.

Yamaguchi et al. [1] used optimized fluorescence in a similar way to measure fatty acid binding proteins (FABP) in rat liver. Fatty acid binding proteins are thought to facilitate the transport of long chain fatty acids in cells and may protect specific enzymes against inhibition by acyl-CoA esters. The authors derivatized the column eluent with dansylundecanoic acid that binds to the FABP to give it fluorescent properties. The reaction was carried out post–column, using the apparatus shown in Figure 7.16

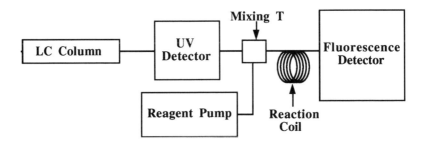

Figure 7.16 Post Column Reactor for the Detection of FABP with Dansylundecanoic Acid (ref.1)

The rat liver cytosolic fraction was injected directly into a Tosoh TSK gel G2000SW XL column (30 cm long, 7.8 mm I.D.) and eluted with 0.1 M potassium phosphate buffer (pH 7.2) at a flow rate of 0.5 ml/min. Subsequent to the column, the eluent was mixed with the dansylundecanoic acid reagent in a low volume mixing T and then passed through a coil for the derivatizing reaction to complete. The derivatized eluent was monitored with an Hitachi F1000 fluorescence spectrometer fitted with a 12 µl flow cell. This cell volume might appear rather large, but it should be recalled that the column was nearly 8 mm I.D and thus the peak volume was also fairly big. Consequently, an oversized cell volume could be tolerated without adversely affecting the separation. The excitation wavelength was

350 nm and the emission wavelength monitored was 500 nm. An example of the separations that were obtained are shown in Figure 7.17

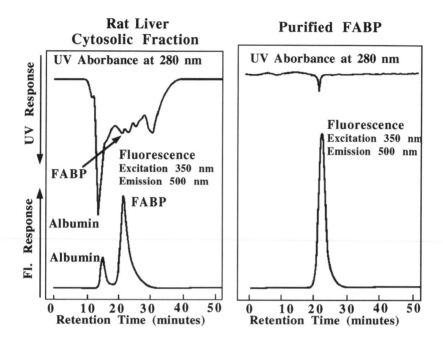

Figure 7.17 Chromatograms Obtained from Rat Liver and Purified FABC Employing a GC/FS Tandem System (ref. 1)

The advantages of the optimized fluorescence detection are clearly demonstrated. On the one hand the sensitivity of the system to the FABC is greatly enhanced but, on the other, and even more impressive, is the improved apparent resolution that is obtained by the selective method of detection.

The Perkin Elmer fluorescence spectrometer was used in tandem with a Waters-based liquid chromatograph assembly, by Adams *et al.* [2], to determine Flurbiprofen and its major metabolite (4'-hydroxyflurbiprofen). The sample matrixes were physiological fluids such as blood serum or urine. The samples of blood serum, (100 µl) were deprotenized with acetonitrile (1 ml) and buffered to a pH of 2.6 with 2 ml 0.05M potassium phosphate. The structural analog, 2-(2-methoxy-4-biphenyl)propionic acid was used as an internal standard. 100 µl aliquots of the supernatant liquid were separated on a Waters µBondepak C18 column, using a mixture of 55% of 0.05M potassium phosphate (pH 2.6), and 45% tetrahydrofuran as the mobile phase. The optimum excitation wavelength was 260 nm and the emission wavelength that was monitored was 320 nm. An excellent separation was obtained and as a result of the selectivity of the spectrometer when operated at the optimum excitation wavelength, and monitoring at the optimum emission wavelength of the substances of interest, the peaks were completely free from undetected contaminant materials. The recoveries of the drug and metabolite ranged from 97.4% to 105.5%. The lower limit of detection for the drug, Flurbiprofen, was approximately 1×10^{-6} g/ml, with a linear dynamic range extending to 50×10^{-6} g/ml. The linear range is small, but no less than would be expected for fluorescence measurements.

The same type of apparatus was used by Soroka *et al.* [3] to determine a number of different metals as their fluorescent 8-hydroxyquinoline-5-sulfonic acid complexes. Nearly 80 different metal species were examined and the optimum excitation wavelengths and emission wavelengths for each were reported. Some examples of which are given in Table 7.3

Table 7.3 Optimum Excitation Wavelengths and Optimum Emission Wavelengths of the 8-hydroxyquinoline-5-sulfonic acid Complexes Some Metal Elements

Metal	λ, Optimum Excitation (nm)	λ, Optimum Emission (nm)
Zinc	393	506
Cadmium	387	522
Magnesium	393	506
Calcium	393	506

The 8-hydroxyquinoline-5-sulfonic acid chelating reagent must be carefully purified before use, to eliminate any trace of fluorescing materials that would contribute background noise to the measurements.

The authors demonstrated the efficacy of their system by monitoring the chromatographic separation of the zinc, cadmium, magnesium and calcium complexes by fluorescence detection using the optimum excitation and emission wavelength given in Table 7.3. It is seen that for the metals examined, the optimum excitation and emission wavelengths are very similar, and seems to be more a function of the characteristics of the chelating agent than those of the metals. The sensitivity of the system indicated a wide variation between metals, the minimum appearing to be about 5×10^{-12} mol.

The Multiwave Fluorescence Detector

Tanabe *et al.* [4] developed a novel instrument that employed interference filters in an attempt to provide multiwave fluorescence detection without using a monochromator. The device operated at four different wavelengths and a diagram of their apparatus is shown in Figure 7.18. The excitation light from a 200 W xenon arc lamp was first reflected by a UV mirror (90% average reflectance for light between 325 and 475 nm) to minimize heat transfer and remove most of the visible light. The excitation light then passed through interference filters that could transmit light at 320 ± 2 nm, 360 ± 2 nm and 400 ± 2 nm, and was then focused onto the rounded shaped end of a quartz optical fiber bundle. This fiber bundle directed the light to the detector module, which consisted of a square quartz tube mounted in a black Teflon cell holder. The fluorescence signal was observed through side windows in the cell. Two detector assemblies, each consisting of a light sensor and an interference filter, were mounted in brass holders on either side of the cell. Consequently light emission at four different wavelengths could be measured simultaneously. Different filters were used for different excitation light frequencies; for example, with an excitation light wavelength of 400 nm, the filters in the four sensor units were chosen to transmit light at 420 ± 2 nm, 440 ± 2 nm, 460 ± 2 nm and 500 ± 2 nm respectively. The output from the sensors was acquired by a computer and the data stored. The separation, monitored by any one of the fluorescence channels, could then be reconstructed. An example of the use of the apparatus to resolve a convoluted peak containing two solutes is shown in Figure 7.19.

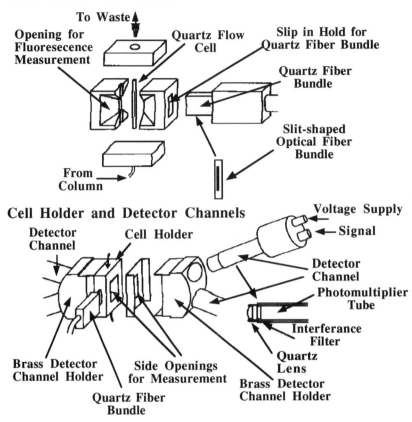

Reprinted with permission from K. Tenabe, M. Glick, B. Smith, E, Volgtman and J. D. Winefordner, *Anal. Chem.*, **59(8)**(1987)1124, Copyright 1987 American Chemical Society

Figure 7.18 Schematic Diagram of the Detector Module

The wavelength of the excitation light was 400 nm and the emission light was monitored at 420 nm, 440 nm, 460 nm and 500 nm respectively. Chromatograms were reconstructed from the output at all four wavelengths, and the combined chromatograms are included in Figure 7.19. It is seen that the two components are clearly differentiated by multi-channel fluorescence detection, whereas the dual nature of the convoluted peak is indiscernible by non-selective detection. Although this apparatus is simpler, and probably less expensive to make than the fluorescence spectrometer, its performance is also severely limited in comparison.

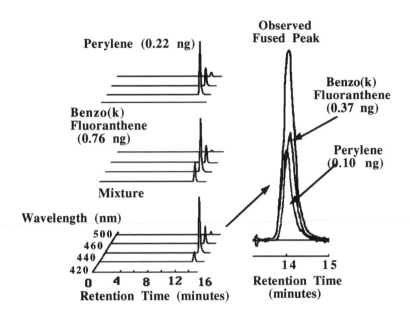

Reprinted with permission from K. Tenabe, M. Glick, B. Smith, E, Volgtman and J. D. Winefordner, *Anal. Chem.*, **59**(8)(1987)1124, Copyright 1987 American Chemical Society

Figure 7.19 The Resolution of Convoluted Peaks

There are distinct disadvantages to the use of the traditional dispersive type monochromators for analyzing the fluorescent light produced from the sample. Either the separation must be arrested leaving the sample stationary in the sample cell, and the fluorescence spectrum obtained by the usual stop–start procedure, or the analyst must be content with measuring the fluorescent light at a single or narrow band of wavelengths. Although it is difficult to select the excitation light by any means other than the dispersion type monochromator, it certainly would be possible to utilize a diode array system to analyze the fluorescent light. Thus the use of a dispersion monochromator to select the wavelength of the excitation light, and a combination of the simple dispersion monochromator and the diode array to measure the fluorescent light, could produce a vastly improved fluorescence spectrometer.

There are however, some problems to be overcome in the use of the diode array, that arise from the low sensitivity of the diode, compared with the normal photomultiplier tube. This low sensitivity is the result of the high background noise level, which consists of 'fixed pattern' noise and integrated 'dark current' (background signal). The fixed pattern noise comes from the many electronic components contained in the entire electronic system, and will include noise from the detector electronics, data acquisition, computer processing and display circuits. This source of noise can be largely removed by digital background subtraction.

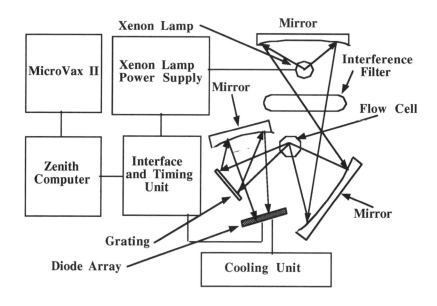

Reprinted with permission from J. Wegrzyn, G. Patonay, M. Ford and I. Warner, *Anal. Chem.*, **62(17)**(1990)1754, Copyright 1990 American Chemical Society

Figure 7.20 The Diode Array Fluorescence Spectrometer

The dark current noise arises from the thermal discharge of reverse bias on each diode. Fluctuations in dark current are directly related to temperature changes in the diode and also to variations in integration time [6]. The noise from integrated dark current can therefore be reduced by cooling the diode array. Thus, by digital background subtraction and the use of a Peltier cooling system to reduce the temperature of the diodes, the use of the diode array for fluorescence spectroscopy becomes possible.

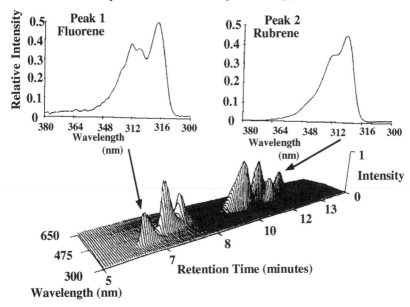

The components of the mixture in order of their elution are as follows,

1. Fluorene	2.2 µg/ml	5. Perylene	0.9 µg/ml
2. Anthracene	1.6 µg/ml	6. Benzo(k)fluoranthene	1.6 µg/ml
3. Fluoranthene	1.3 µg/ml	7. Rubrene	1.9 µg/ml
4. Benzo(b)fluoranthene	2.6 µg/ml		

Reprinted with permission from J. Wegrzyn, G. Patonay, M. Ford and I. Warner, *Anal. Chem.*, **62(17)**(1990)1754, Copyright 1990 American Chemical Society

Figure 7.21 Chromatograms of a Seven-component Mixture of Polycyclic Aromatic Hydrocarbons

Diode arrays were first examined for use in liquid chromatography detection in the mid-1970s. Wegrzyn *et al.* [6] were one of the earlier workers to explore the possibilities of a diode array for fluorescence measurements in liquid chromatography and a diagram of their tandem apparatus is shown in Figure 7.20. Light from a xenon lamp was focused onto the flow cell by two curved mirrors. Between the mirrors was an interference filter which selected the wavelength of the excitation light. To change the excitation light, the filter had to be removed and an appropriate replacement inserted. The fluorescent light was then focused by means of a plane and curved mirror onto the diode array. The diode array was cooled by a Peltier cooling unit. The liquid chromatographic system, that was used in conjunction with the apparatus, consisted of a Perkin Elmer Series 10 isocratic pump, a Rheodyne 7510 injector with a 20 µl sample loop, and an ODS column, 15 cm long, 4.6 mm I.D. packed with particles 10 µm in diameter.

The test sample used was a mixture of polycyclic aromatic hydrocarbons, which was eluted isocratically with a solvent mixture of 80% acetonitrile and 20% water. The separation they obtained is shown in Figure 7.21, and it is seen that the apparatus functions quite well, but the sensitivity compared with the simple fluorescence detector is relatively low. It is also seen from the two-dimensional presentation in Figure 7.21 that good spectra can be obtained from partially resolved peaks and would certainly aid in confirming the identity of a specific solute.

A Tandem Instrument that Monitors UV Absorption, Fluorescence and Luminescence

In early 1989 Yappert *et al.* (7) described a tandem instrument involving a liquid chromatograph and a spectrometric assembly that monitors UV absorption, fluorescence and luminescence simultaneously. A diagram of the spectrometric system is shown in Figure 7.22, and it is seen that it is a fairly complicated arrangement. Pulsed light from a broad wavelength xenon lamp passes through an IR filter to the first half mirror. Light reflected at right angles, passes to a second half mirror, and the light that passes through the second half mirror is focused onto a photoelectric cell. The output from this photoelectric cell is used in a feedback system to control the intensity of the light emitted from the xenon lamp.

The light that passes directly through the first half mirror is focused onto a monochromator that selects the wavelength of the light that is to act as both the excitation light for fluorescence and the light used for absorbance measurements. Light of the selected wavelength passes through a third half mirror, and is focused onto the sample cell. The light reflected at right angles from the third half mirror is focused onto another photoelectric cell. The output from this second photoelectric cell provides a reference for fluorescence measurements. The fluorescent light emitted at right angles to the excitation light is focused onto a monochromator and the dispersed light measured by a diode array. Light reflected at right angles from the second half mirror is focused through an optical fiber into the sample cell, and the transmitted light passes to a second monochromator, and dispersed onto another diode array.

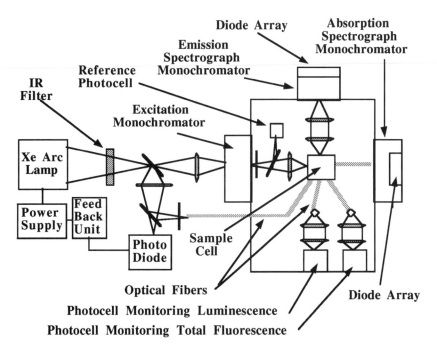

Reprinted with permission from M. C. Yappert, M. W. Schuyler and J. D. Ingle, Jr. *Anal. Chem.* **61**(1989)593, Copyright 1989 American Chemical Society

Figure 7.22 The Spectroscopic System that Measures UV Absorption, Fluorescence and Luminescence

The fluorescence emitted in the sample cell is also focused by means of another optical fiber onto a simple photoelectric cell that provides an output equivalent to the total fluorescent light. Finally another optical fiber, situated downstream in the sample cell and well away from the excitation light, monitors the delayed fluorescence or, as it is more commonly known, the luminescence.

Depending on the method of measurement, the sensitivity of the system to quinine ranged from 1.2 to 40 ng/ml. As one might expect, the highest sensitivity observed was obtained from the fluorescence channel. The amount of fluorescent light emitted by a sample, is a function of the intensity of the excitation light as well as its wavelength. It follows that by increasing the intensity of the excitation light, by the use of a laser light source, the sensitivity of the fluorescent spectrometer would also be increased.

Laser-induced Fluorescence Detection Employing a LC/FL Tandem Combination

Roach and Harmony [8] developed a laser induced fluorescence instrument for the analysis of amino acids, based on an apparatus developed earlier by Sepaniak and Yeung [9]. The modification was carried out to significantly improve the sensitivity. The instrument was basically a laser source, transmitting light through a flow cell, and the fluorescent light so produced was measured by a fluorescence spectrometer arrangement.

A diagram of the apparatus developed by Sepaniak and Yeung is shown in Figure 7.23. The separation apparatus was very simple and comprised a mobile phase supply, pump, sample valve and column. The eluent passed to quartz capillary tube that was mounted on an adjustable stand so that it could be placed directly in the laser light path. The volume of the quartz tube was about 20 µl but, according to the authors, could be reduced to a few microliters if so desired. The mobile phase, subsequent to traversing the quartz tube, passed to waste. An argon laser was used, the light from which was focused by means of a simple lens onto the center of the quartz tube. Fluorescent light generated was collected by a quartz fiber and first passed thorough two broad band interference filters to remove stray light, and then to the monochromator.

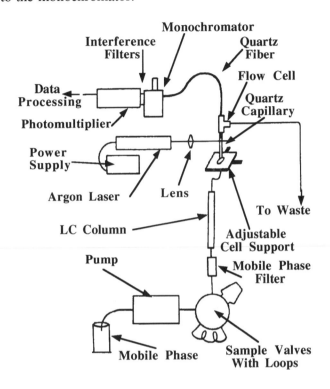

Figure 7.23 The Laser Induced Fluorometric Instrument Combined with the Liquid Chromatograph

Roach and Harmony employed the tandem system for the separation and trace analysis of amino acids. The derivatization process (Carlson *et al.* [9]), employed the reagent 2,3-naphthalenedialdehyde which, when used in the presence of cyanide, yields adducts of excellent thermal stability and gave high quantum yields of fluorescent light. An excellent separation was obtained with the desired high sensitivity. The sensitivity of the tandem system, as defined by the minimum detectable mols ranged from about 0.2 to 0.4 fmol. of the original acid. Assuming an average molecular weight for the amino acid of about 150, then this molar sensitivity corresponds to a mass sensitivity range (minimum detectable mass) of 3×10^{-13} to 6×10^{-13} g (0.3–0.6 pg).

Synopsis

Today the LC/UV tandem systems are basically multi-wavelength detectors. The multi-wavelenght dispersion detector can only monitor a separation at one wavelength, but has the higher resolution. The diode array detector monitors UV absorption simultaneously at all wavelengths, but its resolution is limited by the number of diodes. The tandem LC/UV system is either used to monitor a separation at the optimum wavelength, test the purity of a peak from its spectrum, or compare the spectrum of a peak to that of a reference. Tandem systems employing fluorescence spectroscopy help identify eluted compounds, selectively monitor a specific type of sample, or enhance the sensitivity of the system to a particular compound. Both excitation spectra and emission spectra can be obtained and in modern instruments, the change in excitation wavelength and emission wavelength can be programmed to suit the elution pattern of the eluents. Fluorescent derivatives can be used very effectively to enhance the sensitivity or render substances that do not naturally fluoresce amenable to fluorescent detection. More recently the diode array (appropriately cooled to reduce the dark noise) has also been used to monitor fluorescence spectra.

References

1. M. Yanaguchi, K. Wade, J. Ishida and M. Nakamura, *Analyst*,**117**(1992)1859.
2. W. J. Adams, B. E. Bothwell, W. M. Bothwell, G. J. VanGiesen and D. G. Kaiser, *Anal. Chem.,* **59(11)**(1987)1504.
3. K. Soroka, R. S. Vithanage, D. A. Phillips, B. Walker and P. K. Dasgupta, *Anal. Chem.*, **59(4)**(1987)629.
4. K. Tenabe, M. Glick, B. Smith, E, Volgtman and J. D. Winefordner, *Anal. Chem.*, **59(8)**(1987)1124.
5. Y. Talmi, *Appl. Spectrosc.*, **36**(1982)1.
6. J. Wegrzyn, G. Patonay, M. Ford and I. Warner, *Anal. Chem.*, **62(17)**(1990)1754.
7. M. C. Yappert, M. W. Schuyler and J. D. Ingle, Jr. *Anal. Chem.* **61**(1989)593.
8 . M. C. Roach and M. D. Harmony, *Anal. Chem.*, **59(3)**(1987)411.
9. M. J. Sepaniak and E. S. Yeung, *J Chromatogr.*, **190**(1980)377.
10. R. G. Carlson, K. Scrinivasachar, K. R. S. Givens and B. K. J. Matuszewske,

CHAPTER 8

LIQUID CHROMATOGRAPHY IR SPECTROSCOPY (LC/IR) TANDEM SYSTEMS

Infrared spectra, in general, provide more information on molecular structure than UV spectra, and by comparison with reference spectra confirm solute identity with greater certainty. This is largely because the IR spectra of the majority of compounds contain far more fine structure than there is present in their corresponding UV spectra. However, the association of the IR spectrometer with the liquid chromatograph is far more difficult than the UV spectrometer, due to the inherent IR absorption characteristics of the solvents normally employed in LC development. In fact, the majority of useful solvents absorb light in the infrared range, and, more important, those ranges are frequently the most informative in structure elucidation and sample identity confirmation. This means that, for the practical association of the liquid chromatograph with the IR spectrometer, either the solvent needs to be removed before measurement, or an infrared transparent solvent must be employed. Unfortunately, of the few solvents that are transparent to the IR over the wavelength range that is important, many are not compatible with the mobile phase requirements of the liquid chromatograph.

The IR spectrometer is generally less sensitive than the UV spectrometer and thus the LC/IR tandem system is at a further disadvantage. The introduction of the FTIR instrument has partly compensated for the sensitivity difference but the IR spectrometer is still several orders less sensitive than the mass spectrometer. Consequently, to obtain an adequate signal, the size of the sample that is placed onto the column will need to be increased, which, however, may be unacceptable to the chromatographic system, particularly if small–bore columns are used.

As a result of these problems, prior to about 1975, LC/IR analyses were carried out off–line, the peaks being collected and the solvent removed by appropriate procedures. The eluted solutes were then either redissolved in a suitable IR solvent, or compressed into a potassium bromide disk, and the spectrum obtained employing standard techniques.

The introduction of the Fourier transform IR spectrometer not only partly solved the problem of sensitivity but, as the spectra are stored, the problem of solvent absorption could also be partly solved. In theory, the background spectrum of the solvent can be subtracted from the spectrum obtained for the solute, leaving the actual spectrum of the solute as a difference value. Unfortunately, this involves subtracting two very large signals, and thus the signal to noise of the difference spectrum is rather low. It follows that spectra obtained in this way would be very noisy, causing uncertainty both in spectra interpretation and in the confirmation of substance identity. Nevertheless, in the early days of FTIR, Kizer *et al.* [1] demonstrated that this technique would work but, as would be expected, the spectra were, indeed, very noisy which resulted in very poor sensitivity.

LC/IR Transport Interface

The introduction of the concept of employing transport interfaces to conduct the sample from the chromatograph to the spectrometer was introduced by Scott *et al.* [2] for liquid chromatography/mass spectrometry tandem systems. In due course, the transport concept was also adopted for LC/IR combinations. Probably the most practical LC/FTIR interface that is commercially available is based on the solvent transport concept.

An Early LC/FTIR Interface

Some of the first to attempt to utilize a transport system for LC/IR operation were Kuehl and Griffiths [3]. Initially they tried to employ moving ribbon devices with pre-concentrating techniques, in a similar manner to that of Scott *et al.*, but were rather unsuccessful. Their final system was rather crude but, nevertheless, an effective transport system that utilized the principle of the rotating disk. Their final model took the form of a carousel of cups containing potassium chloride.

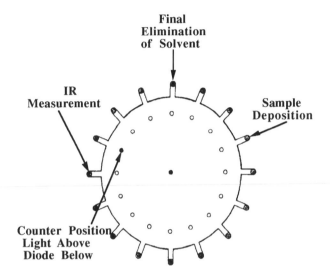

Figure 8.1 Carousel Transport for On-line IR Monitoring of LC Column Eluents

A diagram of their carousel is shown in Figure 8.1. The carousel was very similar to a fraction collector, and consequently the device might be considered more like an off-line auto fraction collector than an in-line interface. In fact, all transport interfaces might be deemed automated off-line monitoring device, depending on the speed of the transport process. Fast transport, as with the wire or belt transport interface used in LC/MS, gives the impression of being an in-line devices. Conversely, the carousel transport interface described here, being relatively slow, appears more like an automated off-line IR sampler.

The LC/IR interface carousel had 32 cups, each fitted with a fine mesh screen and containing potassium chloride powder. The position of the carousel was controlled automatically and there were only three positions where specific sampling activities took place. In the first position the eluent was deposited on the potassium chloride directly from the column, until the powder was wetted and the halide was saturated with mobile phase. The cup was then moved to the second position, where a stream of air was drawn through the potassium chloride to remove the solvent. After drying, the cup the was moved to the third position where infrared light was directed through the halide, and the spectrum was taken. The use of the carousel containing potassium chloride powder, together with the evaporation of the solvent and thus concentrating the solute, certainly increased the sensitivity of the LC/IR combination. However, the intermittent nature of the sample collection made the system unsuitable for modern LC columns, where many peaks can be eluted in a few seconds.

Initially, off line sample collection of chromatographic eluents for subsequent IR examination was employed more as an expedient than as a chosen method. Sample collection provided an opportunity to remove interfering solvents from the sample before the IR spectra were obtained. In fact, the main advantage of the original carousel interface of Kuehl and Griffiths [3] was to eliminate the mobile phase. The carousel, and similar types of interface, act basically as a chromatographic 'memory', collecting all the solutes that are eluted as a 'physical' chromatogram, consisting of localized masses of solutes, deposited on the transport medium. The concept of a chromatographic memory was first introduced by Karmen [4], who used a wire transport detector to collect samples of eluent onto the wire surface, which was then stored on a reel. Subsequently, the wire was unwound from the reel, and passed continuously through the flame of a flame ionization detector, producing a record of the separation. The most effective LC/IR interfaces are based on this principle, but the time interval between collection and measurement varies somewhat between different interfaces.

The Development of LC/IR Transport Interfaces

Jino and Fujimoto [5,6] adopted a similar approach to that of Kuehl and Griffiths, but employed a potassium bromide plate as the transport system.

This modification allowed the transport medium itself to be used as the sample holder for the IR measurement. The rotation of the plate was actuated by a signal from the detector, so that as a peak began to elute, the plate was moved to a new collection position. When the elution of the peak was complete, the disk moved on again, isolating the sample at a specific position at the perimeter of the plate. The authors utilized a small–bore column and the eluent (flow rate 5 µl/ min) was allowed to fall on the plate and evaporate. The eluent was also monitored by a UV detector, the signal from which was used to identify the time the plate should be moved. The solvent was allowed to evaporate under ambient conditions, or the evaporation could be assisted by the use of an infrared heater, directed on the collection area.

After the chromatogram had been developed, and the spots were free from solvent, the light transmitted through the sample and the plate was measured and the spectrum of each solute obtained. It is clear that the system excluded the use of any solvent in which the halide was significantly soluble. Good sensitivity was reported, but the minimum detectable mass or concentration was not precisely defined in chromatographic terms. However the procedure (and, for that matter, the procedure of Kuehl and Griffiths [3]) would probably be considered as little more than novel methods of fraction collecting and, in fact, were really off–line procedures. Nevertheless, the results from these early workers encouraged the transport concept to be examined further.

Raynor *et al.* [7] employed a similar interface to monitor the IR spectra of some polymer additives, using super critical fluid chromatography. They used a heated transfer line that terminated 50 µm above a potassium bromide window in a similar manner to that of Jino and Fujimoto. The sample compartment was purged with nitrogen to reduce interference from water vapor and carbon dioxide. The spectra were collected with the microscope in the IR transmission mode. The IR beam was stopped down to an aperture of about 150 µm, or less, depending on the size of the deposit. The results obtained from the separation of a mixture of different polymer additives are shown in Figure 8.2.

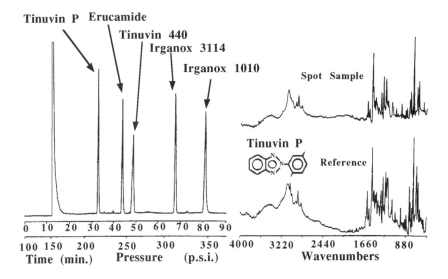

Reprinted with permission from M. W. Raynor, K. D. Bartle, I. L. Davies, A. Williams, A. A. Clifford, J. M. Chalmers and B. W. Cook, *Anal. Chem.*, **60**(5)(1988)427, Copyright 1988 American Chemical Society

Figure 8.2 The Separation of Some Polymer Additives Using a LC/FTIR Tandem System

The sample injected on the column contained 200 ng of each component. On the right-hand side of Figure 8.2, the spectra obtained for Tinuvin P is shown, taken directly from the potassium bromide window together with the reference spectrum for the same compound. It is seen that the spectrum is clear and well resolved, and more than adequate to confirm the identity of the solute. It would appear from the report that the window was not moveable and so each peak was manually collected.

During the 1980s the transport concept was actively developed by a number of workers. Gagel and Bieman [8] described a disk transporter that had a reflective surface, which was used in conjunction with a simple nebulizer to deposit the sample on the surface of the rotating disk. Their basic apparatus is shown diagramatically in Figure 8.3.

They employed an aluminum disk on the surface of which was cemented a circular glass mirror. The disk was rotated continuously as the separation was developed, leaving a trail of solid deposits in the form of a spiral on the surface of the reflective plate. Evaporation was aided by the use of a simple nebulizer. The column eluent was fed into a T junction and one of the other limbs carried a flow of nitrogen gas. The mixture of gas and eluent passed out through though the third limb, *via* a narrow nozzle, which directed the spray onto the surface of the disk. When the separation was complete, the disk was placed into a modified total reflectance accessory.

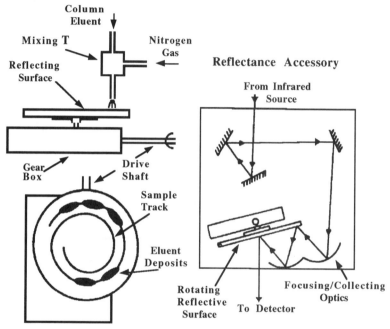

Reprinted with permission from J. J. Gagel and K. Bieman, *Anal. Chem.*, **58(11)**(1986)2184, Copyright 1986 American Chemical Society

Figure 8.3 The Layout of the Transport LC/FTIR Apparatus Developed by Gagel and Bieman.

The surface was continuously monitored by the IR spectrometer, while the disk rotated, and the reflectance–absorbance spectra were continuously collected. The LC/FTIR interface appeared to be a very practical system, and seemed to function without contributing any significant peak dispersion that might cause loss of chromatographic resolution. The minimum mass required to provide a satisfactory spectrum will depend on the characteristic absorbance of the substances being monitored. Nevertheless, the results indicated that between 50 and 100 ng of sample would provide a recognizable spectrum.

Subsequently, Gagel and Bieman [9] modified the design of their nebulizer to improve the deposition procedure, and to make it more suitable for spraying aqueous solvents. The intent was to reduce the spreading of the eluent during deposition and to concentrate the material in a smaller area to improve the sensitivity. The modified jet design involved the use of two nitrogen streams in the nebulizer head very similar to one of the types examined by Lange and Griffiths [10] and a diagram depicting their alternative jet design is shown in Figure 8.4.

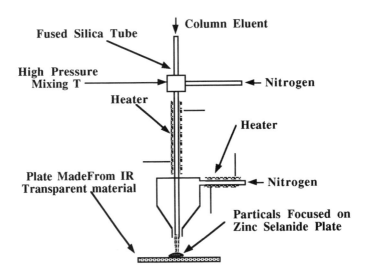

Figure 8.4 The Modified Nebulizer of Gagel and Bieman.

The column eluent was mixed with nitrogen under pressure in a high pressure mixing T. The nitrogen–eluent mixture was directed to the deposition surface through a syringe needle, fitted to the other port of the high–pressure T. The syringe needle itself was situated inside another

The column eluent was mixed with nitrogen under pressure in a high pressure mixing T. The nitrogen–eluent mixture was directed to the deposition surface through a syringe needle, fitted to the other port of the high–pressure T. The syringe needle itself was situated inside another nozzle, through which heated nitrogen was flowing. This arrangement warmed the eluent–nitrogen mixture, and thus increase the evaporation rate, and also heated the receptor plate, which helped to remove the last residues of solvent. The simple arrangement allowed the system to be employed with aqueous solvent mixtures and, at the same time, significantly increased the overall sensitivity of the apparatus. The sensitivity of the modified interface is demonstrated in Figure 8.5.

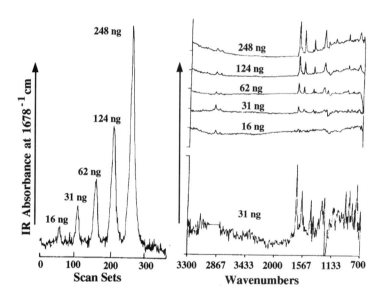

Reprinted with permission from J. J. Gagel and K. Bieman, *Anal. Chem.*, **59(9)**(1987)1267, Copyright 1987 American Chemical Society

Figure 8.5 Results from the Modified LC/FTIR Interface Demonstrating the Overall Sensitivity of the Tandem Instrument

solute that would provide a signal to noise ratio of 2, was about 16 ng. However, it would appear from the spectra on the right that to obtain a spectrum that has sufficient information for sample identification, the mass of solute injected must be greater than 31 ng.

Solvent elimination from transport systems, such as those previously described, has proved to be relatively easy with non–aqueous mobile phases. However, the majority of LC separations are carried out employing reverse phase columns. Such columns require the use of mobile phases with high water contents, and such mixtures do not evaporate easily. Poor volatility causes the solute deposits to be smeared into one another, which can seriously impair the separation. In addition, the presence of water in the mobile phase will also restrict the choice of the transport medium, as it must not be soluble, or affected by water. Another problem associated with the deposition of the solute on the transport medium is the need to keep the area of deposition as small as possible, which is essential to prevent one spot merging into its neighbor.

Lange and Griffiths [10] extended the development of the LC/FTIR interface using concentric flow nebulization to improve the deposition of the solute, and to focus the solute onto a narrow spot. The different types of nebulizers they examined are depicted in Figure 8.6. The first design was merely an emulation of the thermospray employed as an interface in LC/MS. Although satisfactory for interfacing with a mass spectrometer, the thermal spray jet gave a diffuse distribution of eluent, and spots of sufficiently small diameter could not be obtained. The second method, termed by the authors *hydrodynamic focusing*, employed a concentric gas flow which certainly helped reduce the size of the spot due to the Bernoulli effect. However, when the diameter of the jet becomes too small, the jet breaks up again and a diffuse deposition occurs. To aid in the evaporation, and to reduce the jet diameter still further, the jet was heated as in the thermospray method. For optimum performance, the inner jet should protrude about 1 mm beyond the outer tube carrying the nebulizing gas. This provides the required narrow jet and small spot diameter; this form of deposition was termed *concentric flow nebulization*.

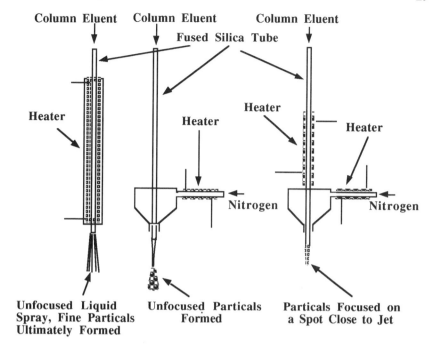

Figure 8.6 Different Types of Nebulizers

The nebulizer was arranged to deposit the solutes on a rotating disk as shown in Figure 8.7. The outer sheath of gas communicates momentum to the liquid center and the stream is broken up into droplets. The heating causes rapid evaporation and the solid material is directed to a very small circular spot on the surface of the stage. It was shown that this nebulizing procedure provided narrow discrete spots of solute, and could be used satisfactorily with aqueous solvent mixtures. In similar systems, the transport medium consisted of a layer of potassium chloride on the surface of a zinc–selenium metallic stage, and the IR spectrum of the deposited material, obtained by diffuse transmission spectrometry. However, it was not possible to use potassium chloride with aqueous solvents. The authors used a zinc–selenium window as the transport material, and then examined the spots after deposition, using an FTIR microscope to measure the transmission spectra.

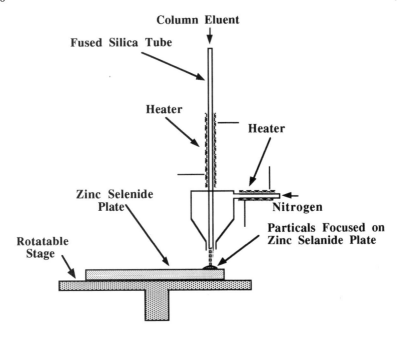

Figure 8.7 Detail of Stage and Concentric Flow Nebulizer

The total system is contained in an evacuated aluminum enclosure, which is fitted with windows to observe the nebulization process. The enclosure is evacuated by means of a standard vacuum system, fitted with a solvent trap to remove solvent vapor. The concentric flow nebulizer projects through the upper plate and electrical connections are made through the plate to the nebulizer heater. The heater consists of a few turns of chromel wire. Helium is fed into the column eluent, exterior to the deposition chamber, and the total nebulizer assembly is shown in the upper part of Figure 8.8. Helium was chosen as the nebulizing gas, as opposed to nitrogen, as helium has the higher thermal conductivity, and allows greater heat transfer to the mobile phase. The average spot diameter of the solutes deposited on the disk was reported as about 100 µm. The transport disk is shown in the lower half of Figure 8.8. The position of the rotatable stage is adjusted by means of a micrometer screw, mounted beneath the stage, and external to the vacuum chamber. The stage consists of a metal disk on the surface of which rests a zinc selenide plate. The position of the plate is adjusted to be between 0.125 and 0.5 mm from the center jet of the nebulizer. The gas flow rate used for nebulization is adjusted to be appropriate for the column flow rate, but should not exceed 80 ml/ min.

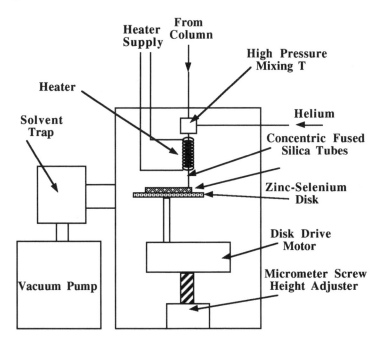

Figure 8.8 An Exploded View of the LC/FTIR Interface

It appeared from the report that although the separation was also monitored by a UV detector, the disk was rotated manually. The disk was moved to a new position just before a peak started to elute, and just after the elution of the peak was complete. After the separation was finished and the individual solutes were deposited on the zinc selenide plate, the disk was removed. The spectra were measured off–line, using a Perkin Elmer Model 1800 FT–IR spectrometer, fitted with a Spectra–Tech IR–Plan microscope. Good spectra were reported to have obtained, and the system exhibited a very useful sensitivity. The spectra were obtained from 60 ng of material with a signal to noise ratio of at least 9. It would also appear that the method of deposition was very efficient, and useful spectra could be obtained without excessively loading the LC column. Nevertheless, the system was basically a very efficient and simple fraction collecting device. However, is should be noted that the spectrometer and collection device could probably be oriented, so that after passing through the coating procedure, as the disk was rotated it would automatically place the sample in the FTIR optical unit. In this case the time interval between collection and measurement would be reduced, and the need for the physical transfer of the plate from one unit to the other eliminated. Under such circumstances the device might be considered an in–line interface, but it would be a moot argument. In addition the disk could easily be automated and either rotated continuously at an appropriate speed, or actuated automatically from the signal produced from the UV detector.

Pentoney and Griffiths [11] examined a number of methods for measuring the spectra from solid samples distributed on a carrier. They considered six different procedures. These included conventional transmission (i) and external reflection spectrometry (ii) of the sample deposited on a zinc selenide plate. They also investigated the reflection adsorption spectrometry (iii) of a sample deposited on a smooth metallic surface. In addition the diffuse reflection spectrometry was examined where the sample was deposited on a thin layer of sodium chloride attached to either a metallic (iv) or an infrared transmitting substrate zinc selenide (v). Finally they also examined spectra taken by diffuse transmission of samples deposited on zinc selenide. From he results it was concluded that conventional transmission spectra of samples, situated on a flat infrared window, gave the best compromise between high sensitivity and the faithful representation of relative band intensities, and adherence to the Beer–Lambert law.

Somsen *et al.* [12] described a jet spray assembly as an interface for reversed phase LC/FTIR, which involved the continuous deposition of the eluent from a narrow bore reversed phase column, onto the surface of a linearly moving substrate. After deposition, the immobilized chromatogram was analyzed by linearly moving the substrate under an FT-IR microscope while collecting the spectra. The arrangement performed as a transport interface where the transport medium was used as a chromatographic memory. Although the eluates were collected and stored on line with the chromatograph, they were analyzed later at some convenient time. A diagram of their apparatus is shown in Figure 8.9.

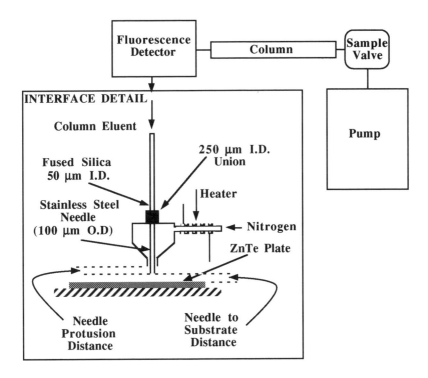

Figure 8.9 The Linear Memory Transport Interface.

The sample valve (Valco) had an internal loop, 1.9 µl in volume and the column was 17 cm long, 1.1 mm I.D., packed with Rosil C18 reversed phase, particle size 5 µm. The mobile phase was a methanol/water mixture (either 95:5 or 80:20) used at a flow rate of 20 µl/min. The IR data was obtained from a Bruker 1FS-85FT-IR spectrometer, equipped with a Bruker A590 FT-IR microscope. The microscope contained a 16x Cassegrainian lens and a narrow-range mercury-cadmium telluride detector. The microscope had an adjustable rectangular aperture and normally 128 scans were co-added.

The column conduit was a fused capillary tube, 40 cm long, 50 μm I.D which was connected, *via* a suitable union, to a stainless steel needle, 100 μm I.D and 475 μm O.D. In the spray unit, a nitrogen flow that could be heated up to 200°C, passed through a concentric outside tube, 600 or 900 μm I.D. This system evaporated the solvent and deposited a solid sample onto the transport medium. The length of inner tube that projected beyond the nitrogen sheath, and the distance between the end of the deposition jet and the transport medium surface, were adjustable. The model compounds used were fluoranthene, pyrene, benzo[a]anthracene and benzo[k]-fluoranthene. A separation of these compounds is shown in Figure 8.10 monitored directly by fluorescence detection and by scanning the transport medium by densitometry.

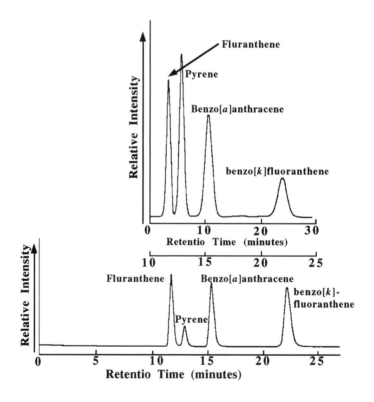

Figure 8.10 Chromatograms of the Test Mixture Taken On-line by a Fluorescence Detector and Off-line by Densitometry.

LC/IR Tandem Systems

The chromatograms demonstrates the effect of the deposition interface on the chromatographic resolution. It is seen that the integrity of the separation is not significantly impaired by the spraying technique. It was found there was an optimum distance for the jet to project beyond the sheath flow jet, and there was an optimum distance between the end of the deposition jet and the transport medium surface. These optimum distances were found to be approximately the same, that is about 0.5 mm and 0.5 mm respectively. It was also found that a zinc selenide window gave better spectra than a aluminized reflective surface when used as transport media. A three-dimensional reversed phase LC/IR plot for the separation of some polynuclear aromatic hydrocarbons is shown in Figure 8.11

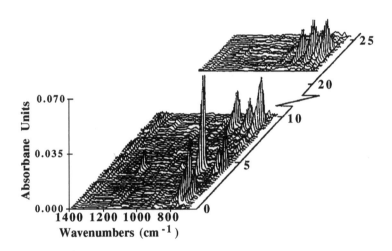

Figure 8.11 A Three-dimensional Reversed Phase LC/IR Plot of the Separation of Some Polynuclear Aromatic Hydrocarbons

The data was measured by scanning the substrate in steps of 200 μm using a rectangular aperture. The spectra of the individual solutes are clearly recognizable in the figure.

The signal to noise ratio of the spectra was measured for the solute pyrene over a range of sample sizes. The sample size that would just provide a identifiable spectrum was taken as the limit of detection. The spectrum taken at the limit of detection together with the spectrum of pyrene taken from Figure 8.11 are shown in Figure 8.12.

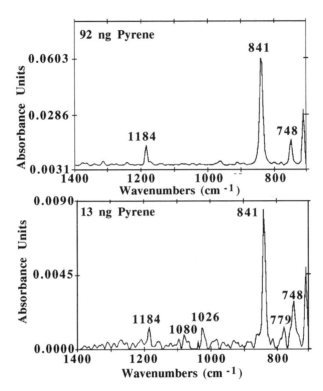

Figure 8.12. Reversed Phase LC/IR Spectra of Pyrene Taken at Different Sample Sizes

It is seen that the spectrum taken from Figure 8.11 is extremely clear and unambiguous and was obtained from about 92 ng of sample. The lower chromatogram in Figure 8.12 shows the limit of detection, which appears to be about 13 ng, an extremely low sample size for this type of interface. This interface system provides high sensitivity and excellent spectra but the authors only claim success with mobile phases carrying up to 20% of water. Although this opens up a significant range of analyses to this particular LC/FTIR tandem system, the limit of 20% water in the mobile phase still excludes the majority of reversed phase applications.

More recently, Turula and Haseth [13] have employed the particle beam atomizing technique, originally described by Browner *et al.* [14–16]. This technique was initially employed as a mass spectrometry interface but has now been adapted for use in LC/FTIR. The system was employed with a standard liquid chromatograph (the Hewlett–Packard 190L binary gradient instrument) and the spectra were obtained on a Perkin Elmer 1725X FTIR spectrometer. The particle beam atomizer is shown in Figure 8.13.

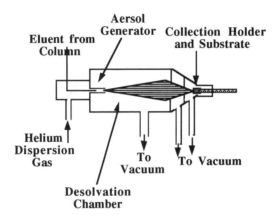

The approximate shape of the aerosol 'plume' in both the desolvation chamber and the momentum separator is shown shaded.

Figure 8.13 The Particle Beam Atomizer.

In normal operation the column eluent is pumped through a small capillary and exits as a liquid jet. The emitted stream is then nebulized with a stream of helium gas. The droplets of liquid, so produced, pass into a desolvation chamber, where the solvent molecules evaporate and are removed. The size of the droplets is controlled by the diameter of the capillary (*ca* 16–25 μm I.D.). The rate of evaporation rises as the sizes of the drops decrease due to the increase in relative surface area. The solute particles then pass into a separator chamber which is evacuated, and then strike an infrared transparent substrate. This substrate is then placed in the FTIR spectrometer and the spectrum obtained by co–adding 1000–2000 single–beam scans. The nebulizer was not used with a transport collector, but there appears no reason why this should not be possible. The main advantage of the device appears to be that proteins can be collected with the minimum amount of structural alteration, and thus true spectra of the protein can be obtained.

The Series 100 LC–Transform™ LC/FTIR Interface

One of the few LC/FTIR interfaces that are commercially available is that provided Lab Connections Inc., the design of which is based on, or similar to, the apparatus described previously by Lang and Griffiths. The instrument is a transport type of interface, but uses a monocrystalline germanium disk as the carrier. The germanium disk is transparent to IR light over the range of wavelengths that provide molecular structure information. A section of the disk is shown in Figure 8.14.

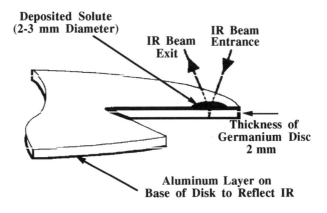

Courtesy of Lab Connections Inc.

Figure 8.14 Section of the Germanium Disk Transport System

The disk is about 1.5 mm thick and the lower surface is coated with aluminum to reflect any IR light that passes through it. The passage of the IR light beam, to and from the optical unit, is illustrated in Figure 8.14. Germanium has a high transmittance to light in the infrared range (6000 to 500 cm^{-1}) and the disk is arranged so that the light beam passes through the deposited solute to the reflective backing, and then back

LC/IR Tandem Systems

through the solute to the optical unit. In this way, the path length of the light through the sample is effectively doubled. The disk is easily cleaned and is completely reusable. It is also easy to remove the disk from one unit and insert it into the other. The interface consists of two modules, the coating module and the optical module. Diagrams of the two modules are shown in Figure 8.15.

The LC column is connected to the chromatography module and eluent from the column passes into an ultrasonic nebulizer. The nebulizer sprays the solvent in a tightly focused jet onto the surface of the sample–collection disk. The temperatures of both the ultrasonic nebulizer and the collection disk are carefully controlled.

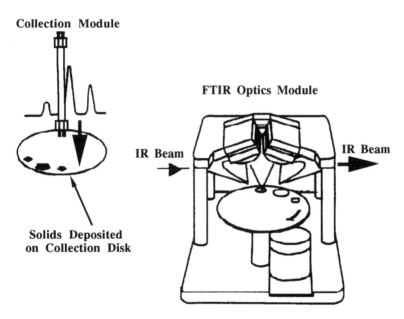

Courtesy of Lab Connections Inc.

Figure 8.15 The Coating and Optical Modules of the Interface

The disk rests on a heated stage in an evacuated compartment that rotates slowly during the elution of the sample. All the solvents evaporate leaving a deposit of solvent–free solutes on the disk surface. As a consequence, localized masses of solutes are located round the perimeter of the disk in the form of a 'physical' chromatogram. After the separation is complete, the disk is removed and placed on the stage of the optical module, which is located in the sample compartment of the FTIR bench. The compartment is purged with carbon dioxide–free air. The stage rotates and the incoming IR beam passes though the sample, and the spectrum is obtained by reflected transmission. As already discussed, this procedure produces a two–pass transmission spectrum of the sample. As the disk is rotated and the spectra obtained, they are stored and so the spectrum of any particular peak can be recalled and printed out as desired. In addition the individual spectra can be integrated and the results plotted against time to produce a chromatogram of the separation. The FTIR spectra obtained from the analysis of an oil–based steroid solution used for intravenous injection containing benzyl benzoate and testosterone cypionate are shown in Figure 8.16.

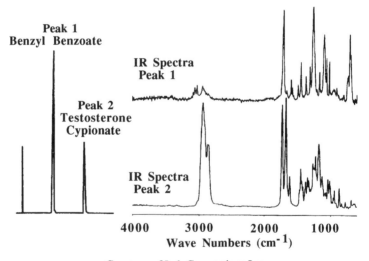

Courtesy of Lab Connections Inc.

Figure 8.16 IR Spectra of Benzyl Benzoate and Testosterone Cypionate Obtained from the LC/IR Interface

A sample of the oil (0.5 ml) was mixed with 15 ml of methanol and shaken for 10 minutes. The mixture was then centrifuged at 1100 g for 5 minutes to obtain the supernatant layer of methanol, which contained the steroids. A sample of the methanol layer was placed on the column. The separation was carried out on a Zorbax column, 25 cm long, 4.6 mm I.D., packed with a reversed phase, ODS 5U C18. The nebulizer temperature was set at 30°C and the transport stage at 35°C. The chromatogram was obtained from a UV monitor. The first peak can be identified from its spectrum as the preservative, benzyl benzoate, and the second as testosterone cypionate. The benzyl benzoate has a large extinction coefficient and produces a disproportionally strong UV absorption peak. The spectra of the active component clearly shows the dual carbonyl peaks at 1676 and 1732 cm^{-1} that correspond to the ketone and ester carbonyl groups. Adequate spectra could be obtained from 100 ng of sample deposited on the disk. However, if the technique were to be used for the analysis of blood or urine, a solid phase extraction stage would be necessary to obtain sufficient sample at a satisfactory concentration. Two examples of some spectra taken with the Lab Connections instrument, for the identification of specific drug precursors or derivatives are shown in Figure 8.17.

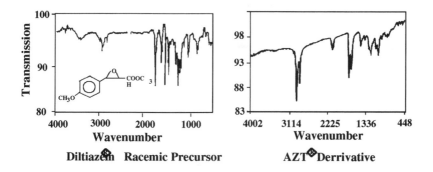

Courtesy of Lab Connections Inc. (The spectra of the AZT derivative was kindly supplied to Lab Connections Inc. by Dr. Phyllis Brown and John Imari of Rhode Island)

Figure 8.17 Spectra of a Drug Precursor and a Drug Derivative Taken from the LC Column Eluent

It is seen that the spectra exhibit significant fine structure and would be quite satisfactory for confirming the identity of a specific drug or derivative. Such spectra, used in conjunction with a mass spectrum, would also be very useful for elucidating the structure of a completely unknown substance. The spectra obtained by solid deposition on an IR transparent surface can differ significantly from those obtained by other sampling methods, *e.g.* by using Nujol mulls or potassium bromide pellets. The difference will be more apparent with certain types of compounds. For example, a sugar solid, formed by the rapid evaporation of water from solution, will tend to form a visibly clear 'glass'. Scattering and fine structure artifacts in spectra observed in solid preparations will be absent in such samples.

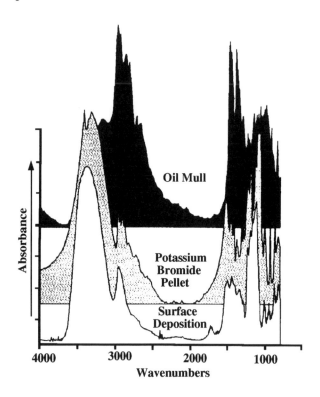

Courtesy of Lab Connections Inc.

Figure 8.18 Spectra of Dextrose Taken by Three Different Sampling Methods.

The differences that can be observed between different methods of sample preparation are demonstrated in Figure 8.18. The foreground spectrum was obtained from a small drop of dextrose sugar solution applied to the surface of the sample collection disk and evaporated. The spot appeared as a round clear deposit. The middle spectrum was obtained from a potassium bromide pellet made from the same dextrose sugar. The third spectrum was obtained from a mineral oil mull of the dextrose sample, which was smeared on the disk surface. The latter two spectra share many of the fine structure artifacts associated with matrix effects. It is apparent that for certain samples this type of sampling give improved spectra over those produced by the more conventional sampling procedures. An example of the use of the tandem instrument for analyzing natural products is the assay of natural orange juice concentrate.

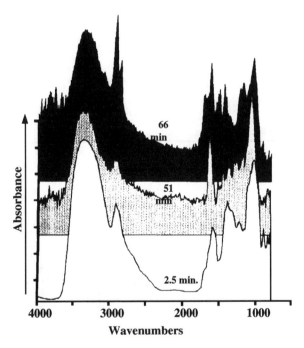

Courtesy of Lab Connections Inc.

Figure 8.19. Spectra of Three Peaks from the Separation of a Sample of Natural Orange Juice

The orange juice was centrifuged and the supernatant liquid filtered. The filtrate was passed through a Waters C18 Sep–Pak®, and the retained solutes eluted with acetonitrile. In this sample preparation procedure, most of the flavenoids are retained but the simple sugars are not. A sample of the acetonitrile extract was placed on a reverse phase column, and the separation developed using gradient elution. The spectra (from 100 scans at 8 cm^{-1} resolution) of an early major peak and two of the later eluting minor peaks are shown in Figure 8.19. It is seen that good spectra are obtained and the signal to noise is more than adequate for reference spectra matching.

The LC/FTIR tandem instrument has not been used frequently in natural product assays, but this has been almost entirely due to relatively poor sensitivity of the early instruments.

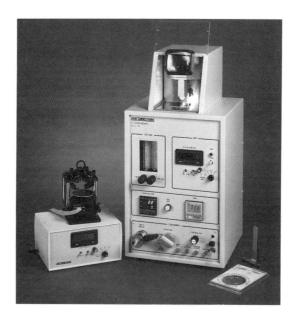

Courtesy of Lab Connections Inc.

Figure 8.20 The Lab Connections, Inc. LC/IR Interface System Alternative LC/FTIR Interfaces

It is clear that the performance of the commercially available instruments has now been much improved, and it is likely that the technique will become far more popular in the future. A photograph of the 100LC–Transform™ LC/FTIR Interface fabricated by Lab Connections Inc. is shown in Figure 8.20.

The obvious alternative to a transport interface is the in–line flow–through cell, and in 1983, Brown and Taylor [17] introduced a micro IR cell, 3.2 µl in volume, that fitted directly into the IR spectrometer. This arrangement constituted a true in–line LC/IR combination. They employed a small–bore column, and claimed an overall increase in mass sensitivity of about 20 orders of magnitude, relative to the standard 4.6 mm I.D. column. They employed an FT/IR spectrometer, but the actual sensitivity improvement was obscured by the fact that the column length of the small bore column, was significantly different from that of the standard column. As a result, the actual sensitivity improvement, in terms of minimum sample mass that would provide an acceptable spectrum, could not be accurately calculated. A simple cell for directly interfacing an LC microbore column to an FTIR spectrometer was described by Johnson and Taylor [18].

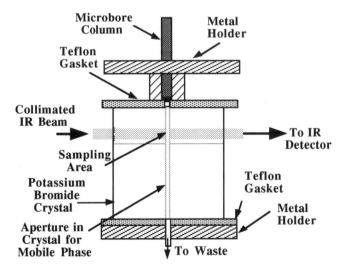

Figure 8.21 Zero Dead Volume Micro IR Cell

They claimed that the use of this cell would reduce the detection limit (that is the minimum mass required to produce a useful IR spectrum) to as little as 50 ng of material. A design of their flow cell is depicted in Figure 8.21. The cell was fabricated from a crystalline block, of either calcium fluoride or potassium bromide, 10 x 10 x 6 mm, with a 0.75 mm hole drilled through it. The cylindrical aperture in the crystal carried the mobile phase from the column through the block and then to waste. The collimated IR beam passes through the block, normal to the flow of the mobile phase, and is arranged to transverse a section of the 'hole' conduit. Since the focal diameter of the FT/IR spectrometer was 3 mm, and the hole in the cell only 0.75 mm, a beam condenser was used to reduce the focal diameter to that of the hole. The authors observed that a maximum IR *signal* could be obtained when the peak maximum was in the light path. Nevertheless, they also noted that the maximum *signal–to–noise* was obtained by summing the spectra from scans taken across the peak, between ± 1.53 σ of the Gaussian profile, as it passed through the cell. They also reported a general point of interest when designing such cells. It was found easier, and more effective, to modify optically the size of the IR beam to suit those of the flow cell, than to try to fabricate the cell to have proportions that will accurately suit the dimensions of the IR beam.

Sabo *et al.* [19] developed an LC/FTIR interface for both normal and reversed phase chromatography, using an attenuated total reflectance cell. The flow–through cell is made from a cylindrical shaped zinc selenide crystal, with cone–shaped ends, mounted in a stainless steel cell. The crystal, blazed at 45°, gives 10 reflections of the IR beam down its length. The cell fits into an optical bench, which focuses the incident beam onto the incident cone–face, and directs the radiation leaving the crystal onto the detector.

The volume of the cell is rather large, *vis.* 24 µl, which could adversely affect the resolution of a small–bore column. On–the–fly spectra of the components from a 100 µl sample of a solution, containing 2% of acetophenone and ethyl benzoate and 1% of nitrobenzene, gave clearly identifiable spectra. Nevertheless, relative to other LC/FTIR systems, this was not a very sensitive device.

Conroy and Griffiths [20] developed a rather involved solvent extraction device that could be employed with an LC/FTIR combination. The device embodied an extraction procedure that continuously extracted the dissolved solute from the column eluent into dichloromethane. The dichloromethane was then concentrated and finally dispersed onto a plug of potassium chloride powder. The solvent was evaporated and a spectrum taken in the usual manner. This device appears a little clumsy but, nevertheless, introduced a new basic concept for fabricating LC/IR interfaces. Following the basic principle of Conroy and Griffiths, Johnson *et al.* [21] developed a rather unique extraction cell for an LC/IR system.

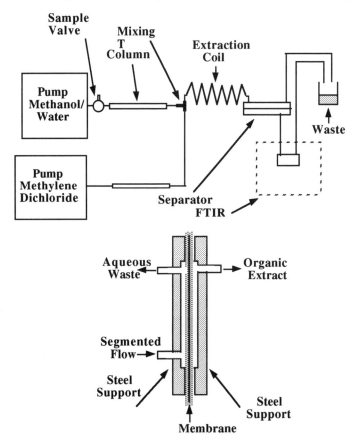

Figure 8.22 An Extraction Interface for LC/IR

They produced a segmented flow, by mixing the aqueous eluent from a reversed phase column with chloroform. The extraction solvent was then separated from the segmented flow by means of a 'hydrophobic' (dispersive) membrane. A diagram of their apparatus is shown in Figure 8.22. There are two pumps, one provides the solvent for the chromatographic development and the other the extraction solvent, which can be either chloroform or carbon tetrachloride. After passage through the column, the two streams are mixed at a T junction and form a segmented flow as both solvents are virtually immiscible. The segmented flow passes through an extraction coil, which allows sufficient time for the extraction to take place, and then to a separator. A diagram of the separator is shown in the lower portion of Figure 8.22.

The separator is made of stainless steel, and the volume on either side of the membrane is about 16 µl, the membrane itself having pores about 0.2 µm in diameter. The membrane is sealed against the steel flanges by compression. The amount of solvent that is passed through the membrane is controlled by adjusting the differential pressure across the membrane. This device could obviously cause serious peak dispersion and, in the form described, would be unsuitable for use with high efficiency or small-bore columns. It was found that samples containing 300 µg of material were necessary to produce a satisfactory spectrum, which indicated a relatively poor sensitivity.

Hellgeth and Taylor [22] developed the segmented flow interface further, improving both the method of producing the segmented flow, and the phase extractor. The segmented flow generator was constructed from a 1/16. in Swagelok T union. The union was drilled out 1/16 in. I.D. to allow the 1/16 in. tubes to be inserted, so that the ends of the tubes were only 0.45 mm apart. The aqueous solvent and the organic extraction solvent passed into the mixing T through tubes 0.020 in. I.D. The segmented flow left the mixing T through a length of Teflon tube, 75 cm long and 0.8 mm I.D., which also acted as the extraction conduit. The membrane separator consisted of two stainless steel plates with grooves in each surface, and a triple-layer membrane of Gore-Tex sheet. The basic system is diagramatically represented in Figure 8.23

LC/IR Tandem Systems

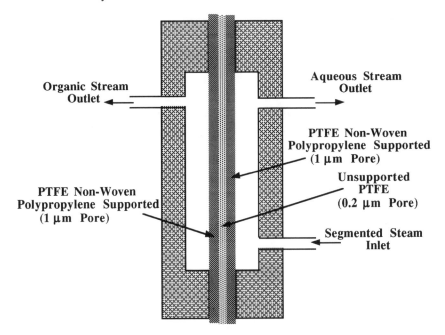

Figure 8.23 Diagrams of the Phase Separator.

The membrane is fabricated from two materials. The inner layer comprises an unsupported 1 μm pore Teflon sheet. This sheet is sandwiched between two outer sheets, each of 1 μm pore Teflon, supported by non-woven polypropylene membranes. The membrane is assembled so that the supporting non-woven polypropylene membranes are on the outer surfaces. The infrared cell is a modified form of the Spectra-Tech Inc. de-mountable flow cell, fitted with windows of either calcium fluoride or zinc selenide. The system appears to function well and satisfactory spectra are shown that had been obtained from 100 μg of material. Although an improvement on the sensitivity obtained by Conroy and Griffiths, it is still relatively poor compared with that obtained with the rotating disk interface.

More recently, the extraction system has been extended further by Somsen *et al.* [23]. These authors have developed a segmented flow concentrator that causes little or no band dispersion, and thus conserves the resolution of the column. The integrity of the chromatographic separation, however, was easily maintained by earlier models. More importantly, they utilize the concentrator in conjunction with the disk transport interface, and as a consequence, achieve a marked improvement in sensitivity. The apparatus consists of a conventional liquid chromatograph, and includes a pump, pulse damper, injection valve and column. Subsequent to the column, the eluent enters a T piece and is joined by a stream of immiscible extraction solvent. The solvent used largely in this work was methylene dichloride. The extraction solvent is supplied from another pump and pulse damper, and the solvent also flows through a column prior to the T piece, to provide more pulse damping. The eluent and solvent pass through an extraction coil as segmented flow, which provides the necessary time for the solutes to diffuse from the aqueous phase into the solvent. Due to the flow being segmented, there is no parabolic velocity profile in the fluid, and thus little or no peak dispersion can occur.

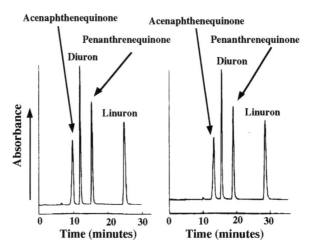

Reprinted with permission from G. W. Somsen, E. W. J. Hooijschuur, C. Goopijer, U. A. Th. Brinkman and N. H. Velthorst, *Anal. Chem.*, **68(5)**(1996)746, Copyright 1996 American Chemical Society

Figure 8.24 Chromatograms Monitored Before and After the Extraction Process

After passing through the extraction coil, the segmented flow enters a phase separator, similar to that described by Johnson *et al.* [21], and the separated solvent then flows through a UV absorption detector. The eluent leaving the UV detector then passes directly into a spray jet assembly, similar to that manufactured by Lab Connections Inc., which has already been described. Nitrogen is passed through a heater to the spray jet assembly, to aid in the nebulization process. In order to confirm that the extraction system did not contribute significantly to peak dispersion, a separation was monitored with a UV detector placed before and after the solvent extraction process. The chromatograms obtained are shown in Figure 8.24. It is seen from Figure 8.24, that very little peak dispersion takes place in the extraction tube with segmented flow, and that the resolution obtained from the chromatographic column is not significantly denigrated. Due to the finite volume of the extraction tube, however, there is a significant retention delay manifested between the two chromatograms. Nevertheless, as all the peaks are displaced by the same amount, this displacement has no effect on resolution. The delay amounted to about 3.5 min, but this will vary with both the flow rate and volume of the extraction system. In general the delay (Δt) will be given by,

$$\Delta t = \frac{V_E}{Q}$$

Where (V_E) is the volume of the extraction system,
and (Q) is the flow rate.

An example of the use of the system for the separation and identification of a series of phenylureas is shown in Figure 8.25.

It is seen that an excellent separation was obtained, and as the chromatogram was monitored after the extraction process, the results again confirm that the integrity of the separation is not impaired by the extraction procedure. The spectrum shown below was obtained as a result of an injection of 150 µl of a solution, containing 180 ng/ml of acenaphthenequinone, and thus represents a total mass of about 27 ng. It is clear that the system has a very useful sensitivity and can handle aqueous mobile phases very easily. However, the solvent extraction technique does not seem to have a great advantage over the system used by Lange and Griffiths [10] and developed by Lab Connections Inc., and is certainly more complicated and cumbersome. It should also be noted that if the mobile phase employed contains any solvents, then these will also be extracted by the methylene dichloride.

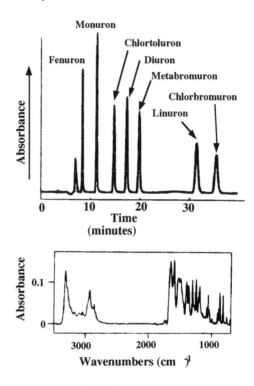

Reprinted with permission from G. W. Somsen, E. W. J. Hooijschuur, C. Goopijer, U. A. Th. Brinkman and N. H. Velthorst, *Anal. Chem.*, **68(5)**(1996)746, Copyright 1996 American Chemical Society

Figure 8.25 The Separation of a Mixture of Phenyl Ureas.

However, providing the solvents were reasonably volatile, they would be removed in the nebulizing process. In addition, there must not be sufficient solvent present to render the mobile phase miscible with the methylene dichloride and prevent the formation of segmented flow.

Consequently, some restrictions are placed on the choice of phase systems that are required to effect the separation.

Interface for the Combination of Liquid Chromatography and Raman Spectroscopy

Tandem systems involving the combination of the liquid chromatograph with the Raman spectrometer are not widespread, largely due to inherent insensitivity of the earlier instruments. However, the use of the laser as the excitation energy source has significantly increased the sensitivity of Raman measurements. Consequently the LC/Raman combination is likely to become as practical and useful as the LC/FTIR tandem system. The advantages of Raman spectroscopy over infrared spectroscopy lies in the fact that Raman spectra can be readily obtained in aqueous media. This renders the LC/Raman system particularly attractive to the biotechnology field. The structural information provided by the Raman spectrum is very similar to that provided by the IR spectrum. The Raman spectrometer has been connected to the liquid chromatograph to provide Raman spectra of eluted solutes in real time by Sheng *et al.* [24]. In order to obtain the necessary sensitivity in a flow through cell, the authors employed the technique of surface enhanced Raman spectroscopy (SERS). They mixed the column eluent with a suspension of fine particles of silver which acted as the enhancing surface. The use of the technique depends critically on the operating conditions, and particularly on the physical properties of fluid entering the cell and the surface characteristics size, and degree of aggregation of the silver particles. The general arrangement of their basic system is depicted in Figure 8.25.

A standard liquid chromatograph is used and the outlet is fed to a mixing T. The silver solution is pumped through a heat exchanger to the opposite limb of the mixing T, and the mixture is taken from the center limb. To realize maximum mixing, the outside limbs of the mixing T are arranged to be at an angle of 30° to the center limb [25]. The conduit between the T and Raman cell must have a minimum length to ensure adequate mixing. However, within this constraint, all connections carrying the column eluent, and the mixture of column eluent and silver dispersion, must be kept as short as possible to minimize peak dispersion.

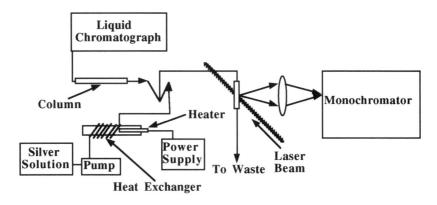

Figure 8.26 The Liquid Chromatography/Raman (LC/RA) Tandem Instrument

The preparation of the silver dispersion must be carried out extremely carefully by the citrate reduction process [25] and is heated, prior to mixing with the column eluent, by means of a coiled heat exchanger. The temperature is adjusted so that the mixture entering the Raman cell is *ca* 50–70°C. This was found essential to achieve the necessary sensitivity. The connection between the mixing T and the Raman cell consists of 12 cm of knotted tube, which helps ensure that the silver dispersion and the column eluent are completely mixed before entering the flow cell. The flow cell consists of a glass capillary, 100 mm long and 1 mm I.D. fitted into an aluminum holder.

The laser is the Coherent Innova 100 Kr^+ and the 476 nm line is used at 40 mW power. The excitation beam is set at 45° to the normal of the surface and the Raman scattered light is collected in a back–scattering geometry. The scattered light is focused onto a Spex Triplemate spectrometer, where it is dispersed onto a diode array, which is operated at –40°C to achieve maximum sensitivity. The solutes used for testing the system were adenine, guanine, xanthine and hypoxanthine. An example of the results that were obtained, is shown in the three dimensional chromatogram of the separation depicted in Figure 8.27.

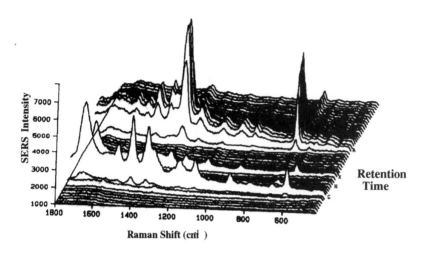

Reprinted with permission from R. Sheng, F. Ni and T. M. Cotton, *Anal. Chem.*, 63(5)(1991)437, Copyright 1991 American Chemical Society

Figure 8.27 A Three-dimensional Chromatogram Obtained from an LC/RAMAN Tandem Instrument Separating Three Purine Bases.

It is seen that a good separation is obtained, although it is difficult to determine exactly the extent to which the separation was degraded in the interface. In any event, it appears to be relatively little, and considering that this type of instrument is still in the relatively early stages of development, these results bode well for the future of this particular tandem technique.

In the past, tandem systems involving the combination of the liquid chromatograph in-line with the infrared spectrometer have not performed well. Most IR spectra of LC eluents have been obtained by what are, in effect, off-line procedures. This is apparent from the many examples given in this chapter. This lack of interest has largely arisen from the poor sensitivity of the early instruments, and also, perhaps, from the clumsy operating procedures that were entailed. However, the problem of sensitivity appears to have been solved by employing the FTIR instrument with the right type of interface. As already stated, as a result of the extensive research that been carried out on the development of interfaces for LC/FTIR tandem instruments, the use of the technique is likely to increase significantly in the future.

Synopsis

Initially LC/IR was carried out off–line, fractions of the eluent being collected, and the IR spectra being taken in a bromide disk, or Nujol mull. The advent of the FTIR instrument improved the IR sensitivity, and as the spectra could be stored, background subtraction was now possible. However, this procedure gave a very low signal to noise, and thus poor overall sensitivity. Transport interfaces were the first type to be developed which were, in effect, fraction collecting devices, that subsequently took the collected sample to the IR measuring system to obtain a spectrum. The transport concept evolved through a potassium bromide plate, a disk with a metallic reflective surface, to a IR transparent disk made of germanium. A number of different nebulizing jets were also developed, to deposit the solute on the disk more efficiently, including a supersonic nebulizer. Transmission and/or reflection spectra were taken by an FTIR spectrometer, and sensitivities of 50–100 ng per spectra were obtained using these methods. Flow–through cells were developed by drilling holes in alkali metal halide crystals, but were found to have very limited sensitivity. Membrane extraction techniques were also investigated, which although cleverly devised and carefully constructed, could not match the sensitivity obtained from the disk transport system. A recent interesting innovation involved the use of a Raman spectrometer to monitor LC column eluents. The eluent from the column was mixed with a silver dispersion which strongly enhances the Raman signal, and the mixture then passed through a cell illuminated by a laser beam. The scattered light was then focused onto a monochromator and a spectrum obtained in the usual manner. This device is relatively new, and the actual sensitivities that were obtained are difficult to determine from the literature, but without doubt this concept holds exciting prospects for the future.

References

1. K. L. Kizer, A. W. Mantz and L. C. Bonar, *Am. Lab.* **May** (1975)
2. R. P. W. Scott, C. G. Scott, M. Munroe and J. Hess. Jr., *The Poisoned Patient: The Role of the Laboratory,* Elsevier, New York (1974)395.
3. D. Kuehl and P. R. Griffiths, *J. Chromatogr. Sci.,* **17**(1979)471.
4. A. Karmen, *Anal. Chem.*, **38**(1966)286.
5. K. Jino and C. Fujimoto, *J. High Resolut. Chromatogr.*, (1981)10277.
6. K. Jino, C. Fujimoto, and Y. Hirata, *Appl. Spectrosc.*, **36,1**(1982)67.
7. M. W. Raynor, K. D. Bartle, I. L. Davies, A. Williams, A. A. Clifford, J. M. Chalmers and B. W. Cook, *Anal. Chem.*, **60(5)**(1988)427.
8. J. J. Gagel and K. Bieman, *Anal. Chem.*, **58(11)**(1986)2184.
9. J. J. Gagel and K. Bieman, *Anal. Chem.*, **59(9)**(1987)1267.
10. A. J. Lange and P. R. Griffiths, *Anal. Chem.*, **63(8)**(1991)782.
11. S. L. Pentoney and P. R. Giffiths, *Anal. Chem.*, **61(19)**(1989)2212.
12. G. W. Somsen, R. J. Van de Nesse, C. Gooijer.U. A. Th. Brinkman, N. V. Velthorst, T. Visser, P. R. Kootstra and A. P. J. N. De Jong, *J. Chromatogr.,* **552**(199)635.
13. V. E. Turula and J. A. de Haseth, *Anal. Chem.*,**68(4)**(1996)629.
14. R. C. Willoughby and R. F. Browner, *Anal. Chem.*,**56**(1984)2626.
15. R. F. Browner, P. C. Winkler, D. D. Perkins and L. E. Abbey, *Michrochem. J.* **34**(1986)15
16. P. C. Winkler, D. D. Perkins, K. W. Williams and R. F. Browner, *Anal. Chem.*, **60**(1988)489.
17. R. S. Brown and L. T. Taylor, *Anal. Chem.*, **55**(1983)1492.
18. C. C. Johnson and L. T. Taylor, *Anal. Chem.*, **56**(1984)2642.
19. M. Sabo, J. Gross, J. Wang and I. E. Rosenberg, *Anal. Chem.*, **57**(1985)1822.
20. C. M . Conroy and P. R. Griffiths, *Anal. Chem.*, **56**(1984)2636.
21. C. C. Johnson, J. W. Hellgeth and L. T. Taylor, *Anal. Chem.*, **57**(1985)610.
22. J. W. Helgeth and L. T. Taylor, *Anal. Chem.*, **59(2)**(1987)295.
23. G. W. Somsen, E. W. J. Hooijschuur, C. Goopijer, U. A. Th. Brinkman and N. H. Velthorst, *Anal. Chem.*, **68(5)**(1996)746
24. R. Sheng, F. Ni and T. M. Cotton, *Anal. Chem.*, **63(5)**(1991)437.
25. G. D. Clark, J. M. Hungerford and G. D. Christian, *Anal. Chem.*, **61**((1989)973.
26. R. S. Cheng, L. Zhu and M. D. Morris, *Anal. Chem.,* **58**(1986)1116.

CHAPTER 9

LIQUID CHROMATOGRAPHY/MASS SPECTROSCOPY (LC/MS) TANDEM SYSTEMS

Tandem instruments involving the combination of the liquid chromatograph and the mass spectrometer are generally far more complicated than their GC/MS counterparts. There are two reasons for this; first the substances separated by liquid chromatography are, in the main, relatively involatile; second, as a result of their poor volatility, the simple electron impact and chemical ionization sources are no longer adequate. This situation has evoked the development of a number of unique ionization procedures and interfaces, exclusively for use with LC/MS tandem instruments.

The problem of ionizing involatile materials, however, is not restricted to the successful use of the mass spectrometer with the liquid chromatograph, the difficulty also arises with the normal use of the mass spectrometer. As the temperature of a sample is raised its volatility increases, but so does the rate of thermal decomposition. Small molecular weight substances, and those of low polarity, can usually be rendered sufficiently volatile by merely raising the temperature. Strongly polar and high molecular weight materials, on the other hand, start to thermally decompose, before the parent material has sufficient vapor pressure to provide ions by electron impact or chemical ionization. It follows that alternative methods of ionization are necessary, in order to obtain mass spectra of such compounds. A number of these alternative ionization methods can be easily incorporated into an interface that will operate with a liquid chromatograph. Other types, however, require complicated transport interfaces for them to function satisfactorily.

An early solution to the problem of sample involatility was the use of a solid probe, which was coated with the sample and inserted into the electron beam of the ionizing source. Subsequently, the probe was heated, and this increased the range of substances that could be ionized. However, the majority of high molecular weight or strongly polar materials started to decompose, long before they were sufficiently volatile to provide a vapor pressure that was adequate to produce ions, particularly parent ions. Many high molecular weight samples, however, do have a significant vapor pressure before the decomposition temperature is reached, but it is very low. It follows that if the ionization process could act directly on the solid surface, sufficient ions should be produced to permit mass spectra to be obtained. The process is a form of desorption ionization.

In general, desorption ionization is carried out by coating the sample on the surface of a suitable metal, and then bombarding the sample with particles of high energy. On collision with the sample, some of the energy is communicated to surface molecules and ejects them from the surface, and some of the energy is used for their ionization. There are a number of different ways of communicating the energy to the surface molecules, some of which will now be briefly described.

Secondary Ion Mass Spectrometry (SIMS)

In secondary ion mass spectrometry the excitation beam that produces the ionization consists of a stream of ions having a kinetic energy in the kilo-electron volt range. These ions are generated in a specially designed ion-gun. Those most commonly used are the Ar^+, O_2^+ and Cs^+ ions. If the sample, or sample support, is conducting (*i.e.* a metal) then the charge received by the target can leak away. If, however, the target is not conducting, an electric charge can build up on the sample surface and interfere with the focusing of the ions that are produced. This charge build up can be eliminated by flooding the sample with low–energy electrons from a separate electron source. Abundant molecular ions are usually formed by the secondary ion process, such as (M+H)$^+$ and (M+Ag)$^+$, from thin layers of sample deposited on a metal substrate such as silver. The use of other materials as a substrate, such as nitrocellulose, can improve the ionization efficiency for very high molecular weight samples. To employ this method of ionization in an LC/MS tandem instrument, the sample must be presented to the ionization source as a dry film on a solid carrier, and therefore requires the use of a transport interface.

Fast Atom Bombardment (FAB)

Fast atom bombardment spectrometry is an extension of the secondary ion mass spectrometry, and the principle of the ionization process is shown in Figure 9.1.

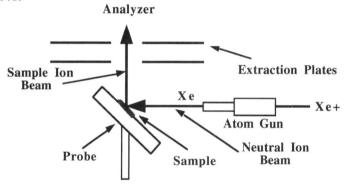

Figure 9.1 The Fast Atom Bombardment Ionization Source

A beam of energetic particles is generated and directed onto the sample, which is carried as a thin film on a clean metal support. The secondary ions that are produced are extracted by a suitable ion optical arrangement into the mass spectrometer analyzer. The impact of an ion striking the surface produces an intense thermal spike. The energy from this thermal spike is then dissipated through the outer layers of the sample. In the original device developed by Barber [1], neutral atoms of argon and xenon were used, but eventually these were replaced by charged ions such as Ce$^+$ and Xe$^+$. If a dry sample is used, the surface becomes damaged by the intense incident beam, and the yield of secondary ions rapidly decreases.

However, if the sample is dissolved or dispersed in a liquid matrix such as glycerol, the liquid surface is continuously renewed and an intense primary beam can be safely used. This method of ionization has become a standard process for obtaining the mass spectra of very polar or labile substances. The temperature of the glycerol can be quite critical. At -20°C the glycerol becomes too viscous, and cannot dissipate the energy sufficiently rapidly, and at 40°C the vapor pressure is too high and the glycerol evaporates in the high vacuum of the source. Optimum sensitivity appears to be realized when the glycerol is about 25°C. It is clear that this form of ionization would need to be employed with a transport interface if used in a tandem instrument associated with a liquid chromatograph.

Plasma Desorption Mass Spectrometry

Plasma desorption is achieved by the use of a radioactive source, the fission particles being used to ionize the sample. Employing ^{252}Cf as the source the technique can be used with the time of flight mass spectrometer in a rather clever manner. ^{252}Cf has a half life of 2.7 years and decays giving an alpha particle and two charged fission fragments simultaneously emitted in opposite directions. Typically a pair of fission fragments might be ^{106}Te and ^{142}Ba with energies of 104 and 79 MeV respectively. A diagram of the plasma desorption apparatus is shown in Figure 9.2.

The sample is placed on a thin aluminum sheet or a sheet of aluminized polyester film, and the aluminum is connected to a high positive potential (assuming the sample ions will be positively charged). When fission occurs, one particle strikes the sample and produces ions, while the other, emitted in the opposite direction, is sensed by the trigger detector and starts the time of flight measurement. The ionized sample molecule or fragment is accelerated to its characteristic velocity, passes through the drift region of the spectrometer and is finally sensed by the stop sensor which arrests the time measurement.

The yield of sample ions is very low and so a large number of spectra needs to be collected, which may take several minutes. It is apparent that this long sampling time would prohibit the use of the plasma desorption technique as a tandem instrument in real time.

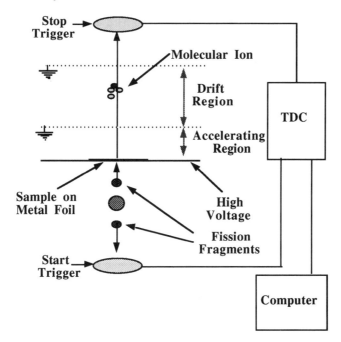

Figure 9.2 The Plasma Desorption Ionization System

It might be possible however, to use an aluminum sheet as a transport medium, and store the eluted solutes on it in the form of a chromatographic memory. The sheet could then be scanned in the plasma desorption interface off-line.

The Inductively Coupled Plasma Ionization Source

The inductively coupled plasma ionization source is very similar to the ICP torch that has been described previously. The eluent is first nebulized and the atomized eluent passes through a spray chamber and then enters the ICP torch. The gas and ions leaving the jet first enter an expansion chamber through a cooled sampling cone with an orifice, about 1 mm in diameter, where the gas expands and is pumped away. A portion of the jet from the torch then passes through a second orifice called the skimmer cone. The ICP produces singly-charged positive ions for most elements which makes it a very efficient source for coupling to a mass spectrometer. A series of electrostatic lenses focus the ions into the quadrupole mass spectrometer analyzer. After dispersion the ions are detected in the normal way with an electron multiplier tube. A diagram of the ICP interface and the quadrupole mass spectrometer is shown in Figure 9.3.

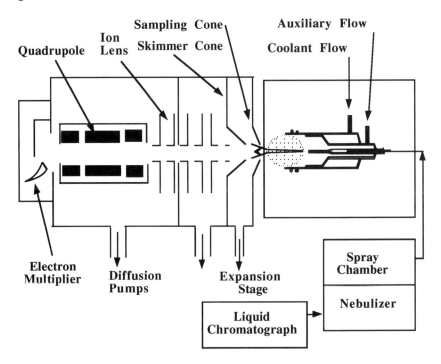

Figure 9.3 The Liquid Chromatography/Inductively Coupled Plasma Mass Spectrometer (LC/ICPMS) Tandem Instrument

The quadrupole mass spectrometer separates the ions on a basis of mass-to-charge ratios (m/z). Unfortunately, the ICP is not a very efficient excitation/ionization source for nonmetals such as the halogens and for elements such as arsenic and selenium. Helium plasma has an ionization potential of 24.5 eV compared with that of argon, 5.75 eV, and is consequently a more efficient excitation/ionization source. It follows that the microwave induced helium plasma is likely to be more efficient. The arrangement for the microwave induced helium plasma is very similar to that of the ICP torch as depicted in Figure 9.3, except that the plasma is induced by a resonant microwave cavity surrounding the body of the torch, and not by a cooled coil. In addition, with the ICP, the isotopes of argon, oxygen, nitrogen and hydrogen can combine with themselves, or with other elements, to produce isobaric interferences. The use of helium, which is essentially mono-isotopic, significantly reduces the number of interferences compared with the argon plasmas.

Laser Desorption Mass Spectrometry

The laser can be used for two purposes, first to desorb the sample as vapor into the ionizing system, and second to actually produce the sample ions. However, ions that are created by pulsed lasers produce bursts of ions and are obviously unsuitable for use with the scanning spectrometer. Intermittent ion production, however, is compatible with the time of flight mass spectrometer, which can record all the ions produced by each laser pulse. Another advantage of this combination is the nearly unlimited mass range of this particular type of mass spectrometer. A diagram of the laser desorption/time of flight mass spectrometer system is shown in Figure 9.4.

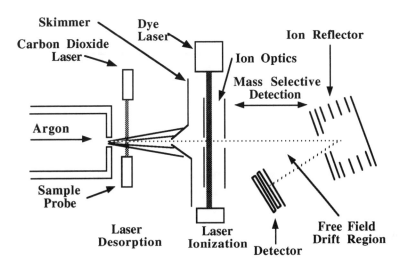

Figure 9.4 The Laser Desorption Mass Spectrometer Combination

The sample is placed on a probe that receives high-energy laser pulses from a carbon dioxide laser. The bursts of vaporized sample are propelled by an expanding argon stream from a 100 μm orifice, through a skimmer, into the accelerating section of the mass spectrometer. In this section, the molecules are again subjected to a high–energy laser light beam, this time from a dye laser, which produces ions by photo–ionization. The ions formed are immediately accelerated through the region into a drift section, where they are deflected by an ion reflector, to an electron multiplier detector. It is seen that to desorb and ionize a sample from a liquid chromatograph, a fairly complicated transport system would again be necessary.

Laser desorption can also be used with conventional electron impact and high–pressure chemical ionization sources, and also in conjunction with other ionization modes. A diagram of a sample probe that can be used with an electron impact ion source is shown in Figure 9.5.

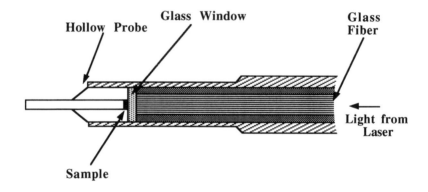

Figure 9.5 The Laser Desorption Sample Probe for Use with Electron Impact Ionization Sources

The laser light is focused by means of suitable lenses, and a optical fiber pipe, though the probe and onto a glass window at the end of the probe. The sample is placed next to the glass window and the desorbed molecules pass along the hollow tube to be ionized by electron impact. This type of desorption causes very little thermal degradation, and has been used satisfactorily for measuring the spectra of labile materials, such as certain antibiotics.

Matrix Assisted Laser Desorption/Ionization (MALDI)

This type of ionization system can provide spectra of substances of extremely high molecular weight. The laser usually employed, is the Nd-Yg-Laser, having a wavelength of 286 nm, and a pulse width of about 10 ns. The sample is dispersed in a suitable involatile liquid to prevent decomposition and allow the surface to be continually renewed. A number of different liquids have been employed, including glycerol and nicotinic acid. Nicotinic acid is particularly useful as it absorbs very strongly at 286 nm the wavelength of the laser light. A diagram showing the basic layout of the matrix assisted laser desorption/ionization source and mass spectrometer is shown in Figure 9.6.

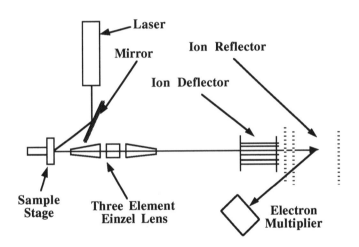

Figure 9.6 Diagram of the Basic Matrix Assisted Laser Desorption Ionization Source

Light from the laser is arranged to strike the surface at a 45° angle, and the ions emitted are collimated, by a three element Einzel lens, through an ion deflector, to the time of flight mass spectrometer. The ions are accelerated by an appropriate voltage, and then allowed to drift to an ion reflector and then back to the electron multiplier. This system was employed by Karas and Hillenkamp [2] to obtain the mass spectrum of a number of proteins including bovine albumin. An example of a spectrum they obtained for bovine albumin is shown in Figure 9.7.

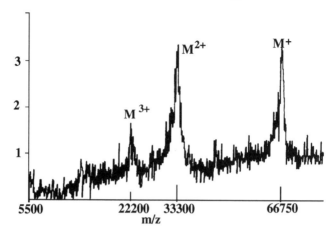

Reprinted with permission from M. Karas and F. Hillenkamp, *Anal. Chem.*, **60(20)**(1988)2299., Copyright 1988 American Chemical Society

Figure 9.7 The Matrix-UV-Laser Desorption Spectrum of Bovine Albumin

The spectrum shows the single, double and triple charged parent ions, the molecular weight of the parent ion being about 66750 daltons. The standard deviation of the parent ion peak is about 300 daltons. It is seen that the technique can help provide mass spectrum of very large molecules, apparently with little or no decomposition. This type of system would be ideal to combine with the liquid chromatograph, but for satisfactory operation it must again evoke the need for a transport interface.

Field Desorption Ionization

The process of field desorption involves the extraction of ions from a sample (deposited on a specially prepared surface) by the use of extremely high electrical fields. The preparation of the emitting surface can be quite involved and appears to be somewhat of an art as well as a science. A diagram of the field desorption source is shown in Figure 9.8. The emitter is mounted on an insulated probe, which can be introduced into the ion source by means of the standard vacuum lock. About 2 mm from the probe tip is a counter electrode, which is held at a potential of about 10,000 V relative to the probe.

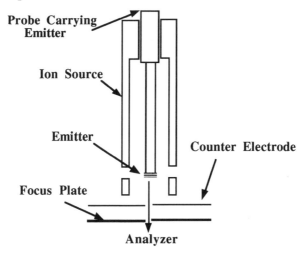

Figure 9.8 Diagram of a Field Desorption Source

The ions produced from the emitter are accelerated toward the counter electrode, pass through a hole into the focusing section, and then into the mass spectrometer analyzer. The preparation of the emitter surface is both tedious and time consuming. The most commonly used surface consists of carbon micro needles that are formed on tungsten wire in a vacuum. One substance that is used for this purpose is benzonitrile. The material is placed on the wire and the temperature adjusted to gently pyrolyse the material, producing carbon spots along the wire. A high voltage is then applied and the pyrolysis continued, and carbon needles are formed that grow from the original carbonaceous deposits on the wire. The sample is then coated on the carbon surface and in an electric field, the high potential gradient close to the needle points causes ions to be emitted from the sample. Field desorption ionization also provides spectra of high molecular weight materials but, due to the nature of the probe surface, the technique would be very difficult to use with tandem instruments. Even with the use of a transport system, the need for such a carefully prepared emitter surface would make it rather impractical. However, the introduction of a different emitter surface that could be part of a transport interface might render it more feasible.

Liquid Chromatography/Mass Spectrometry (LC/MS) Techniques

The liquid chromatograph/mass spectrometer combination can only be successful if the mobile phase is eliminated, either before entering or from inside the mass spectrometer. This is necessary to ensure that the production of ions is not adversely affected or that the spectrum is not swamped with fragment ions from the solvent. The evaporation of the mobile phase generates a considerable volume of solvent vapor and if allowed to enter the ion source must be rapidly pumped away. The problem can be reduced by using a splitting system, in much the same way as the early GC/MS tandem systems operated, or by employing the solvent vapor as a chemical ionization agent. In fact the first LC/MS system developed was that by McLafferty [3], who used the direct sampling technique passing the eluent directly into the mass spectrometer. A small portion of the column eluent was split from the main stream, and fed by means of capillary tube, directly into the source of a high-resolution mass spectrometer. The solvent was used as a chemical ionization agent and thus produced chemical ionization spectra of the eluted solutes.

There are two solutions to the solvent problem in LC/MS tandem instruments. The first is to remove the solvent exterior to the mass spectrometer while contained on a suitable transport medium. The dry solutes on the surface of the transport medium are then moved into the ion source for ionization. The transport concept is not used extensively as an interface for LC/MS in contemporary tandem instruments, but has been found to be extremely useful as an interface for LC/FTIR, as already discussed. The second solution to the elimination of the solvent is to utilize the direct inlet approach of McLafferty, but with suitable interfaces that will cope with the solvent and provide a variety of ionization techniques. The transport interfaces will first be discussed, largely because historically these were the first to become commercially available. In addition, as alternative transport media were identified, and evoked the development

LC/MS Tandem Systems

of efficient LC/FTIR transport interfaces, so may the same occur with LC/MS interfaces, particularly with the advent of the new ionizing techniques.

Transport Interfaces: The Wire Transport Detector

In 1974 Scott *et al.* [4] developed a moving wire transport system for transferring the solute from the LC column to the ion source of the mass spectrometer. This interface allowed the use of electron impact ionization, and so all the expected ion fragments were produced to facilitate interpretation of the spectra. The basic principle was similar to that of the moving wire LC transport detector [5,6], and in the original model, the wire train from the detector was modified for use with the LC/MS interface. A diagram of the LC/MS transport interface is shown in Figure 9.9.

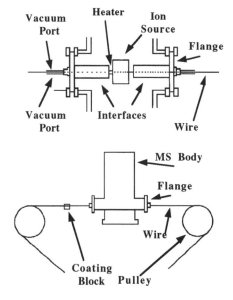

Figure 9.9 The Transport Interface for LC/MS

The column eluent passes over a moving stainless steel wire, 0.005 in O.D., coating it with a thin film of column eluent. In order to enter the ion source, the wire passes through three orifices and two chambers connected in series. Each chamber is connected to a vacuum pump that reduces the pressure in the first chamber to a few microns, and in the second chamber to about 10^{-5} mm of mercury. During passage through these chambers, the solvent evaporates from the wire, leaving the solute as a thin film coated on the wire surface. A current of a few milliamperes is passed continuously through the wire, which only becomes hot when it enters the high vacuum of the ion source, where it can no longer lose heat to its surroundings. The temperature of the wire rises rapidly to about 250°C in the ion source, vaporizing the sample directly into the electron beam. The wire leaves the ion source through the same type of vacuum lock as it entered, *i.e. via* a second pair of differentially pumped chambers. A detailed diagram of the vacuum locks is shown in Figure 9.10.

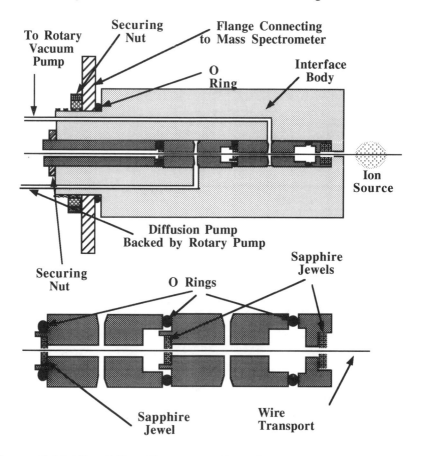

Figure 9.10 The Wire Transport Interfaces

LC/MS Tandem Systems

The main body of the interface is constructed of stainless steel and is fitted to the side flanges of a Finnigan quadrupole mass spectrometer, such that the interfaces are re-entrant to the ion source, and terminate a few millimeters from the electron beam. The two chambers are separated and terminated by sapphire jewels 0.1 in. O.D. and 0.018 in. thick. The jewels in the left-hand interface, where the sample is introduced have apertures 0.010 in. I.D. The jewels in the right-hand interface, where the wire transport leaves the mass spectrometer to the winding spool, have apertures 0.007 in. I.D. The larger diameter apertures on the feed side are employed to reduce the surface of the wire being 'scuffed', which might result in possible solute loss. Sapphire jewels are necessary to prevent frictional erosion of the apertures by the stainless steel wire. The first chamber of each interface is connected to a 150 l/min. rotary pump which reduces the pressure in the first chamber to about 0.1 mm of mercury. The second chamber of each interface is connected to an oil diffusion pump backed by a 150 l/min. rotary pump. The pressure in the second chambers is reduced to about 5-10 μm of mercury. The entrance and the exit of each interface is fitted with a helium purge that passes over the aperture through which the wire is entering or leaving and ensures that only helium is drawn into the interfaces. In this way background signals from air contaminants are greatly reduced. The purge also allows the use of methane as a chemical ionization agent if so required. The pressure in the source was maintained at about 1×10^{-6} mm of mercury.

The sensitivity of the device to diazepam, evaluated by monitoring the total ion current, was found to be about 4×10^{-6} g/ml. Bearing in mind that only a small proportion of the eluent is taken on the wire, this was calculated to be equivalent to about 7×10^{-10} g/spectrum (0.7 ng). A total ion current chromatogram, of a sample of the mother liquor from some vitamin A acetate crystallization, is shown in Figure 9.11. The total ion current is taken as proportional to the sum of all the mass peaks in each spectrum. The separation in Figure 9.11 was carried out employing incremental gradient elution using 12 different solvents. It is seen that a good separation is obtained and the integrity of the separation is maintained after passing through the interface. It also shows the complete independence of the interface to the nature of the solvents used for the mobile phase

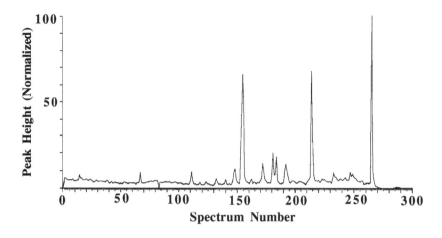

Figure 9.11 The Total Ion Current Chromatogram of a Sample of Mother Liquor from a Vitamin A Acetate Crystallization

It is also seen that the system is entirely independent of the solvent used in the separation, which ranged in polarity from the very dispersive *n*-paraffins to highly polar aliphatic alcohols and included chlorinated hydrocarbons, nitroparaffins, esters and ketones.

The Belt Transport Detector

The wire transport interface was modified by McFadden *et al.* [7], who replaced the wire with a continuous belt made either from stainless steel ribbon or a high-temperature plastic band. The belt could be thermally cleaned after passing through the ion source, and prior to re-entering the coating block. In a similar manner to the wire transport interface, the belt is actuated by a motor-driven pulley with manual speed control. A diagram of the McFadden belt interface is shown in Figure 9.12.

The column eluent is taken up as a thin liquid film on the surface of a stainless steel or high-temperature plastic ribbon (about 3.2 mm wide and

0.05 mm thick). The belt then transports the film to the vacuum locks and on the way is heated by an infrared heater to facilitate evaporation of the solvent. The remaining solvent is removed during passage through the vacuum-locks so that only about 10^{-7} g/sec of the solvent actually enters the mass spectrometer.

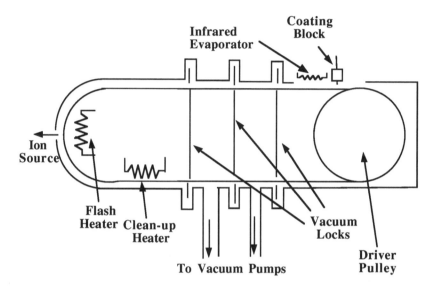

Figure 9.12 The Belt GC/MS Interface

The first interface pump removes air at a rate of 500 l/min and the second at about 300 l/min. The pressure in the first vacuum-lock is maintained at between 1 to 20 torr, and that in the second vacuum-lock between 0.1 to 0.5 torr. As a consequence the mass spectrometer source can be easily maintained at a pressure of about 10^{-6} torr. Flash vaporization of the solute is achieved by radiant heating, and occurs in a small chamber that butts directly onto the solid probe entrance to the ionization chamber. The sample vapor passes through a small hole in the chamber wall directly into the ion source. The flash heater is either a Nichrom coil or a quartz heater tube. In a similar manner to the wire transport interface, the slots in the vacuum-locks are made of sapphire strips to prevent abrasion. An example of the use of the belt interface to monitor the separation of a pesticide mixture is shown in Figure 9.13.

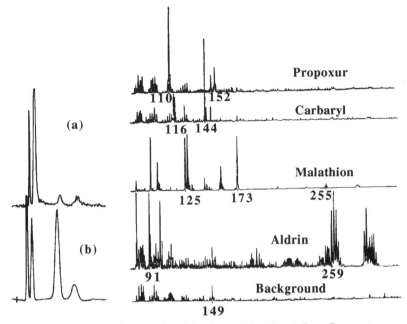

(a) Separation Monitored by Total Ion Current
(b) Separation Monitored by UV Detector

Figure 9.13 The Separation of Some Pesticides Using the Belt Interface

It is seen that there is little loss of resolution in the interface, and very little cross-contamination occurs between the peaks, resulting from incomplete removal of each solute from the belt, prior to the next coating. The sensitivity of the system to carbaryl was claimed to be about 1 ng per spectrum, but the actual sensitivity in g/ml was not possible to calculate from the data available.

Other Transport Systems

In 1982 Alcock et al. [8] reviewed the various types of transport interfaces that were available at that time, with particular reference to their use with small-bore columns.

Games et al. [9] examined the peak dispersion that could take place in the transport interface itself, together with the electronics associated with the mass spectrometer. They concluded that the LC/MS system, incorporating the transport interface, behaved as a low-dispersion LC detector and consequently, could be used very effectively with small-bore columns. In fact, once the eluent is deposited on the wire, and evaporation begins, the longitudinal diffusion of the solutes along the surface becomes exceedingly small. Consequently the peak widths, and the position of the peaks relative to one another, are 'frozen' on the transport medium, just as they were when they left the column.

Hayes *et al.* [10] conducted a systematic study on the effect of the method of solvent deposition on the transport medium on the overall performance of the LC/MS system. They designed a nebulizer to spray the column eluent onto the moving belt, a diagram of which is shown in Figure 9.14.

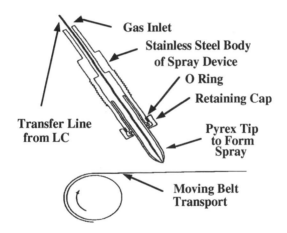

Reprinted with permission from M. J. Hayes, E. P. Lanksmeyer, P. Vouroo and B. L. Karger, *Anal. Chem*, **55**(1983)1745, Copyright 1983 American Chemical Society

Figure 9.14 The Spray Deposition Device for Belt Transport Interfaces

A stainless steel tube, 0.0625 in. O.D. and 0.007 in. I.D., is placed concentrically inside a Pyrex tube that carries the nebulizing gas. The Pyrex tube is held inside an outer steel tube by means of a screw cap and an O ring. The effects of nebulizer temperature and gas flow rate, on the integrity of the elution curve produced by the interface, were investigated.

From the results of their studies, the authors also claimed that small-bore columns could be used satisfactorily with the interface without degrading the resolution of the column and, at the same time, provide good quality mass spectra. They also claimed detection limits of 40 pg and linear dynamic ranges extending over four orders of magnitude.

Fan *et al.* [11] developed a belt interface system for LC/MS that could provide both secondary ion mass spectra and laser desorption mass spectra. A diagram of the dual ionization source is shown in Figure 9.15.

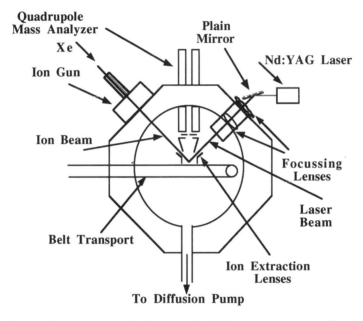

T. P. Fan, A. E. Schoem, R. G. Cooks and P. H. Hemberger, *J. Am. Chem. Soc.*, **103**(1981)1295. Copyright 1981 American Chemical Society

Figure 9.15 The Dual Ionization Transport Interface

The apparatus consists of a quadrupole mass spectrometer with a moving belt which is used in conjunction with a Finnigan/ INCOS data system. Samples are deposited on the belt by a type of thermospray process. Both the ion beam and the laser light are so aligned as to strike the belt at 45° to the surface, and normal to each other. Xenon gas, ionized by electron impact, is used as the source of primary ions, which are focused by an

Einzel lens and steering plates, through an aperture, onto the belt forming a spot covering an area of about 5 mm^2. The beam strikes the belt with an energy of about 3 keV. The laser light can be focused on the same spot by a suitable set of mirrors and lenses. The transport medium, ultimately employed by the authors, is a carbon steel belt that has a blackened surface, and consequently absorbed more energy from the laser beam. Providing laser energies are kept reasonably low, there is no degradation of the surface. Comparing the results obtained from the two methods of ionization, laser desorption appears to have the advantage, in that it provides more reliable molecular weight information. Furthermore, by varying the power of the laser, different degrees of fragmentation can be achieved and consequently more structural information can be obtained. However, as it is difficult to control the laser energy absorbed, the mass spectra are less reproducible. In contrast, continuous ionization by primary ion bombardment provides more characteristic fragment peaks, without losing molecular weight information and both the mass spectra and the chromatograms are less noisy and more reproducible.

Direct Inlet LC/MS Interfaces

Direct inlet interfaces allow a portion of the column eluent to be injected straight into ion source *via* a suitable restriction without any prior concentration procedure. A diagram of a simple direct inlet interface is shown in Figure 9.16.

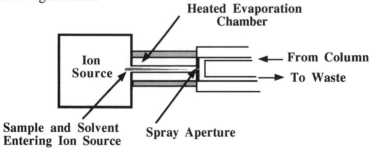

Figure 9.16 The Direct Inlet LC/MS Interface

The restriction is usually composed of an orifice 2-5 μm in diameter through which the column element is forced. A liquid jet is first formed that rapidly breaks up into droplets. The droplets then pass into a heated chamber, where they are vaporized and the sample and solvent then enter the ionization chamber. The tip of the jet can be suitably cooled to prevent premature evaporation of the eluent drops. The spectrometer pumping system cannot usually cope with much more that a few microliters of liquid solvent as vapor, so the column eluent must be split, and this is accomplished with the use of a down-stream needle valve. The interface is usually employed with a chemical ionization interface and the solvent vapor is used as the reagent gas. The use of the solvent as the reagent gas can either place restrictions on the choice of mobile phase that can be employed with the chromatograph or control the nature of the ions that are produced in the chemical ionization source. This interface has the disadvantage that a very small orifice is necessary for vaporization which can easily become blocked. This type of interface has been most successfully used for solutes that have a reasonable vapor pressure at the source temperature

The Thermospray Interface

The thermospray interface evolved from the simple inlet system, by the rather straightforward modification of heating the tip of the entry tube (or column) to a relative high temperature. One of the first reports of the successful use of the thermospray interface for LC/MS was that by Covey and Henion [12]. As a result of the heated tip, the solvent is vaporized right at the tip, and not somewhere inside the delivery tube or column. Under these conditions, the vapor has sufficient energy to change the remainder of the liquid to a mist of small droplets This results in much better control of both the nebulizing and the ionizing process. Due to the low pressure, an electric field cannot be applied (as in electrospray), as an a electric discharge would form. Instead, if there are no ions present in the mobile phase an electrolyte such as a 0.1 M solution of ammonium acetate may be added prior to volatilization. Due to the presence of the ions, the droplets assume charges approximately half of which will be negative and half positive. Either can be used for spectroscopic examination depending on the electrode configuration of the mass spectrometer. In this way the thermospray process differs considerably

LC/MS Tandem Systems

from electrospray where the ions assume a charge determined by the direction of the ionizing field. The consensus of opinion appears to be that as these charged droplets evaporate, the diameter of the droplets are reduced and the charge density increases. There comes a point when the electrical forces become equal to the surface tension forces which contain the drop, and the droplet explodes. This is called Raleigh's limit and the maximum permissible charge (q_r) that can be carried by a drop radius (r) can be mathematically expressed by the following equation.

$$q_r = \left(\frac{8\pi}{e}\right)\sqrt{(\gamma e_0)r^3}$$

where (g) is the surface tension of the liquid
and (e_0) is the permittivity of free space.

The thermospray produces larger drops (*ca* 10 µm) than electrospray (1-2 µm). The small size of the droplets produced by electrospray probably accounts for the production of multiple ions, which as will be seen later, allows very high masses to be measured.

An example of the thermospray Ionizer is shown in Figure 8.17. The device consists of a stainless steel tube, containing in one end a metal cap made from a high-conductivity metal such as copper. Through the center of the stainless steel tube and copper cap passes, either the column itself if a capillary column, or a conduit of small internal diameter carrying the column eluent. The conduit or column projects slightly out from the end of the heater cap. In the copper cap is placed a cartridge heater and a thermocouple, which measures the temperature of the probe tip, and also provides the controlling signal to maintain the cap at a selected temperature. As the pumping rate of the mass spectrometer vacuum system is limited, either a microbore column can be used, or if a normal packed column is employed, the eluent is split. The flow rate range that can be accepted by the thermospray interface is about 100-1000 µl/min. In contrast, the electrospray, which will be described later, can only accept a flow rate that lies between about 1 to 10 µl/min. The flow capacity of the thermospray is similar to that of the particle beam interface which will also be described later.

The properties of the thermospray system were investigated by Voyksner *et al.* [13]. They noted that the LC/MS thermal spray system frequently produced molecular weight information (parent ions), and exhibited lower detection limits than the other forms of LC/MS interfaces.

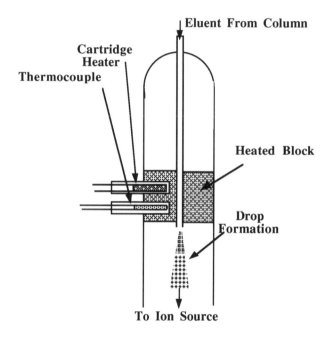

Figure 9.17   The Thermospray LC/MS Interface

They found that when they used a thermal spray system with a 0.1 M ammonia acetate buffer, and a mobile phase that contained a high proportion of water, high sensitivities were achieved. The optimum interface temperature varied with solvent composition and could be determined by maximizing the solvent buffer ion intensities. The spectra obtained from thermospray ionization resembles chemical ionization using ammonia as the chemical ionization reagent. The ions formed appear to be the result of protenation, resulting in ammonium addition and proton-bound solvent molecular clusters. The thermospray ionizing procedure

LC/MS Tandem Systems

was reported to be very soft producing very few molecular fragment ions. The system has been used very successfully in the analysis of triazine herbicides and organo-phosphorus pesticides and excellent specificity and sensitivity was achieved. A number of different forms of this device were described, but they are all basically similar to that shown in Figure 9.17. An interesting form of the thermospray interface has been described by De Wit *et al.* [14] in 1987, for use with open tubular LC columns.

Blakely and Vestal [15] employed the thermospray system with the quadrupole mass spectrometer, and demonstrated that it could provide stable vaporization and ionization at flow rates up to 2 ml/min, with an aqueous mobile phase. If the mobile phase contained a significant amount of ions in solution (ca 10^{-4} to 1.0 M), no extra thermal ionization source is required to achieve detection of many non-volatile solutes at the sub-microgram level. They found that with weakly ionized mobile phases, a conventional electron beam needs to be used to provide gas-phase reagent ions for the chemical ionization of the solute.

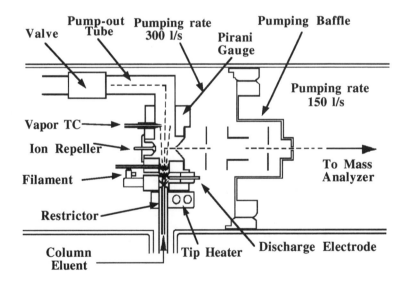

Figure 9.18 The Vestec Model 210 Thermospray Interface

The design of a more recent form of thermospray interface, the Vestec Model 201 used by Via and Taylor [16], is shown in Figure 9.18. This interface is designed to cope with column flow rates of up to 1.5 ml/min, and requires the use of two oil diffusion pumps backed by a single mechanical pump. The two pumps differentially exhaust the vacuum manifold, the source, and the analyzer regions of a quadrupole mass spectrometer. In addition, a further two diffusion pumps, backed by a single mechanical pump, are also coupled directly to the source, opposite the sample inlet. This pump removes about 99% of the vaporized mobile phase, whereas the heavier molecules pass through an ion aperture in the sampling cone, and into the mass spectrometer. The ions are formed immediately after the nebulization, and pass alongside a repeller plate, held at a high potential, that impels the ions through a hole into the ion source. Once in the ion source, the ion optical system directs them into the analyzer section of the spectrometer. The reagent gas is methane and its flow is controlled by separate needle valves.

An example of the use of the thermospray ionization source for the measurement of furazolidone in some pharmacokinetic studies is afforded by he work of McCracken *et al.* [17]. Furazolidone (N-(5-nitro-3-furfurylidene-3-amino)-2-oxazolodinone is a 5-nitrofuran antibiotic that is added to animal feeds to help prevent such infections as *Escherichia coli* and *Salmonella* in cattle, pigs and poultry. Consequently, it is important to determine the amount of furazolidone residues (if any) that might remain in the meat after slaughter for human consumption. In Europe the maximum level that is tolerated is 5 µg/kg of animal foodstuff.

Furazolidone is light sensitive and so operations must be carried out under artificial yellow light. Liver and muscle tissue were examined and 2 g samples were homogenized with 40 ml of a mixture of McLlvaine buffer and methanol (7+3) and then centrifuged for 15 minutes. The supernatant liquid was evaporated to 15 ml at 40°C and 25 ml of dichloromethane was then added and the mixture shaken for about 1 minute. The upper aqueous layer was discarded and the solvent extract evaporated dryness and taken up in 2 ml of dichloromethane and 6 ml of hexane. The sample was passed through a prepared Bond-Elut NH$_2$ extraction cartridge, washed with 5 ml hexane/ dichloromethane mixture (1+1) and 2 ml of hexane/chloroform mixture (1+1). The cartridge was extracted with 5 ml of chloroform/ methanol mixture (7+3), the extract evaporated to dryness and redissolved in 100 µl of mobile phase. The sample was injected onto a reversed phase column, 12.5 cm long, 4 mm I.D. containing RP18 stationary phase bonded to 5 µm particles. The tandem instrument was a Vestec thermospray LC/MS Model 201A. Chromatograms showing the elution of the Furazolidone by single ion monitoring is shown in Figure 9.19

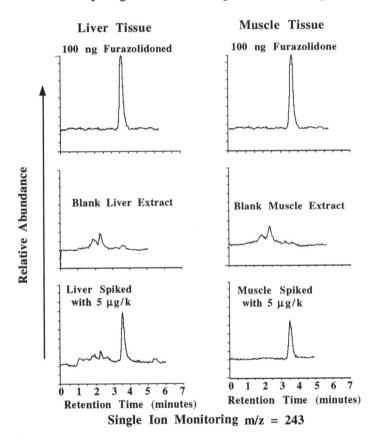

Figure 9.19 Single Ion Chromatograms (m/z=243) Furazolidone Demonstrating the Sensitivity Levels Obtainable (ref.17)

It is seen that as a result of the use of single ion monitoring, the procedure is highly selective. The recovery of the drug ranged between 65 and 70%. The minimum detectable level of contamination was about 1 μm/kg of tissue which was quite adequate to confirm that a sample did not contain the drug in excess of the tolerance level. A common analytical problem of importance in the explosives industry, is the identification of stabilizer derivatives in an explosive material, the presence of which are an indication of its age or stability. Many nitrocellulose propellants are stabilized with diphenylamine. This stabilizer is thought to react with any NO2 that is released during aging or decomposition, to produce nitrated derivatives of the stabilizer.

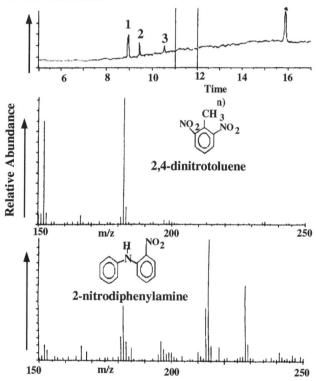

1. 2,6-dinitrotoluene, 2. impurity, 3. 2-nitrodiphenylamine, 4. 4-nitrodiphenylamine.

Reprinted with permission from J. Via and L. T. Taylor, *Anal. Chem.*, **66(9)**(1994)1385, Copyright 1994 American Chemical Society

Figure 9.20 The Separation of a Propellant Test Mixture

Consequently, an analytical method that will identify and measure the presence of nitro-diphenylamines in an explosive would be very useful. LC/MS employing a thermospray interface is an obvious solution to this problem. Via and Taylor [16] developed a method, using their SFC/MS tandem system, to separate, identify and assay the amount of nitrated diphenyl-amines present in a nitrocellulose explosive sample. The results from a test sample, containing the stabilizer and its derivatives, that was separated by SFC and analyzed by the mass spectrometer are shown in Figure 9.20.

The reconstructed total ion current chromatogram of the separation is shown at the top of Figure 9.20. It is seen that a good separation of the solutes of interest was obtained. It is also seen that the spectra for 2,4 dinitrotoluene and 2-nitrodiphenylamine are clean, and would allow the substance to be identified with little or no ambiguity. The authors claimed that the sensitivity of the analytical procedure was about 1 ng. However, according to Via and Taylor, in a private communication, Wilkes of the Vestec Corp. claimed that satisfactory spectra could be obtained from samples present at the picogram level. For example, 60 pg of 2-nitrodiphenylamine injected on the column could provide an identifiable mass spectrum.

Cannavan, *et al.* [18] used the high sensitivity available from the thermospray interface LC/MS, in a similar manner to the previous authors, to determine the amount of levamisole in animal tissues. Levamisole [1]-(–)-2,3,5,6-tetrahydro-6-phenylimidazole(1,1-b)thiazole, the laevorotatory isomer of tetramisole, is an anthelmintic drug used to control gastrointestinal parasites in cattle pigs and sheep. As in the assay of furazolidone in animal tissue, a somewhat complicated sample preparation procedure was necessary. 3 g of the tissue sample was mixed with 2 g of anhydrous sodium sulfate, 9 ml of ethyl acetate and 0.5 ml of 50%w/v caustic potash solution. and the mixture homogenized. To 6 ml of the supernatant liquid, 6 ml of n-hexane was added, and the mixture passed through a Baker Bond CN cartridge column. The column was washed with 5 ml of chloroform/n-hexane mixture (1+1) and air-dried. The levamisol was eluted with two 5 ml aliquots of methanol, evaporated to dryness, and the residue taken up in 200 µl of the mobile phase. This solution was used for the LC/MS analysis. 50 µl samples were placed on a Li-Chrospher 60RP-select B column 12.5 cm long, 4 mm I.D. packed with 5 µm particles. The mass spectrometer was the Hewlett-Packard 5989A MS engine with thermospray. Single ion chromatograms obtained from the analysis are shown in Figure 9.21.

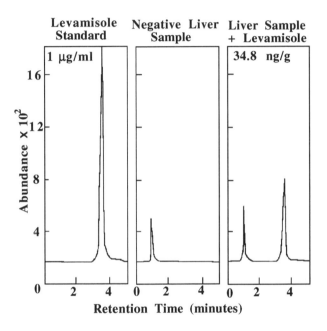

Figure 9.21 Single Ion Chromatograms of Levamisole Extracted form Liver Tissue (ref.18)

It is seen that again by single ion monitoring the peak of interest is picked out with only one other appearing in the chromatogram. The limit of detection was found to be about 5 ng/g and it was calculated that the mean recovery for liver, kidney and muscle tissue was 93%, 85% and 79% respectively. This method is now used as the standard assay for levamisole in eats.

Blanchflower and Kennedy [19], developed a similar procedure utilizing LC/MS with a thermospray interface for the assay of Nitroxynil in meats. A modified extraction procedure for the specific material was developed

LC/MS Tandem Systems

but the LC/MS analysis was almost identical. Nitroxynil is used in animals to control liver fluke and consequently the amount ingested when consuming meats from such animals needs to be continuously monitored. Single ion monitoring was again used and a low detection limit of 2 ng/g was reported. The average recover of the drug from a range of different tissues was found to be 82%

The choice of the most effective LC interface is complicated by the fact that the majority of LC separations are carried out employing reversed phase chromatographic distribution systems. This means that an aqueous mixture of polar solvents must be used as the mobile phase. Unfortunately, this renders the electrospray interface (which will be discussed next) more useful than the thermospray. It would appear from contemporary literature that the majority of LC/MS tandem systems now utilize the electrospray interface, and far fewer applications employing the thermospray interface are now being reported.

The Electrospray Interface

One of the earlier reports describing the *electrospray* interface was that of Whitehouse *et al.* [20] and this device is probably the most commonly used interface in modern LC/MS tandem instruments today. A diagram of the elctrospray interface is shown in Figure 20.

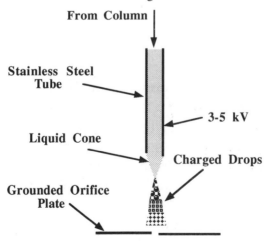

Figure 9.20 The Electrospray Interface

The electrospray interface differs from the thermospray in that it is operated at atmospheric pressure whereas the thermospray usually functions at a reduced pressure, *ca* 1 to 10 torr. Typically the analyte solution is sprayed from a stainless steel capillary usually situated about 1 cm form the ion sampling orifice. A potential of 3-5 kV is applied between the jet and the orifice plate and the gradient can be positive or negative in nature. The ions are formed by the potential between the capillary jet and the plate The electrospray effect changes with different ionizing potentials. As the potential difference is increased, the drop size increases, but initially there are no ions formed. As the voltage increases further a liquid cone is formed which generates an expanding mist of droplets. In this form the droplets are charged and ions are formed. The flow rate is limited to a few microliters /min, as the volume that can be drawn from the jet by electrical sheer forces is limited. In most instruments solvent evaporation is aided by a warm stream of nitrogen flowing counter current to the spray and any uncharged droplets are swept away from the orifice. At still higher voltages a corona discharge is formed which interferes with the ion productions and is to be avoided.

A diagram of the Hewlett-Packard electrospray ionization LC/MS interface is shown in Figure 9.21. The column eluent is mixed with a nebulizing gas and the spray jet directed onto a disk target which is at a high potential relative to that of the spray nozzle. In the center of the target is a pinhole entry into the interface and the jet is directed slightly to the side of this aperture. The fine droplets at the periphery of the spray are drawn into a chamber (held at a reduced pressure) through a pinhole aperture. Inside the chamber, the droplets are entrained in a stream of hot nitrogen gas, that rapidly evaporates the solvent, producing ions in the manner described above. The core of the jet is skimmed by conical screens to remove the drying gas, and the ions then pass directly into an ion-optical arrangement, that directs them into the mass analyzer. Ions with multiple charges are produced as a result of a molecule being associated with more than one proton. Consequently, an ion of molecular weight 1000, carrying three charges, will appear at an m/z value of 333.3 on the spectrum where an ion of mass 333.3 and unit charge would normally appear.

LC/MS Tandem Systems

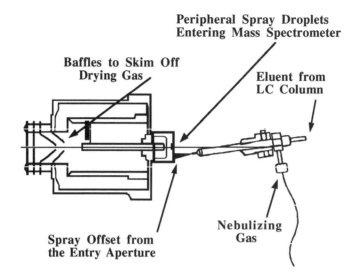

Courtesy of the Hewlett-Packard Company.

Figure 9.21   The Hewlett-Packard Electrospray Ionization LC/MS Interface

As already discussed, it is seen that the production of multiple charged ions, in effect, increases the mass range of the spectrometer. As an example of the use of the ion spray LC/MS interface, the total ion current chromatogram of a sample from the tryptic digest of lysozyme is shown in Figure 9.22. A Vydac C-18 reversed phase column 250 mm long and 2.1 mm I.D. was used to separate the components of the mixture, and was thermostatted at 50°C. The flow rate was 200 µl per min and the separation was developed by gradient elution. The mobile phase composition was programmed from pure solvent A (0.1% trifluoro-acetic acid (TFA) in water) to 60% solvent B (0.1% TFA in acetonitrile) over a period of 60 minutes. The total column eluent was passed into the interface but, as seen from the diagram of the interface and its mode of operation, only a portion of the solute actually enters the interface. Furthermore, of that portion that does enter, only a fraction of it is carried into the mass analyzer. The mass spectrometer employed was the Hewlett-Packard MS Engine Quadrupole Mass Spectrometer. The mass spectrum of the peak eluted at 30.35 minutes is shown in Figure 9.23.

362   Tandem Techniques

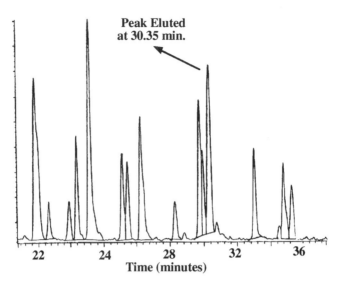

Courtesy of the Hewlett Packard Company.

Figure 9.22 The Total Ion Current Chromatogram of a Sample from the Tryptic Digest of Lysozyme

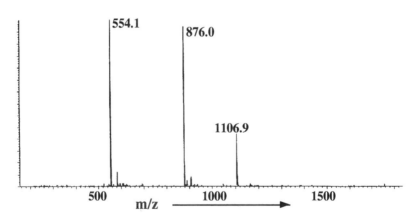

Courtesy of the Hewlett-Packard Company.

Figure 9.23 The Mass Spectrum of the Peak Eluted at 30.35 Minutes in the Tryptic Digest Chromatogram of Lysozyme

LC/MS Tandem Systems

The Figure 9.23 shows three peaks at m/z values of 554.1, 876.0 and 1106.8. It is seen from Figure 9.22, that the peak was not completely resolved from its neighbors, and thus it was necessary to decide whether all three peaks originated from the same solute, or whether any pair of the peaks resulted from multiple charges on the same molecule. A simple program is included in the Hewlett-Packard data handling software that tests the interrelationship between the individual mass peaks of the mass spectrum. The software determines whether the heights of any pair or group of peaks are linearly related to each other. If the respective peaks increase or decrease proportionally with one another, then they must originate from the same parent molecule, and probably represent multiple charges on the same molecule. The peaks in Figure 9.23 were tested in this way, and the results from the correlation program are shown in Figure 9.24.

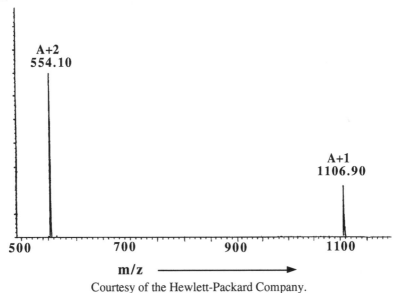

Courtesy of the Hewlett-Packard Company.

Figure 9.24 Mass Spectrum Showing Multiple Charged Peaks of a Parent Ion

The data processing demonstrated that the peak at 876.0 m/z was not related to the other two and, furthermore, the peaks at 554.1 and 1106.9 were doubly and singly charged species of the same ion.

e.g. $\dfrac{1106.9 - 1(H+)}{2} + 2(H+) = 554.95$ c.f.   554.1 (from spectrum)

The electrospray interface has been widely used in LC/MS tandem instruments, and there are many applications reported in the literature. Nevertheless, the device in some forms does have some drawbacks. The electrospray system may not work well with mobile phases having high water content, and as the majority of chromatographic separations are carried out employing reverse phase distribution systems, this limitation can be a serious problem. However, certain interface designs work better than others with aqueous mixtures, and some electrospray nebulizers are designed specifically to cope with solvents of high water content. It follows that in purchasing an electrospray interface, its efficacy for the particular analyses with which it is to be used, must be ascertained. Its areas of use have increased rapidly over the past five years, and some particular applications that illustrate its versatile capabilities will be described.

Applications of the Electrospray LC/MS Interface

Davis *et al.* [21], used an electrospray interface to couple a microbore column to a Finnigan MAT TSQ 700 triple sector quadrupole mass spectrometer. The basic layout of the tandem system is shown in Figure 9.25. The microbore column was 15 cm long, 0.25 mm I.D. and packed with a C18 reversed phase support. The flow rate was only 1-2 µl /min and thus a special electrospray assembly was designed to accommodate these low flow rates. The gradient was performed and stored in the manner of Snyder and Saunders [22] and Katz and Scott [23]. The eluent was monitored by an on-column UV detector, and the eluent passed directly to the micro electro-jet assembly. The micro-spray was constructed from a flame-drawn length of fused silica tubing, 5 cm long, 350 µm O.D. and 150 µm I.D. The aperture diameter of the drawn jet ranged from 1-5 µm. Two lengths of 150 mm O.D. and 25 mm I.D. tubing were placed inside the tube, and sandwiched between them was a hydrophilic PVDF frit. These insert tubes reduced the dead volume of the spray jet, and provided a support for the filter, which was necessary to avoid the jet becoming blocked.

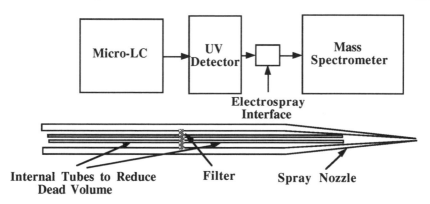

Figure 9.25 The On-line Micro-electrospray Interface

Even with this precaution, the tube life was limited, and the jet was liable to become blocked after a few hours of use. The jet was positioned accurately by means of a micrometer-driven optical rail assembly. Spray potentials ranged from 500-1000 V, and the optimum operating potential appeared to be about 100 V in excess of that at which spray droplets started to form.

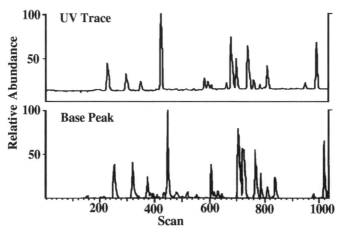

Reprinted with permission from J. Via and L. T. Taylor, *Anal. Chem.*, **66(9)**(1994)1385, Copyright 1994 American Chemical Society

Figure 9.26 Comparative Chromatograms of a Cytochrome c Standard Digest Mixture Monitored by UV Absorption and by the Base Peak Intensity

The micro-electrospray interface was used to monitor the separation of a number of polypeptides and an example of the separation obtained from 2 pmol of sample is shown in Figure 9.26.

It is seen that very similar chromatograms are obtained, and little or no loss of chromatographic resolution takes place in the micro-electrospray interface. The sample size was 2 pmol which, if a mean molecular weight of 1000 is assumed, will represent a sample mass of about 2 ng. However, a better example of the sensitivity obtainable from the system is afforded by the reconstructed chromatograms shown in Figure 9.27. The sample size was only 40 fmole, which, again assuming a mean molecular weight of about 1000, would be equivalent to a sample mass of 40 pg.

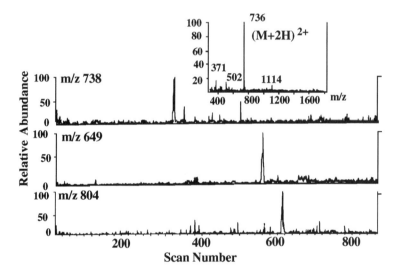

Reprinted with permission from J. Via and L. T. Taylor, *Anal. Chem.*, **66(9)**(1994)1385, Copyright 1994 American Chemical Society

Figure 9.27 Selected Mass Chromatograms for Three of the Peptides from a Sample of Lys C Digest of Cytochrome

It is seen that the combination of the microbore column and the micro-electrospray provides a very high sensitivity without compromising the performance of either the liquid chromatograph or the mass spectrometer.

Another interesting application of the electrospray interface not only demonstrates its efficacy for natural product investigation but also shows some of the foibles that can be associated with the liquid chromatography column. As a result of the contemporary popularity of herbal remedies in the United States, there has been a demand for analytical techniques to monitor the substances imported, to ensure their integrity and safety. Van Breeman *et al.* [24], developed an analytical method for measuring the ginsenoside content of ginseng products, marketed as roots, capsules, tablets and liquid extracts. Ginsenosides are made up of a series of triterpine saponins in proportions that are characteristic of their country of origin. The individual ginsenosides have been separated by reverse phase chromatography and ion exchange chromatography, but the use of specialized carbohydrate analysis columns, containing aminopropyl functional groups, have also proved useful.

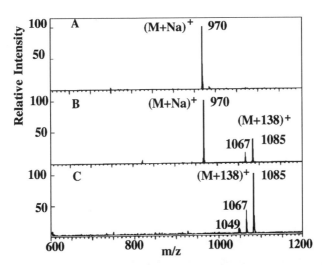

A. Ginsenoside standard added by direct infusion. B. Ginsenoside standard introduced from liquid chromatograph C. Ginsenoside extract introduced from liquid chromatograph.

Reprinted with permission from R. B. van Breeman, C.R Huang, Z. Z. Lu, A. Rimando, H. H. S. Fong and J. F. Fitzloff, *Anal. Chem.*, **67**(21)(1995)3985., Copyright 1995 American Chemical Society

Figure 9.28 Electrospray Mass Spectra of *ca* 100 pmol of Ginsenoside

Van Breeman *et al.*, examined the ginseng root powder by extracting the powdered root with aqueous methanol, evaporating to dryness, dissolving the residue in water, and passing the solution through a solid-phase extraction tube. The retained material was displaced with butanol, evaporated to dryness, and redissolved in methanol. Samples of the methanol solution were placed on a carbohydrate analysis column (aminopropyl bonded to silica) and separated using a water/acetonitrile gradient. As the ion monitored was an adduct of sodium $(M+Na)^+$, the mobile phase also carried 100 μM sodium chloride. The column eluent was monitored by a standard Hewlett-Packard 5989B MS engine quadrupole mass spectrometer, and the spectra that were obtained for a specific ginsenoside are shown in Figure 9.28.

It is seen that the standard sample, injected directly into the mass spectrometer, gave the expected sodium adduct ion, $(M+Na)^+$ at an m/z value of 970. The same peak appears when the standard is chromatographed and passed through the electrospray interface. However, it is seen that two new peaks appears at m/z values of 1067 and 1085. When the root extract is separated and monitored under the same conditions, there is no peak at 907 m/z and the peaks at m/z values of 1067 and 1085 have become much bigger, and that at 1085 is now the major peak. Although all standard ginsenoside sample produced both the sodium adduct ion and the $[M+138]^+$ ion, the natural products only gave the $[M+138]^+$. It was also found, that the 138 adduct to the ginsenoside was most likely to be the protonated adduct,(3-aminopropyl)-trihydroxysilane, $[NH_3(CH_2)_3Si(OH)_3]^+$ which was present either as a reagent contaminant in the bonded phase or was produced by the decomposition of the stationary phase. The 1067 ion appeared to be produced by the removal of water from the $[M+138]^+$ ion. The spectra represented only 100 pmol of ginsenoside and so the appearance of the (3-aminopropyl)-trihydroxysilane adducts might be a function of the sample size and if larger charges were employed the sodium adduct might again appear.

The various penicillins, particularly penicillin G, are widely used in veterinary medicine and to prevent antibiotic residues from entering the human food chain, their levels in animal products for human consumption need to be carefully monitored. The antibiotics of interest included, oxacillin, cloxacillin and dicloxicillin Blanchflower *et al.* [25] developed a procedure for simultaneously monitoring five penicillins in muscle and kidney tissue and milk. The authors employed an LC/MS tandem instrument fitted with an electrospray interface, and utilized single ion monitoring to selectively locate and measure each penicillin. Most biological samples require complex sample preparation and the measurement of antibiotics in animal tissue and animal products is no exception. Tissue sample were pulverized, spiked with the standard and homogenized. Acetonitrile was added, sonically mixed and then centrifuged. A portion of the supernatant liquid was treated with phosphoric acid, mixed with dichloromethane and again centrifuged. Acetonitrile and n-hexane were added, the mixture shaken and centrifuged. The lower layer was washed with water. The solvent mixture was then extracted with phosphate buffer, centrifuged and the lower layer treated with tetrabutylammonium hydrogen sulfate. The solution was then extracted with dichloromethane, the extract evaporated to dryness and dissolved an acetonitrile water mixture. The liquid chromatograph was a Merck-Hitachi Model L6000 pump, a Rheodyne 7125 injector and an Intersil ODS-2 reversed phase column 15 cm long and 4.6 mm I.D. The outlet from the column was coupled to a Megabore probe of a VG Platform ES-MS which was operated in the negative ion mode. The source was maintained at 120°C and the flow rates of the drying and nebulizing gas was 10 l/hr. The extraction and focus voltages were about 17 and 24 V respectively

It was found that the fragmentation pattern could be significantly changed by adjusting the voltage on the extraction cone. As the voltage was increased the degree of fragmentation increased. This effect is shown in Figure 9.29. It is seen that 5 V on the extraction cone produced just two ions above m/z of 160, *i.e.* m/z = 434 and 436. Increasing the potential to 10 V produced another peak at m/z 293. At a potential of 20 V the peak at m/z of 293 has markedly increased and has been joined by a significant peak at m/z=295. At the same time the original major peak at m/z=434 had shrunk to a minor peak and the peak at m/z=436 was barely visible. It is seen that the operating conditions of the interface can be critical in order to achieve the desired results. This is typical of the performance of electrospray interfaces.

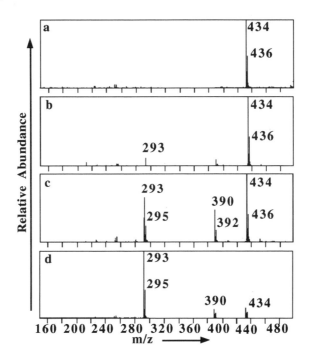

a. Extraction Potential 5 V, b. Extraction Potential 10 V, c. Extraction Potential 15 V, d. Extraction Potential 20 V

Figure 9.29 The Effect of Changing the Extraction Voltage on the Fragmentation Pattern (ref.25)

In the assays the extraction potential was kept between 15 and 17 V. Examples of the chromatograms obtained by single ion monitoring at different m/z values, for Cloroxacillin and Penicillin G contained in muscle tissue, are shown in Figure 9.30. It is seen that monitoring at m/z values of 293, 390 and 434 clearly, and almost exclusively, selects the Cloxacillin from the accompanying unresolved substances. It would also appear that the best signal-to-nose ratio was achieved employing a m/z values of 293 and 434. From the apparent magnitude of the signal-to-noise ratio it would seem that the assay could detect levels of Cloxacillin as low as 40 ng/g.

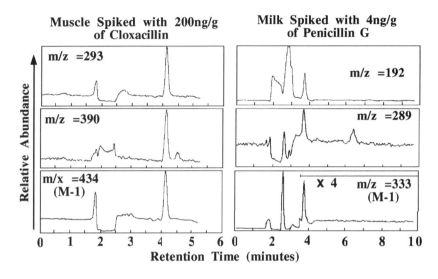

Figure 9.30 Chromatogram from the Assay of Cloxacillin and Penicillin G Monitored at Different m/z Values (ref. 25)

The optimum extraction voltage for assaying Penicillin G in milk appears to be far more critical, the best selectivity being obtained at a m/z value of 333. The lower limit of detection, however, is much smaller and it looks as if less than 1 ng/g might be detectable.

Research by Elliot *et al.* [26], involving the correlation of stereochemical structure with insecticidal activity led to the discovery of permethrin, cypermathrin and many other important pyrethroids. The new pyrethroids lack the cyclopropanecarboxylic ester bond, common to other chemically similar insecticides, and are mainly achiral. They are broadly insecticidal, have low mammalian toxicity's, and as a result are applied to crops, forests, soils, animal feeds and in household use. It follows, that an assay, capable of detecting very low concentrations of such substances is necessary for environmental testing. Fleet *et al.* [27] developed such a technique using positive-ion electrospray mass spectrometry. A VG tandem quadrupole mass spectrometer fitted with an electrospray ionization source interfaced to a tri-axial probe held at 4.3 kV was used for these experiments. A diagram of their apparatus is shown in Figure 9.31 together with the details of the tri-axial probe.

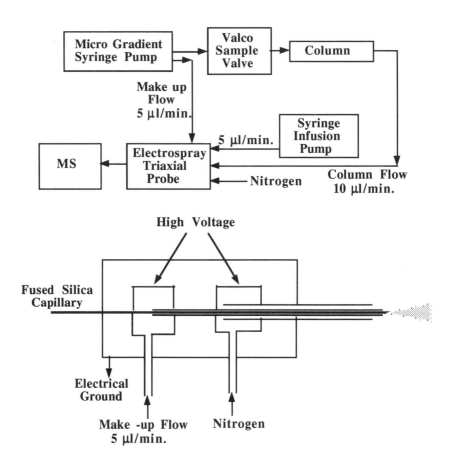

Figure 9.31 The LC/ES-MS Tandem Apparatus and details of the Tri-axial Probe (ref. 27)

The tri-axial probe comprised of two concentric stainless steel capillaries and an inner uncoated fused silica capillary 70 cm long, 75 μm I.D. and 375 μm O.D., that terminated approximately 0.5 mm beyond the end of the inner stainless steel capillary. The make up flow rate of 70% propan-2-ol and 30% water containing 10 mmol of ammonium acetate and 22 mmol of formic acid was 5 ml/min. A Valco valve (capacity 0.06 ml) was used to inject the sample and the column was 25 cm long, 0.5 mm I.D. packed with Spherisorb ODS 2. The positive electrospray spectra of Cypermethrin taken a different cone voltages are shown in Figure 9.32.

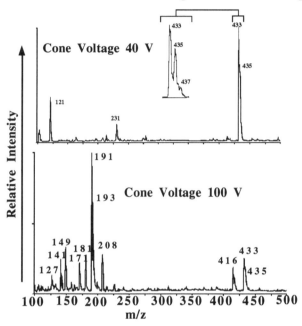

Figure 9.32 Positive Ion Electrospray Spectra of Cypermethrin at Different Cone Voltages

The dependence of fragmentation pattern on cone voltage is similar to that shown by Blanchflower [25] and emphasizes the importance of using the correct cone voltage for the specific molecules being examined. The technique worked well, all the solutes of interest were separated and their identity could be confirmed from their spectra. The technique also proved to be extremely sensitive and for a signal-to-noise ratio of 3, using the full scan mode, the limit of detection was 120-300 pg. If single ion monitoring was employed the limit of detection was reduced to 16-60 pg.

In addition to the cone voltage, the size of the droplets formed in electrospray interfaces is extremely important. Wilm and Mann [28] pointed out, that in general, the lower the flow rate through the electrospray, the smaller the droplets that are produced. Small droplets have high surface to volume ratios, and thus make a large proportion of the analyte molecules available for desorption.

The authors fabricated capillaries with spraying orifices having only 1-2 µm I.D. The capillary tips allowed reduced flow rates of 20 nl/min to be realized. The droplet diameter was claimed to be about 200 nm, in comparison with 1-2 µm, the diameter of droplets generated from conventional electrospray sources. As a consequence, the droplet volume was also 100 to 1000 times smaller. In addition, the miniaturized electrospray inlet could be operated *without* a sheath flow or pneumatic assistance, which made the design of the interface much simpler. Spectra were obtained from a 1 µl sample of an aqueous solution of ovalbumin containing 5 pmol/µl, and so it is clear that the interface functioned well with aqueous solvent mixtures. Assuming a molecular weight of about 44,000 for ovalbumin, this sample volume contains a mass of about 0.2 µg of protein at a sample concentration of about 0.02%.

An interesting extension of the utilization of this type of interface is its use with a MS/MS tandem instrument. The results published by Wilm and Mann [28], are shown in Figure 9.33. The top curve shows an MS analysis of the peptide over the m/z range of 600 to 850. Two ion adducts from the preliminary ionization at about m/z values 670 and 810 were chosen for subsequent examination, and the results obtained are shown by the two spectra in the lower part of the Figure 9.33. It is seen that whereas the sample adduct ions had m/z values of 670 and 810, the m/z range of the MS/MS spectra extend to m/z values of 1100 and 1500. One explanation for this difference might be that the original adduct ions having m/z values of 670 and 810 carried multiple charges.

From the results discussed so far, it would appear that the sensitivity limits of the electrospray interface can vary widely from one design to another, and also, perhaps, between one type of sample and another. In any event, the ultimate practical sensitivity obtainable from the electrospray principle may well be much greater than that which has been so far achieved.

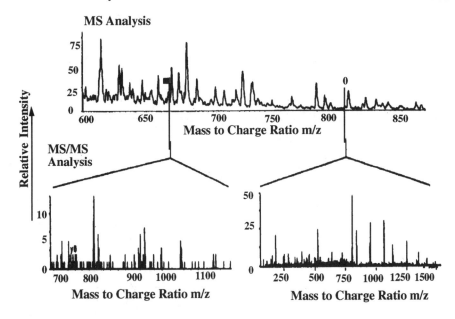

Reprinted with permission from M. Wilm and M. Mann, *Anal. Chem.*, **68**(1)(1996)1, Copyright 1996 American Chemical Society

Figure 9.33 The MS/MS Spectra from a Mass Separation of a Mixture of Peptides.

Post Column Additives in LC/MS using Electrospray Interfaces

A technique using on-line post-column adduct formation in an electrospray interface was developed by Kohler and Leary [29] for the analysis of carbohydrate mixtures. A tri-axial electrospray probe was designed, very similar to that described in Figure 9.31, that would allow the introduction of the eluent, the nebulizing gas and a metal chloride reagent solution simultaneously into the probe jet.

Two LC pumps provided gradient elution facilities and the sample was placed on the column with a low dispersion sample valve, having a 5 µl loop. After injection, the sample contained in the mobile phase passed through a 0.5 µm filter onto the column. The column was the Spherisorb s5-NH_2 microbore column, 25 cm long and 1 mm I.D., operated at a flow rate of 20 µl/min. The center tube (see Figure 9.31 carried the column eluent, the next coaxial tube carried the reagent, and the outside coaxial tube provided nitrogen to assist the nebulization. The post-column addition of metallic chlorides to an eluent carrying carbohydrates provides greater sensitivity, and assists in structural analysis. The relative abundance of protonated carbohydrate and the metal complexes with lithium, sodium, potassium, rubidium and cesium are different for each metal. Complexes with alkali metals significantly increase the sensitivity of the system to the carbohydrate, and reflects the enhanced ionization in the electrospray interface by the presence of the complexing reagents.

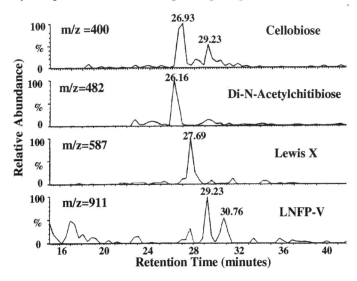

Reprinted with permission from M. Kohler and J. A. Leary, **67(19)**(1995)3508, Copyright 1995 American Chemical Society.

Figure 9.34 The Separation of Four Oligosaccharides Using Metal Chloride Post-column Reagents

It was shown that the lithium complex is over seventy times more abundant than the protonated species. Although the lithium complex appears to provide the maximum sensitivity of the group examined, it was also shown that cobalt complexes provided even greater sensitivity enhancement. Employing the tri-axial electrospray with cobalt chloride as the complexing reagent, the chromatogram of 1 nmol of a mixture of different carbohydrates, produced by single ion monitoring, is shown in

Figure 9.34. Although the chromatographic separation was not complete, the single ion monitoring of the cobalt complex ions clearly shows the presence of each carbohydrate.

The use of different reagents to enhance the production of ions in electrospray ionization has been the subject of a number of investigations, and was employed by Van Breeman [30] in the analysis of carotenoids. Carotenoids are the metabolic precursors of vitamin A, and are also thought to have anticancer activity, and to act as *in vivo* antioxidants. Van Breeman used an LC/MS tandem instrument in conjunction with the electrospray interface, together with the post-column injection of halogenated compounds, to enhance the ionization efficiency. The addition of several different halogen compounds were examined, including chloroform, 2,2,3,3,4,4,4-heptafluoro-1-butanol, 2,2,3,3,4,4,4-heptafluoro-1-butyric acid, 1,1,1,3,3,3-hexafluoro-2-propanol and trifluoroacetic acid. The chromatography column was 12.5 cm long and 4.6 mm I.D., packed with a C18 stationary phase that has been specially prepared for carotenoid separations. The separation was developed employing a 60 min. linear gradient, from 85:15 to 10:90 methanol/methyl *tert*-butyl ether, containing 1.0 mM ammonium acetate, at a flow rate of 1 ml/min. The solute used for testing the different antioxidants was β-carotene and 0.25 µg was injected onto the column. It was shown that the presence of the halogen antioxidants did, indeed, significantly affect the efficiency of ion production. However, in addition, it was evident that an excess of the additive can also reduce the enhancement significantly. The 1,1,1,3,3,3-hexafluoro-2-propanol additive provided the greatest sensitivity. The limits of detection were determined for both β-carotene (at an m/z value of 536) and lutien (at an m/z value of 568) and the results obtained indicated sensitivities of about 2 pmol and about 0.6 pmol for β-carotene and lutein, respectively were easily obtainable.

An example of the application of the technique to the determination of carotenoids in heat processed canned sweet potatoes with the post-column addition of heptafluorobutanol is shown in Figure 9.35, (A) to (E). (A) is the computer reconstructed mass chromatogram of the β-carotene molecular ion, at an m/z value of 536, from the injection of *ca* 20 ng of extract. (B) is the computer reconstructed mass chromatogram of the ion at an m/z value of 568, which corresponds to lutein. (C) is the computer reconstructed mass chromatogram of the ion at an m/z value of 552, showing the isomers of β-cryptoxanthin. (D) is the chromatogram provided by the output of the diode array UV detector at 450 nm, recorded during the analysis shown in (A). (E) is the chromatogram provided by the output of the diode array UV detector at 450 nm, recorded during the separation of 2 µg of sweet potato extract.

It is clear from the results obtained that the addition of 2,2,3,3,4,4,4-heptafluoro-1-butanol, post-column, as an ionizing enhancer when using an electrospray interface, can increase the overall sensitivity of the tandem instrument to carotenoids by about two orders of magnitude.

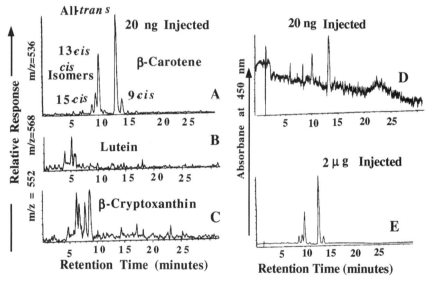

Reprinted with permission from R. B. van Breeman, *Anal. Chem.*, **67**(13)(1995)2004., Copyright 1995 American Chemical Society.

Figure 9.37 Positive Ion Electrospray LC/MS Analysis of an Extract of Heat Processed Canned Sweet Potatoes

The use of sodium replacement ions can also be used to help identify the charged state of the molecular ion. This possibility was investigated by Neubauer and Anderegg [31] to determine the charged states of peptide ions, when using an LC/MS tandem system with an electrospray interface. The addition of submillimolar levels of sodium acetate to the mobile phase, encourages the formation of sodium replacement ions in addition to the normally observed protonated species. The m/z spacing of the sodium adducts allows unambiguous identification of the charged state of the ions and hence their actual mass. It was also found that at the necessary sodium concentration (*ca.* 250 µM) the sodium salt did neither interfere with the chromatographic process, nor cause undue fouling in the mass spectrometer ion source. In general, as the quantity of sodium salt in the mobile phase is increased, the number of sodium replacement ions also increases. The spectra exhibit a number of replacement ions. The m/z values for the sodium replacement ions can be plotted against the number of sodium atoms in the respective replacement ion, which results in linear graph. the slope of the line indicates the number of charges on the sodium adduct ion, *e.g.* a slope of 11.5 would indicate that the ions carried two charges. Consequently from the slope and intercept of the graph the molecular weight of the substance can be calculated

In an attempt to eliminate the restrictions imposed by the nature of the mobile phase on the function of the electrospray interface, Banks *et al.* [32] developed an ultrasonically assisted nebulizer, to allow aqueous mixtures of nucleosides to be separated and monitored by a LC/MS tandem instrument, using the electrospray interface. The major interest of the authors was to employ the electrospray interface in the analysis of nucleosides, and the modified electrospray interface they developed is depicted diagramatically in Figure 9.38. Except for the ultrasonic nebulizer, the interface was similar to those already described. It consisted of a 0.005 in. I.D., 1/16 in. O.D., stainless steel tube, which had been ground to a sharp point at one end fitted into a stainless steel body which was in two parts. A pair of piezoelectric crystals were fitted between the two stainless steel parts, and were driven by a function generator and a power amplifier. The device was designed so that there was independent control of the needle potential, the cylindrical electrode potential, the nosepiece potential, and the capillary entrance potential. In the configuration shown, the needle was always maintained at ground potential and the temperature of the drying gas was held at 82°C.

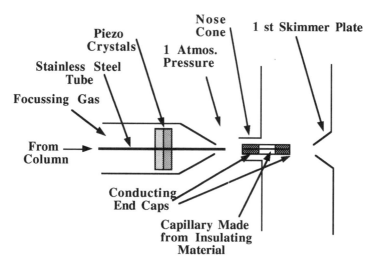

Figure 9.38 The Electrospray Interface Fitted with an Ultrasonic Nebulizer

In other respects, the system closely resembled the standard type of electrospray interface, typically described by Yamashita and Fenn [33]. It was found essential that, for effective nebulization, the frequency of the ultrasonic vibrator must be adjusted to the resonant frequency of the device, and this critical frequency was identified by experiment. Furthermore, the optimum frequency had to be carefully controlled, as a deviation of 0.1 kHz at 180 kHz was found to reduce the spray efficiency by nearly two thirds. The sample used for optimizing the conditions of ion production was adenosine dissolved in pure water (100 pmol/µl) which was considered a 'worst case' example. The interface was employed in the separation of some nucleosides, using a microbore column and a Hewlett-Packard mass spectrometer model HP88A. The nucleosides were derived from a tRNA digest, reduced to its substituent nucleosides through the combined action of nuclease P1, and bacterial alkaline phosphatase. The samples were separated on a C18 column, 1 m long and 0.25 mm I.D., using a gradient from 5% to 20% aqueous methanol solution. Good spectra were obtained from as little as 1.5 pmol of sample. The interface operated well with liquids having high water contents, and in fact, operated well when nebulizing samples in solution in pure water. It was noted, however, that very small samples exhibited a disproportional loss of minor components, which might make quantitative assays uncertain when operating at maximum sensitivity.

The electrospray interface has rendered the LC/MS tandem instrument available to a wide range of applications in biology and biochemistry, which hitherto, were precluded from the technique due to its inability to function efficiently with samples in aqueous solutions. Hua *et al.* [34] employed electrospray ionization in their examination of brevetoxins. Brevetoxins are produced by the dinoflagellate, *Gymnodinium breve*, and are responsible for killing fish and also pose certain health risks to humans.

The crude brevetoxins were isolated from an extract of cultured material using a reversed phase solid-state extraction procedure. A microbore column was employed for the separation, 10 cm long and 1 mm I.D., packed with a C18 reversed phase, having a particle of 3 µm. The column eluent was split at a T junction, part passing to a UV detector operating at 215 nm, and the remainder passing to the mass spectrometer. The mass spectrometer was the Vestec Model 201 fitted with an electrospray inlet system. However, the electrospray source was modified by replacing the nozzle used for the counter electrode by a flat stainless steel plate with a hole in it, 0.44 mm in diameter. In addition, the 200 l /min pump, normally used to evacuate the first stage of the ion source, was replaced with a 500 l /min. pump. These modifications helped reduce the lower limit of detection by a factor of four.

The separation was developed isocratically with a mobile phase containing 15% of water in methanol, at a flow rate of 8 µl/min. The sample consisted of 1 µl of a solution, containing 20 ng/µl (*i.e.* a mass of 20 ng) and the column eluent was split in the ratio 3:1, to the UV detector and the spectrometer respectively. Good spectra were obtained and single ion monitoring produced peaks for each component with a signal-to-noise ratio of about ten, indicating a lower sensitivity limit of less than 3 ng/ml. The significant peaks in the spectra, in fact, were sodium adduct ions of the respective brevetoxins.

An electrospray interface has been designed specifically for use with liquid chromatography micro columns. Micro columns employ very low flow rates and the peaks are eluted in very small volumes so, although the mass placed on column is very small, the concentration of the solute in the eluent is still relatively high. Their design and use for the analysis of biological samples has been reviewed by Yates *et al.* [35] and a diagram depicting the general properties of the micro column electrospray interface is shown in Figure 9.39.

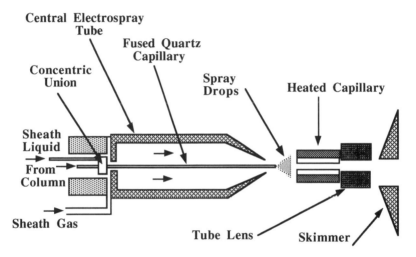

Figure 9.39 Electrospray Interface for Liquid chromatography Micro-columns (ref. 35)

The column flow passes through a fused quartz capillary, which may be a conduit from the column, or the column itself, and is joined by the sheath liquid flow at a T junction. The two streams pass down a jacket tube through which flows a warm current of sheath gas, the different streams meeting at the jet tip. The quartz capillary projects a little beyond the sheath gas nozzle and the mist of drops are formed at the liquid cone beyond the jet inducted by a potential of 2-4.5 kV. This potential is set up between the end of the jet and the end of the heated capillary situated ahead of the electrospray jet. The solvent evaporates and the ions that are formed pass into the heated capillary where the last traces of solvent are removed. At the end of the capillary the ions are focused by a tube lens through the skimmer plates into the mass spectrometer source. This interface system can operate at flow rates of a few microliters a minute and are thus ideally suited for micro column chromatography. An example of the use of the interface to monitor the micro column separation of cytochrome c and to obtain its mass spectrum is shown in Figure 9.40.

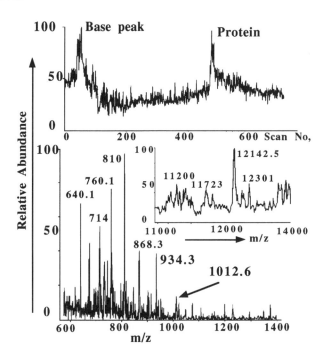

Figure 9.40. A Chromatogram of Cytochrome c Monitored by the Micro Column Electrospray Interface (ref. 35)

The micro column was 15 cm long, 50 μm I.D. and the protein peak represents 10 fmol of cytochrome c. The unprocessed mass spectrum is that obtained directly from the peak. The insert depicts the masses calculated by deconvolution algorithms that are applied to the multiply charged ion peaks in the actual mass spectrum. It is seen that as a result of the production of multiply charged ions, the m/z range for measurement is considerably extended and from very small samples the molecular mass of a relatively large protein molecule can be evaluated.

The electrospray ionization source has become one of the more popular LC/MS interfaces that are commercially available. However, more recently, another type of ionization source, that operates at atmospheric pressure, has become a strong competitor, and provides very similar ionization characteristics, but functions somewhat differently.

The Atmospheric Ionization Interface (API)

In atmospheric ionization the liquid flow is nebulized in the source region, but the ions are formed at atmospheric pressure. Ionization at atmospheric pressure has a number of advantages. It avoids the problems that arise when a liquid flows directly into a vacuum, and it allows the separation system to operate under ambient conditions. This is particularly advantageous where low flow rates are employed, such as with microbore columns and capillary electrophoresis. A diagram illustrating the basic design of the atmospheric ionization interface is shown in Figure 9.41.

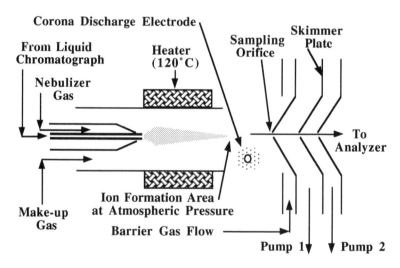

Figure 9.41 The Atmospheric Ionization Interface.

The atmospheric pressure ionization (API) process has some similarity to the electrospray ionization source, and can cope with a wide range of column flow rates, up to a maximum of about 2 ml/min. Consequently,

LC/MS Tandem Systems

the total column eluent can be utilized without splitting the flow. However, even though all the solute may enter the interface, not all the solute molecules are ionized, and not all the ions that are formed enter the mass spectrometer. There are three alternative forms of the API source: one that uses a heated nebulizer with a corona discharge, one that employs an atmospheric electrospray and one that uses an ion spray. The interface that employs a heated nebulizer and a corona discharge ionization process is that depicted in Figure 9.43. The liquid flow is nebulized by means of the gas flow which is then swept by another stream of gas (the sheath gas or make-up gas) through a quartz tube heater that vaporizes the solvent. The sample then drifts through a chamber containing a corona discharge, which is set up by a potential difference of about 2000 volts, applied between a simple electrode arrangement. The charged molecules of the solvent vapor are used as the ionizing agents.

The reactant ions that are formed in the corona discharge collide with the sample molecules and largely give *sample molecule plus a proton* (hydrogen positive ions), *i.e.* $[M+H]^+$. This is a process very similar to chemical ionization but occurs at atmospheric pressure. The ions are drawn by means of an electric field to a plate with an orifice, over which passes another flow of gas called the curtain or barrier gas. The flow of barrier gas helps to prevent uncharged molecules from entering the ion source but the charged ions pass through into the next chamber. The ions then pass through an aperture in the next plate, the skimmer plate, and the space between the sampling plate and the skimmer plate is connected to the first vacuum pump. After passing through the skimmer plate, the ions pass through an aperture in a third plate into the mass spectrometer analyzer. The space between the skimmer plate and the final plate is also connected to a second vacuum pump. By means of the differential pumping the necessary pressure can be maintained in the mass spectrometer. As the sample entering the mass spectrometer is virtually a parent ion already, the fragmentation pattern is very similar to that obtained in MS/MS. In this way, the system differs fundamentally from the electrospray interface. The ionization process is soft, very sensitive and gives good characteristic spectra, that can be used for both sample identification and structural elucidation.

A recent modification of the atmospheric pressure ionization technique, involving a special low dead volume interface for use with microboe columns The packed microbore columns, (170 μm, 320 μm, and 500 μm I.D., with lengths ranging from 5 to 15 cm) were used in conjunction with a low-volume, wall-coated capillary column as an interface. The total ion current chromatogram of a tryptic digest sample, comprising 1 picomole of human growth hormone, is shown in Figure 9.42. The column was packed with an octadecyl bonded phase, having a mean pore size of 300 Å, and a particle diameter of 7 μm. A gradient was employed which extended from 20% solvent (A), (0.1% TFA in water) to 80% solvent (B) (75% 0.1% TFA in water and 25% acetonitrile) over a period of about 1 hour. Flow rates of about 80 to 100 μl per minute were used, with about 3 μl passing to the capillary column and entering the interface.

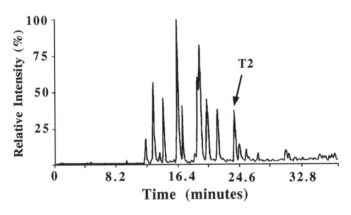

Courtesy of the Perkin Elmer SCIEX Corporation

Figure 9.42 Total Ion Current Chromatogram of a Tryptic Digest Sample of Human Growth Hormone

It is seen that an excellent separation is obtained and apparently little resolution is lost in the capillary interface. The mass spectrum of the peak marked T2 in the chromatogram is shown in Figure 9.43. It is clear that good quality spectra can be obtained up to ion masses of at least 900. Such a combination of techniques can be invaluable for the structure elucidation of compounds generated in biochemical research.

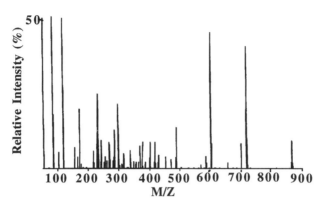

Courtesy of the Perkin Elmer SCIEX Corporation

Figure 9.43 Spectrum of a Product from the Tryptic Digest of Human Growth Hormone Obtained from a Low Dead Volume Atmospheric Ionization Interface

Cai and Henion [37] used a complex combination of sampling techniques to assay LSD and its analogs in urine. First, the authors employed affinity chromatography to extract the substances of interest from the urine. The sample was then displaced from the affinity column and collected in a trap, from which the materials of interest were then displaced onto an LC column for separation. The column eluent was then passed through an atmospheric pressure ionization interface to the mass spectrometer. A diagram of their apparatus is shown in Figure 9.44.

The analytical procedure was complicated. First the immuno–affinity column (a HiPac protein G column, 3.3 mm x 2.1 mm packed with 30 μm particles) was equilibrated with the phosphate buffered saline (PBS). Then 30 μl of PBS-diluted antibody solution (10% antibody–90% PBS) was injected onto the column. Then, humane urine diluted with PBS (50% urine–50 % PBS) was pumped through the protein G column, which was immediately flushed with PBS to remove the weakly bound impurities. During this process, the trapping column (1.5 cm long, 1 mm I.D. packed with 5 μm, C18 particles) and the LC column (15 cm long, 0.3 mm I.D., packed with 3 μm C18 particles) were equilibrated with the mobile phase. The PBS was then pumped through the affinity column, and the trap, desorbing the materials from the affinity column and re-adsorbing them on the trap.

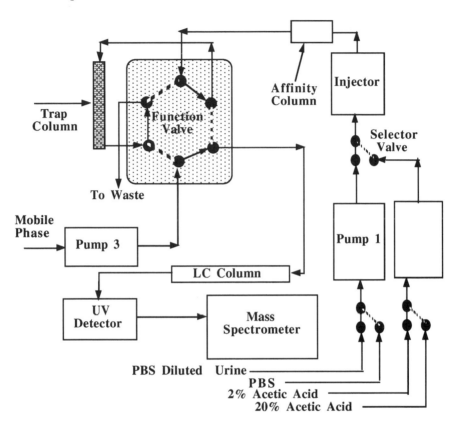

Figure 9.44 Diagram of the Sampling Arrangement for the Analysis of LSD in Urine by an LC/MS Tandem Employing an Electrospray Interface

The trap was then back–flushed, and the desorbed materials eluted through the LC column, through an API interface and into the mass spectrometer. Four metabolic products in addition to the unchanged LSD itself were separated and identified from their mass spectra. The original measured concentration of LSD in the urine was 0.9 ng/ml. These results demonstrated the very high sensitivity that could be obtained by utilizing the selective extraction, that can be provided by affinity chromatography.

LC/MS Tandem Systems

Huang *et al.*[38], in their article on atmospheric pressure ionization, demonstrated the use of the API interface in the LC/MS analysis of some benzodiazepines. They employed a Zorbax-Rx column, 25 cm long, 4.6 mm I.D, which was connected by a standard API interface to the mass spectrometer. The mass spectrometer was scanned between m/z values of 100 and 350 at a scan rate of 3 s/scan.

The separation was developed isocratically using a mobile phase composed of 40% v/v acetonitrile, 25% v/v methanol, and 35 % v/v water, containing 10 nM ammonium formate buffer. 25 ng of each benzodiazepine was present in the sample mixture. All the benzodiazepines were well separated and the analysis was complete in less than seven minutes. The positive ion mass spectra for each component exhibited clear and unambiguous $(M+H)^+$ ions for each benzodiazepine and that there was little fragmentation of the parent ions. This lack of smaller fragments confirmed the gentle nature of the API ionizing process. The API interface offers great promise for extended use in LC/MS tandem instruments, and may well in time become as popular, if not more popular, than the electrospray interfaces. Its great advantages are that it can operate at ambient pressures and temperature, and the gentle nature of the ionizing process. Furthermore, the device does not need the extensive pumping support that is required by the electrospray interface, which simplifies the apparatus and reduces its cost.

Inductively Coupled Plasma LC/MS Interfaces

The most common application of LC/MS tandem instruments with the ICP interface is in the determination of element speciation. In principle, the liquid chromatograph separates the different compounds present in a given mixture, and the mass spectrometer picks out those separated components that contain the specific element or elements of interest. The IC ionization interface is particularly useful for this type of analysis, as it will provide charged atoms of all heavy elements present, and it is the speciation of such elements that is important. The details of the ICP torch, that is used to produce the charged atoms, has already been described and so some examples of its use will now be given.

Shum and Houk [39] employed both size exclusion chromatography and ion exchange chromatography to separate some metal proteins, and two species of selenium, SeO_3^{2-} and SeO_4^{3-}. A detection limit of about 15 pg of selenium was easily achieved. Atomic spectroscopy has become the preferred method for the determination of trace elements due to its high sensitivity and selectivity. However, atomic spectroscopy cannot differentiate between the chemical forms and oxidation states of the element, and these are important, as they determine the toxicity of the substance, and the role the element plays in biosynthesis. If a separation technique is used to differentiate between the different species of the element or its oxidation state, the mass spectrometer can then unambiguously identify the peaks that contain the element of interest. In addition, if isotope ratioing is employed, a quantitative assay can also be accomplished.

The column used for the separation of the metal proteins was 25 cm long, 2 mm I.D., (a GPC column) and that used for the separation of the selenium species was an ion exchange column, 10 cm long and 1.6 mm I.D. The eluent from the columns passed through a fused quartz capillary, 40 cm long, 50 µm I.D., to a direct-injection nebulizer; the distance between the inner capillary and the nebulizer tip was about 25 µm. The ICP/MS tandem instrument was the Elan Model 250 (Perkin Elmer Sciex). A sample of blood serum containing metal proteins of lead, cadmium, zinc, barium, copper, iron and sodium was separated on the size exclusion column, and the elements monitored by, the mass spectrometer. It should be noted that results for all seven metals were obtained from a single injection.

The detection limits for iron, copper, zinc, cadmium and lead were reported to be 3 pg, 0.7 pg, 1 pg, 0.5 pg and 0.5 pg, respectively. It is clear that, although the elements were unambiguously identified, the chemical nature of the eluents can only be assessed from the chromatographic data. It would be possible to employ an electrospray interface to another mass spectrometer, in parallel with the ICP/Ms instrument, and this could provide further information on the structure of the associated protein.

Powell *et al.* [40], developed a sensitive technique, employing an LC/MS tandem instrument, fitted with an API interface, to determine the speciation and the quantitative assay of Cr(III) and Cr(VI). The API interface used a direct injection nebulizer, and they demonstrated a sensitivity of 30, 60 and 180 ng/l for the total chromium, Cr(III) and Cr(VI) respectively. The sample volume injected was 10 µl, and thus the actual mass sensitivity appears to be 0.3, 0.6, and 1.8 pg of the total chromium, Cr(III) and Cr(VI) respectively. A diagram of the sampling arrangement is shown in Figure 9.45.

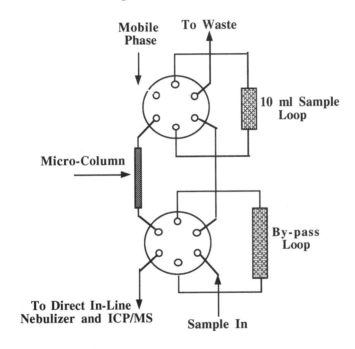

Figure 9.45 Valve Configuration for the Simultaneous Monitoring of Total and Speciated Chromium

The valves were connected so that the 10 µl sample could either be directed through the LC column, which would separate the different chromium species, before the eluent passed to the direct inlet nebulizer, or to a by-pass loop, which would allow the total sample to be passed directly to the nebulizer. The same procedure was used as described previously, the liquid chromatograph separated the different species of chromium, and the ICP/MS identified those peaks that contained the chromium. The two oxidation states of chromium were well separated and both the 52 and 53 isotopes could be clearly monitored and differentiated.

Dissolved organic materials in natural waters play an important role in the normal metabolic processes that take place in aquatic ecosystems. Such material can act as a substrate in heterotrophic processes (metabolic assimilation of decaying vegetable or animal tissue) or as enzymes, vitamins or toxins in different organisms. In natural waters, most of the dissolved organic material comprises humic substances. Humic substances contain a range of polar and dispersive functional groups and can, therefore, interact strongly with heavy metal pollutants as well as PCBs and PAHs and pesticides. Thus heavy metals can be transported by the humic substances over large distances, alternatively, they can trap the heavy metals and act as a detoxification agent. It follows that the speciation of the different heavy metals in water has very important ramifications. On the one hand, a particular species can contribute to water toxicity, and on the other, it can help in the purification of water for drinking purposes.

Rottman and Heumann [41] developed an apparatus to investigate the heavy metal interactions with dissolved organic substances in natural waters, and employed a LC/MS tandem instrument with an ICP interface and a special sample system. A diagram of their sampling arrangement is shown in Figure 9.53. The sample volume was 500 µl, and on injection, the sample passed through a guard column packed with TopOffGel 3PW, and then through a TSK 3000 PW, glass, size exclusion, analytical column. The eluent was monitored by a UV detector and the exit flow passed through a second sample valve (used for calibrating the spike flow) to a Y junction where it was mixed with the spike flow. The mixture then passed directly to the nebulizer of the ICP interface and was then monitored by the mass spectrometer. Water from a bog, river and lake were examined. All the samples indicated clearly fractionated organic substances. The components from the bog water are eluted fairly early in the chromatogram, and as the separation was carried out on an exclusion column, this means that the organic materials were of significantly higher molecular weight (having larger molecular volume) than those from the river and lake water.

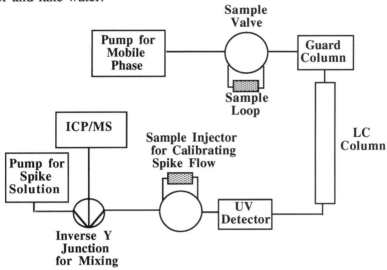

Figure 9.52 The LC/MS Tandem Apparatus for Measuring the Speciation of Heavy Metals in Natural Water

It was also apparent that the distribution of the different heavy metal elements, within the range of humic compounds present, differs quite considerably between the different water sources. The technique is obviously ideal for examining the speciation of the heavy elements, even at the low concentrations normally found in natural waters.

Pergantis *et al.* [42] developed a microscale analytical system, using microbore columns and an ICP interface, with an LC/MS instrument for the detection and estimation of arsenic compounds at the fentogram level. A microscale flow injection system, similar to that already described, was devised to minimize band dispersion that could take place between the injection valve, or column exit, and the nebulizer

The micro-flow nebulizer was designed to operate at low flow rates, *i.e.* in the range between 10 and 150 µl/min. The nebulizer was placed inside the ICP torch, and was designed to give a very fine droplet spray, at the low flow rates demanded by microbore columns. The system was used to separate and identify a range of arsenical, animal feedstock additives, from naturally occurring organic and inorganic arsenic compounds. The microbore column was 1 m long, 1 mm I.D., packed with a reversed phase. Two distribution systems were examined one where the separation was achieved by employing dispersive interactions only, in conjunction with a mobile phase consisting of 0.1% trifluoroacetic acid and 5% methanol in water at a flow rate of 40 ml/min. The other was obtained by exploiting a mixture of dispersive and ionic interactions with the stationary phase, by using an ion pair reagent contained in the mobile phase, which consisted of 1 mM tetrabutyl-ammonium hydroxide and 5% methanol in deionized water. The lower limit of detection was reported as 4 ng/l, which would be equivalent to 4 pg/ml. The volume of sample placed on the column was not clear, but if it were 1 µl, then this would be equivalent to a mass of 4 fentograms. The high resolution of the liquid chromatograph, coupled with the high sensitivity of ICP-MS, makes the tandem combination a very powerful tool for use in many contemporary analytical applications such as this.

The Particle Beam Interface

The particle beam interface involves nebulizing the column eluent, and the solvent free particles of solute, so produced, are then passed into the ionization chamber of the mass spectrometer. Electron impact or chemical ionization spectra can be produced, and the system has been given the somewhat pretentious name of monodisperse aerosol generation interface for chromatography. In addition it has been endowed with an even more pretentious acronym (MAGIC). Willoughby and Browner [43] and Winkler et al. [44] were two of the early groups working on this interface which consists of two parts, the aerosol generator and the momentum separator. A diagram of the aerosol generator described by Creaser and Stygall [45] is shown in Figure 9.53. Nebulization takes place at the end of a fused silica tube, 25 µm I.D., made from the same type of tubing employed routinely in capillary gas chromatography.The liquid jet is formed at the end of the small-diameter tubing, and although simple in design, is very efficient and seldom clogs. Nevertheless precautions must be taken to ensure that no solid particles are carried into the conduit from the column.

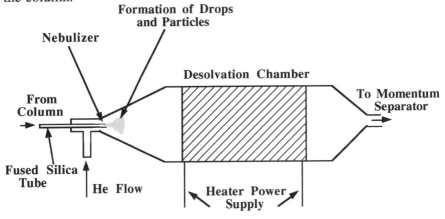

Figure 9.53 Cross-Sectional View of the Aerosol Generator (ref.45)

A diagram of the second part of the interface, the momentum separator, is shown in Figure 9.54.

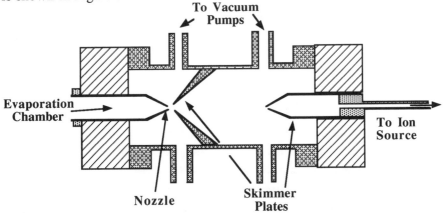

Figure 9.54 A Cross-sectional View of the Momentum Separator (ref.45)

In previous work [46,47]] the most important sources of analyte loss from this type of interface were found to be particle sedimentation, poor nozzle/skimmer alignment and loss due to turbulence. Consequently, the skimmers were designed to provide an undisturbed path from their point of generation, until they were in the ion source of the mass spectrometer. The body of the momentum separator was made of stainless steel and the nozzles and skimmers were machined from 6010 grade aluminum. The separation between the nozzle and the first skimmer was about 10 mm and was adjustable by the use of shims. The first skimmer had a 100° exterior angle and a 95° interior angle with a 0.5 mm orifice. The second skimmer had a 45° exterior angle and a 30° interior angle and a 1 mm orifice. The distance between the two skimmers was also about 10 mm and was also adjustable. Both interlock chambers were pumped with mechanical vane (hot oil) pumps, having a capacity of 21.6 m^3/h. The interface was found to be easy to operate, the skimmer provided little turbulence and it had an overall high transport efficiency. The device was tested by using the normal phase separation of the *cis* and *trans* isomers of retinol acetate. The column was packed with 5 µm silica gel and the solutes eluted isocratically with a mixture of 95% *n*-hexane and 5% diethyl ether at a flow rate of 1 ml/min. The total mass of sample injected was 50 ng and the separation did not appear to be impaired by the interface and the sensitivity that is achieved, compares well with other types of interface. Another example of the use of the interface was for the separation of aliphatic fatty acids using a reversed phase column with gradient development from 80% methanol and 20% water (containing 3% acetic acid) and 100% methanol. There is some indication of peak tailing, although whether that arises from the column or the interface was not clear. This particular example demonstrated that the nebulizer functioned well with aqueous solution containing up to 20% of water. It was not reported whether or not higher water contents work equally as well. The EI spectra exhibited a significant molecular ion for all solutes and was in complete agreement with reference spectra for the respective compound. There was no evidence of thermal degradation occurring in the interface, despite the relatively high temperature of the ion source (*viz.* 240°C).

Cappiello and Famiglini [48] developed the technique further, and designed a micro-flow particle beam interface. They started with the particle beam interface manufactured by Hewlett-Packard, which was similar in form to

LC/MS Tandem Systems

the device described previously, and changed the actual nebulizer to accommodate lower flow rates. The basic design is shown in Figure 9.55, and is a modification of a nebulizer described previously by Cappiello and Bruner [49].

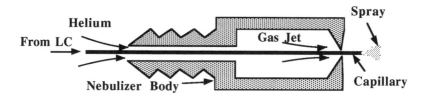

Figure 9.55 The Micro-flow Nebulizer

In this modification the end portion of the coaxial tubing, carrying the helium, has been widened to keep any slow-growing liquid droplets away from the internal gas conduit wall. This enlargement is sharply reduced at the capillary tip, thus keeping the contact surface between the capillary and the wall of the gas conduit to a minimum.

The unique character of this nebulizer is its capacity to nebulize very small liquid flows. In the absence of a helium flow, the liquid is not forced out by the following liquid but remains as an expanding droplet. When the helium flow is commenced, as soon as the droplet size exceeds the diameter of the tip, the energy of the gas breaks the surface tension forces and droplets are formed. The performance of the interface was demonstrated in separating and identification of low levels of caffeine and testosterone.

Clear and unambiguous peaks were produced for 40 pg of caffeine (m/z of monitoring ion of 194) and 600 pg of testosterone (m/z of monitoring ion 124). The flow rate was 2 µl/min. In general, the authors claimed very little nebulizer contamination over long periods of time, better overall performance over a wide range of mobile phase composition, efficient nebulization over a wide range of flow rates and more simple operational procedures.

Permeable Membrane Interface

Before closing this chapter on interfaces for LC/MS tandem instruments, the possibilities of the permeable membrane as a MS inlet system should be mentioned. The membrane interface is strictly only suitable for relatively volatile materials, which would usually be separated by gas chromatography. However, as some environmental samples are contained in aqueous matrixes (*e.g.* the aromatic hydrocarbons in water, or certain pesticides in soil extracts *etc*.) the present stage of the interface development will be briefly considered for possible future use in LC/MS equipment. One of the more recent reports on the design and use of membrane inlet systems was that of Maden and Hayward, [43], who investigated a wide range of different materials for permeability and selectivity.

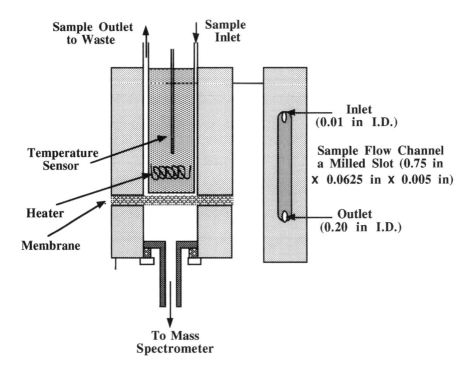

Figure 9.56 The Membrane Interface for the Mass Spectrometer

The materials examined were silicone, two types of latex, polyethylene, polyurethane (polyether), polyurethane (polyester), a copolymer of acrylonitrile and butadiene and polyvinyl chloride. The design of the interface in which they tested the membranes is shown in Figure 9.56. The membrane was positioned between two stainless steel blocks, and the entire interface was heated by a four element heater. The temperature of the interface was monitored and controlled by means of a platinum resistance thermometer. The heaters were carefully located to ensure that the whole block was kept at an even temperature. The sample column eluent or flow injection sample, entered by a narrow channel 0.01 in. diameter, and then passed over the membrane and out through a larger channel, 0.02 in. diameter. The larger exit conduit helped reduce the pressure on the membrane, and prevented mechanical breakdown. The solutes diffused through the membrane under a concentration gradient that was naturally set up, and on the other side of the membrane, evaporated into the high vacuum of the mass spectrometer. The vapor then passed down a heated tube into the ion source.

The results indicated that as one might expect, some membranes were more efficient in the transfer of certain types of materials that others. If the membrane was made of a polymer that was naturally dispersive in nature (*e.g.* silicone membranes) then it would transfer dispersive type compounds such as hydrocarbons efficiently. However, in contrast, the silicone membrane would probably not be so effective for polar compounds such as methanol. Polar membranes such as the polyurethane or polyester would be likely transport polar materials such as alcohols more efficiently than hydrocarbons. It would appear that the membrane interface needs some more development work to be carried out before it might be considered as a competitor for the electrospray or API interfaces. Nevertheless, it is an alternative approach which deserves further consideration and, perhaps with further development, may find use in specific types of analyses.

Finally, in the review of particle beam ionizing techniques, Creaser and Stygall [45] gave a an interesting chart relating the efficacy of the different types of interfaces employed in LC/MS tandem systems and is reproduced in Figure 9.57.

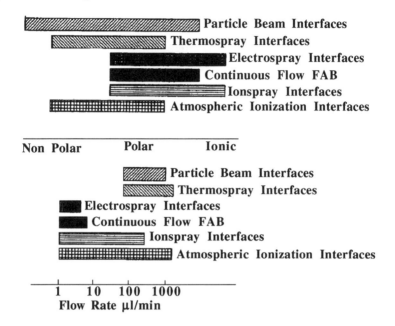

Figure 9.57 The Comparative Performance of the Different Types of LC/MS Interfaces (ref. 50)

It is seen that the particle beam and the thermospray interfaces can cope with the widest range of solvent polarities, whereas the electrospray and ionspray interfaces are not suitable for use with dispersive solvents but are ideal for ionic solutions. It is also seen that the particle beam interface and the thermospray interface operate well with high column flow rates. In contrast the electrospray and continuous flow FAB interfaces are only effective at very low flow rates. The most versatile interfaces with respect to handling a wide range of flow rates are the ion spray and atmospheric ionization interfaces.

Synopsis

The LC/MS tandem systems are more complicated than their GC/MS counterparts, as the eluted substances are mostly involatile, and are not efficiently ionized by electron impact or chemical ionization processes. If volatility is enhanced by raising the temperature, the solutes start to decompose, and the spectra obtained are mixed and generally meaningless. Surface ionization is one solution to the problem, where the material, dispersed on a metal surface, is bombarded with high-energy particles. In secondary ion mass spectrometry (SIMS), Ar^+, O_2^+ and Cs^+ ions, produced by a special ion-gun, can be used to bombard the sample providing the sample is dispersed on a conducting sheet, so that the charge leaks away, and does not affect the subsequent ion focusing. Fast atom bombardment (FAB) is a similar process, in which the colliding atoms are uncharged, but their kinetic energy is sufficient to produce sample desorption. Plasma desorption ionization, promoted by either a radio-active substance such as ^{252}Cf or an inductively coupled plasma can also be effective for ionizing substances on a surface. Laser desorption ionization (LDI) is another method of producing ions of substances of high molecular weight and this can be assisted (Matrix Assisted Laser Desorption Ionization, MALDI) by dispersing the material in substances such as glycerol which help the desorption process. Field desorption ionization involves the ejection of ions by a very strong electric field from a sample deposited on a surface containing points having a very small radius of curvature. The first LC/MS interfaces involved either a wire or belt transport system, that were cumbersome and difficult to operate, but could provide both electron impact spectra and chemical ionization spectra of substances having m/z values of several hundred. Other transport interfaces were developed with special nebulizing jets, and one device had the propensity of providing both laser desorption and secondary ion mass spectra from LC column eluents. In fact, the most important types of contemporary LC/MS interfaces are the direct inlet systems. There are a number of different types of direct inlet system. There is the thermal interface where the column eluent passes through a heated jet into the ion source. Another direct inlet system, probably the most popular today, is the electrospray where a strong electric field acts on the surface of a sample solution as it is sprayed into a dry gas such as nitrogen. This process produces a cloud of charged droplets that rapidly evaporate and, as a consequence, shrink and become smaller in diameter. The accompanying increase in charge density that results from the decrease in volume, surface area, and radius of curvature of the droplets causes very strong electric fields to be formed. As each drop continues to shrink, the electric fields become sufficiently strong to cause the droplets to explode, producing ions. Due to the strength of the electric field, and the large number of ions that are produced, many of the ions that are formed contain multiple charges. As the mass spectrometer measures the m/z values of the ions, this, in effect, increases the mass range of the spectrometer. The device has been the subject of much research and development, including the introduction of novel nebulizing devices, special operating conditions, and modified ion producing techniques. The atmospheric ionization interface operates in a similar way to the electrospray, except it functions at atmospheric pressure, and the evaporated droplets are ionized by a corona discharge, produced by a separate electrode system. Both the electrospray interface and the atmospheric ionization interface have been used extensively in a wide range of analytical applications. The inductively coupled plasma interface and the microwave coupled interface, have also been extensively employed in elemental analysis, particularly in the determination and identification of trace metals. The permeable membrane interface, similar to that described in the chapter on GC/MS, appears to have a limited field of application in LC. The particle beam interface, another direct inlet interface, involves the ionization of dry particles by electron impact, and has been found to function effectively, and appears useful in some specific application areas.

References

1. M. Barber, R. S. Bordoli, G. J. Elliott, R. D. Sedgwick and A. N. Tyler, *Anal. Chem.*, **54**(1982)645A.
2. M. Karas and F. Hillenkamp, *Anal. Chem.*, **60(20)**(1988)2299.
3. M. A. Baldwin, and F. W. McLafferty, *Biomed. Mass Spectrom.*, **1**(1974)80.
4. R. P. W. Scott, C. G. Scott, M. Munroe and J. Hess. Jr., *The Poisoned Patient: The Role of the Laboratory,* Elsevier, New York (1974)395.
5. A. T. James, J. R. Ravenhill and R. P. W. Scott, *Chem. Ind.*, (1964)746.
6. R. P. W. Scott and J. F. Lawrence, *J. Chromatogr. Sci.*, **8**(1970)65.
7. W. M. McFadden, H. L. Schwartz and S. Evens, *J. Chromatogr.*, **122**(1976)389.
8 N.J. Alcock, C. Eckers, D. E. James, M. P. L. Games, M. S. Lant, M. A. McDowall, M. Rossiter, R. W. Smith, S. A. Westwood and H.-Y. Wong, *J. Chromatogr.*, **251**(1982)165.
9. D. E. Games, M. J. Hewlins, S. A. Westwood and D. J. Morgan, *J. Chromatogr.*, **250**(1982)62.

10. M. J. Hayes, E. P. Lanksmeyer, P. Vouroo and B. L. Karger, *Anal. Chem.*, **55**(1983)1745.
11. T. P. Fan, A. E. Schoem, R. G. Cooks and P. H. Hemberger, *J. Am. Chem. Soc.*, **103**(1981)1295.
12. T. Covey and J. Henion, *Anal. Chem.*, **55**(1983)2275.
13. R. D. Voyksner, J. T. Bussey and J. W. Hines, *J. Chromatogr.*, **323**(1985)383.
14. J. S. M. De Wit, C. E. Parker, K. B. Tomer and J. W. Jorgenson, *Anal. Chem.*, **59**(1987)2400.
15. C. R. Blakely and M. L. Vestal, *Anal. Chem.*, **55**(1983)750.
16. J. Via and L. T. Taylor, *Anal. Chem.*, **66(9)**(1994)1385.
17. R. J. McCracken, W. J. Blanchflower, C. Rowen M. A. McCoy and D. G. Kennedy, *Analyst*, **120**(1995)2347.
18. A. Cannavan, W. J. Blanchflower and D. G. Kennedy, *Analyst*, **120**(1995)331.
19. W. J. Blanchflower and D. G. Kennedy, *Analyst*, **114**(1989)1013.
20. C. M. Whitehouse, R. N. Dreger, M. Yamashita and J. B. Fenn, *Anal. Chem.*, **573**(1985)675.
21. M. T. Davis, D. C. Stahl, S. A. Hefta and T. D. Lee, *Anal. Chem.*, **67(24)**(1995)4549.
22. L. R. Snyder and D. L. Saunders, *J. Chromatogr. Sci.*, **7**(1969)195.
23. E. Katz and R. P. W. Scott, *J. Chromatogr.*, **253**(1982)159.
24. R. B. van Breeman, C.R Huang, Z. Z. Lu, A. Rimando, H. H. S. Fong and J. F. Fitzloff, *Anal. Chem.*, **67(21)**(1995)3985.
25. W. J. Blancflower, S. A. Hewitt and D. G. Kennedy, *Analyst*, **119**(1994)2595.
26. M. Elliot, *Recent Advances in the Chemistry of Insect Control, Special Publication No. 53* (Ed. N. F. James) The Royal Society of Chemistry, London, (1985)73
27. I. A. Fleet, J. J. Monaghan, D. B. Gordon and G. A. Lord, <u>Analyst</u>, **121**(1996)55,
28. M. Wilm and M. Mann, *Anal. Chem.*, **68(1)**(1996)1.
29. M. Kohler and J. A. Leary, **67(19)**(1995)3501.
30. R. B. van Breeman, *Anal. Chem.*, **67(13)**(1995)2004.
31. G. Neubauer and R. Anderegg, *Anal. Chem.*,66(7)(1994)1056.
32. J. F. Banks, S. Shen, C. M. Whitehouse and J. B. Fenn, *Anal. Chem.*, **66(3)**(1994)406.
33. M. Yamashita and J. B. Fenn, *J. Phys. Chem.*, **88**(1984)4671.
34. Y. Hua, W. Lu, M. S. Henry, R. H. Pierce and R. B. Cole, *Anal. Chem.*, **67(11)**(1996)1815.
35. J. R. Yates, A. L. MaCormack, A. J. link, D. Schieltz, J. Eng and L. Hays, *Analyst*, **121**(1996)65R.
36. B. Thomson, Tom Covey, B. Shushanm M. Allen, and Takeo Sakuma, Perkin Elmer Corporation, Private Communication.
37. J. Cai and J. Henion, *Anal. Chem.*, **68(1)**(1996)72.
38. E. C. Huang, T. Wachs, J. J. Conby and J. D. Henion, *Anal. Chem.*, **62(13)**(1990)713A.
39. S. C. K. Schum and R. S. Houk, *Anal. Chem.*, **65(21)**(1993)2972.
40. M. J. Powell, D. W. Boomer and D. R. Wiederin, *Anal. Chem.*, **67(14**(1995)2474.
41. L. Rottman and K. G. Heumann, *Anal. Chem.*, 66(21)(1994)3709.
42. S. A. Pergantis, E. M. Heithmar and T. Hinners, *Anal. Chem.*, **67(24)**(1995)4530.
43. R. C. Willoughby, and R. F. Browner, *Anal. Chem.*, **56**(1984)2626.
44. P. C. Winkler, D. D. Perkins, W. K. Wilner and R. F. Browner, *Anal. Chem.*, **60(5)**(1988)489.
45. C. S. Creaser and J. W. Stygall, Analyst, 118(1993)1467.

46. R. F. Browner, A. W. Boorn and D. D. Smith, *Anal. Chem.*, **54**(1982)1411.
47. W. C. Hinds, *Aerosol Technology*, Wiley-Interscience, New York, (1982).
48. A. Cappiello and G. Famiglini, *Anal. Chem.*, **66(22)**(1994)3970.
49. A. Cappiello and F. Bruner, Anal. *Chem.*, **65(9)**(1993)1281.
50. A. J. Maden and M. J. Hayward, *Anal. Chem.*, **68(10)**(1996)1805.

CHAPTER 10

LIQUID CHROMATOGRAPHY/ATOMIC SPECTROSCOPY (LC/AS) TANDEM SYSTEMS

The association of a spectrometer with a liquid chromatograph is usually to aid in structure elucidation or the confirmation of substance identity. In contrast, the various forms of atomic spectroscopy are almost exclusively employed in elemental analysis. The different atomic spectroscopic techniques offer high element sensitivity, unambiguous element identification and, if all the different forms of the technique are included, can accommodate samples contained in almost any type of matrix. Although atomic spectroscopy can easily and accurately identify the element, it can neither determine the form in which the element is occurring, nor can it determine its valency state. As the majority of analyses involving element identification, include identifying the chemical form of the element, the simple atomic spectroscopic analysis is inadequate and some other additional technique is necessary. A preliminary separation process allows the different forms of the element to be isolated, and the individual components, that contain the elements of interest, can then be identified by the appropriate form of atomic spectroscopy. Ideally the two techniques are joined, and the separation and identification carried out sequentially as with (LC/AS or GC/AS) tandem instruments.

Sample introduction in atomic spectroscopy, whether it be to a flame atomic absorption spectrometer, a graphite furnace absorption spectrometer, or an ICP/MIP atomic emission spectrometer, usually involves the nebulization of the sample dissolved in some liquid, and the spray is then passed into the ionization source of the spectrometer. It follows that the eluent from a liquid chromatography column is ideally suited for direct injection into the atomic spectrometer by means of a simple nebulizer. The liquid chromatograph has been coupled with different forms of atomic spectrometer in a variety of ways, and some of these will now be discussed.

Liquid Chromatography Flame Atomic Absorption Spectroscopy (LC/AAS) Systems

The flame AAS is highly element-specific, far more so than the electrochemical detector, but it is not as sensitive. An atomic emission spectrometer, or an atomic fluorescence spectrometer, will readily provide simultaneous multi-element detection, but this is more difficult with the flame AAS. It follows that most LC/ flame AAS combinations are usually set to monitor one element only, throughout the total chromatographic separation.

As the main applications of LC/ Flame AAS is to help determine metal speciation in samples and not merely to identify the presence of a particular element, it is not sufficient to detect the presence of lead, mercury or chromium. One must also to be able to identify the form in which they are present. Depending on the chemical form of a mercury compound, it may or may not be highly toxic. It is also well-known that if chromium is present in the tertiary form it is not particularly dangerous; conversely, in its sixth valency state it is strongly carcinogenic. In order to successfully utilize the LC/AAS combination, both the chromatograph and the spectrometer must be optimized, which has been discussed in some detail in a number of publications [1–3]. It has been claimed [1] that the poor sensitivity that has been obtained from the LC/AAS system, relative to that from the atomic spectrometer alone, was due to the dispersion that takes place in the column. Although substantially true, this misunderstanding arises from the fact that the spectroscopist views the chromatograph as just another sampling device and not as a separation system. The point of interfacing a liquid chromatograph with an absorption spectrometer is to achieve a separation before detection.

LC/AS Tandem Systems 407

Consequently, the important dispersion characteristics are *not* those that take place in the column, but, as has already been discussed in some detail, those that occurs in the interface between the chromatograph and the spectrometer and in the spectrometer itself. These two sources of dispersion, can not only reduce element sensitivity, but also destroy the separation originally achieved in the column. The magnitude of the extra column dispersion is particularly important if high-speed columns, packed with very small particles, are being used since such columns produce very narrow peaks a few microliters in volume. High-speed columns seem ideally suited for use with flame AAS instruments as they can be operated at the flow rates necessary for efficient solvent aspiration into the spectrometer nebulizer. Unfortunately, due to the basic design of most AAS instruments, a significant length of tubing is necessary in the interface if normal operation of the spectrometer is not to be impeded. Katz and Scott [4], solved this problem by the use of low dispersion serpentine tubing as the interface between the exit from the UV detector of the liquid chromatograph, and the spectrometer. A diagram of their interface is shown in Figure 10.1.

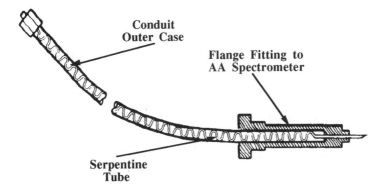

Figure 10.1 The LC/Flame AAS Serpentine Tubing Interface

The principle of low-dispersion tubing has already been discussed, and it is sufficient to say that the outer interface tube was 49 cm long, 0.25 cm I.D. and merely protected the serpentine tube contained inside. The inner serpentine tube had a peak-to-peak amplitude of 1 mm. An example of the chromatograms obtained from a blood sample, monitored by both a UV detector and the flame AAS detector employing the serpentine interface, is shown in Figure 10.2.

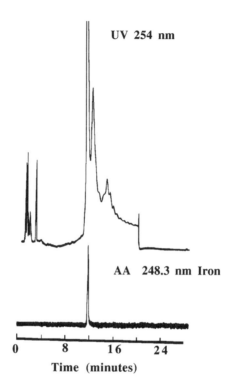

Figure 10.2 Chromatograms of a Blood Sample Monitored by the UV Detector and the Atomic Absorption Spectrometer with the Serpentine Tube Interface

It is seen from the chromatogram monitored by UV absorption that the mixture is complex, and the resolution obtained from the column is rather poor. The number and nature of the peaks shown are almost impossible to identify. In contrast, the flame AAS unambiguously picks out the peak containing iron, and the width of the iron peak clearly substantiates the efficacy of the serpentine interface.

The LC/flame AAS has been employed for a number of years and Holak [5] used it to monitor the separation of a number of mercury-containing drugs, mersalyl, thimerosal and phenyl mercuric borate. Suzuki *et al.* [6] used the technique to identify the heavy metals bound to isoproteins extracted from liver tissue. Robinson and Boothe [7] used the selectivity of the LC/AA system to monitor the alkyl lead compounds in sea water and Messman and Rains [8] separated five alkyl leads, tetramethyl lead (TML), trimethylethyl lead (TMEL), dimethyldiethyl lead (DMDEL), methyltriethyl lead (MTEL) and tetraethyl lead (TEL) in gasoline. An example of their separation is shown in Figure 10.3.

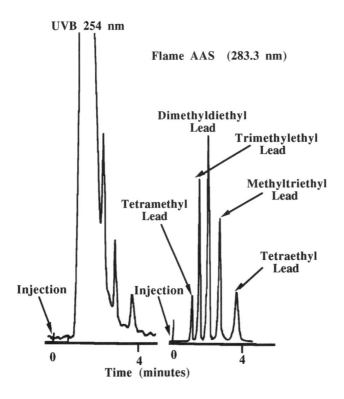

Reprinted with permission from J. D. Messman and T. C. Rains, *Anal. Chem.*, 53(1981)1632, Copyright 1981 American Chemical Society

Figure 10.3 An Example of a Separation of Lead Compounds Demonstrating the Selectivity of the LC/AAS System

It is seen that excellent selectivity can be obtained, eliminating the need for high resolving power columns, as the multitude of interfering hydrocarbons present in gasoline are not detected.

Ajilec and Stupar [9] developed a very similar LC/FAA system for monitoring Fe^{++} and Fe^{+++} in wines. The apparatus they used is shown in Figure 10.4.

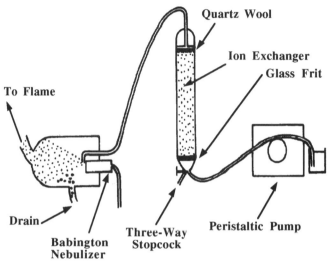

Figure 10.4 An In Exchange LC Coupled to a Babington Nebulizer for the Determination of Iron in Wine (ref. 9)

The simplicity of the tandem system is quite remarkable. The separation of the ion species was achieved with a Dowex 50-X8 (50-100 mesh) column 7.5 cm long and 10 mm I.D. The peristaltic pump provided flow rates between 1 and 4 ml/min. The outlet from the column was fed directly to a Babington nebulizer and the resulting spray passed directly to a flame AA spectrometer. The curves monitored by the flame AA were somewhat asymmetrical but the two iron species were well separated. The limit of detection for iron in wine was found to be about 15 µg/ml with a relative standard deviation of 3%. These limits of detection are not as low as that produced by direct sampling but the LC/MS provides information on speciation as well as the total iron content of the wine

Van Loon and Barefoot {10} reported the development of a transport system [11] that conveyed a sample of the column eluent from the column to the flame for absorption measurements. The basic transport system is diagramatically depicted in Figure 10.5.

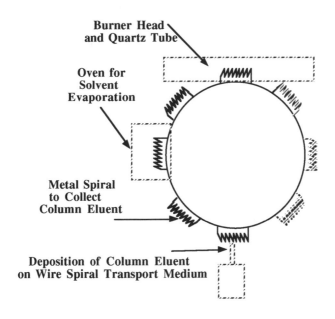

Figure 10.5 The Rotating Spiral Interface for LC/Flame AAS (ref. 10)

It is seen that the rotary transport interface is strongly reminiscent of the carousel interface developed for LC/IR which was described on page 291. The main difference is that the spirals of platinum wire, that hold a sample of the column eluent by surface tension between the coils, are used in place of cups that carried an absorbent in the carousel interface. In fact, the rotating spiral interface also operates in almost exactly the same way. The column eluent passes over a platinum coil, the coil taking up a certain amount of the eluent by surface tension forces. The coil is then automatically moved on, being replaced by the next coil. The solvent evaporates from the first coil and any residual solvent is finally eliminated when it is sequentially moved into the first heated zone. On the next movement the coil is placed in the second heater zone, the burner flame, for absorption measurements. The efficacy of the device was demonstrated in the analysis of proteins containing cadmium, copper and zinc. It was claimed that the sensitivity was two orders of magnitude better than that obtained by employing the conventional nebulizer. However, this would depend somewhat on the design of nebulizer that was taken as the reference interface.

Laser Enhanced Ionization for Measuring Organotin Compounds in Liquid Chromatography Column Eluents

The increasing demands of environmental control have challenged many analytical techniques, and there is a continual search for instruments and devices that will provide higher sensitivity and better resolution. One method for improving sensitivity would be to employ a flame as an atom reservoir, and ionize the atoms by high-energy laser light. As only specific ions of the selected element would be produced, the ionic current would be directly related to the amount of that element present. This procedure has been successfully employed by a number of workers, [12-15]. Epler *et al.* [16] used a laser excitation technique to detect tin compounds and the general layout of their apparatus is depicted in Figure 10.6.

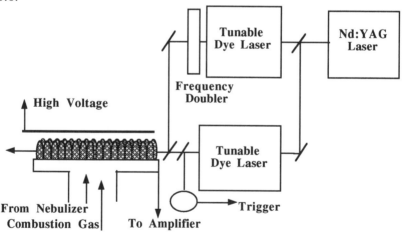

Figure 10.6 Diagram of a Laser Enhanced Ionization Apparatus

Basically, the system consisted of a frequency-doubled YAG laser, operated at 10 Hz, which pumped two dye lasers whose beams were directed into the analytical flame. The first dye laser contained Rhodamine 6G and was tuned to 568 nm. This light was frequency doubled to give ultraviolet light at 284 nm, the wavelength necessary to excite a tin atom from the 5p level at 3428 cm^{-1} to the 6s level at 38629 cm^{-1}, with an energy of 0.086 mJ/pulse.

The second dye laser was tuned to a tin transition, that further excited the tin atoms from the 6s level to a level just below the ionization limit, from which rapid collisional ionization occurs. The 603.77 nm line, with a laser energy of 1.8 mJ/pulse, connects the 6s level with the 7p level at 55187 cm^{-1} which is only 4045 cm^{-1} below the tin ionization limit. The dye used was Rhodamine 640 with 0.1% sodium hydroxide in ethanol. A water cooled cathode, set at a potential of 1000 V, was immersed 2 cm into a premixed air–acetylene flame which was burnt at a 5 cm slot. The two laser beams were co-linearly aligned, 3 mm below the cooled cathode. The ionic current was processed electronically with standard amplifiers.

The assay used to test the method was the determination of tin in drinking water. The separation was carried out on an ion exchange column (Whatman Partisil-10ZSCX cation exchanger) using a mobile phase that consisted of 75% methanol and 25% water, containing 0.05M ammonium acetate buffer (pH, 5.1). The flow rate was 2 ml/min and the sample volume was 20 µl. A normal nebulizer was employed for spraying the sample into the gas stream.

The sample contained tributyltin, tripropyltin and triethyltin, at a concentrations of 260 ng/ml, 310 ng/ml and 300 ng/ml respectively. Peaks were obtained that corresponded to masses of 5.2 ng, 6.2 ng and 6.0 ng of tributyltin, tripropyltin and triethyltin respectively. Sodium and calcium peaks were also disclosed that arose from some naturally occurring impurities in the tap water. The authors reported a detection limit of 3 ng/ml of tin, as tributyltin, which was claimed to be equivalent to a mass detection limit of 0.06 ng of metallic tin.

Laser Excited Atomic Fluorescence for Measuring Organo-manganese and Organo-tin Compounds in Liquid Chromatography Column Eluents

Walton et al. [17] extended the method further, and incorporated a monochromator and photomultiplier as the sensor, so that the device was truly a tandem system, incorporating a liquid chromatograph and a spectrometer. Three different modes were investigated, two utilizing pulsed laser excitation (one measuring fluorescent light emitted from the flame at right angles to the laser beam and the other measuring light reflected back along the line of the laser beam) the third using a high-energy UV light source the emitted light being measured normal to the UV incident beam. The excimer pump laser was the Tachistom Model 800XR, with an output at 308 nm. The dye laser was the Molectron DL 19P (frequency doubled output beam), the output energy/pulse for manganese at 280 nm (Rhodamine dye) was 8 µJ, and for tin at 300 nm (Rhodamine 610 dye) was 5 µJ. The continuum light source was a xenon arc lamp, Model VIXX 300 UV, having an output of 300 W. The flame was an air–acetylene premixed burner, having a 10 mm path length with a pneumatic nebulizer uptake at 1 ml/min for manganese solutions, and 4.0 ml/min for tin solutions.

The separation of the organomanganese species was carried out on a C8 reversed phase column, using a mixture of 65% methanol and 35% aqueous 0.05 M ammonium acetate buffer (pH 4.0), at a flow rate of 1 ml/min. The organotin mixtures were separated on a strong cation exchange column, with a mobile phase consisting of 80% methanol and 20% aqueous 0.2 M ammonium acetate buffer (pH 4.0), at a flow rate of 4 ml/min. Both analytical columns were protected by small guard columns, and the volume of sample placed on either column was 20 µl. The minimum level of detection for organomanganese species was claimed to range from 8 to 22 pg.

The laser-excited atomic fluorescence spectrometer certainly increased the overall sensitivity of the apparatus, and at the same time identified, unambiguously, the metal element that is present. The organomanganese compounds were used extensively as additives to increase the octane rating of gasolines to prevent mixture detonation (as opposed to explosion) but their use was banned some years ago. However, they are still used as boiler and turbine fuel additives and as they are considered environmentally hazardous, they need to be monitored carefully in the various areas in which they are used. The LC/AS tandem instrument incorporating the laser-excited atomic fluorescence spectrometer would appear ideal for this type of assay.

The same tandem instrument was also used for the determination of some organotin compounds. The tin measurements were carried out by resonance (300 nm excitation wavelength/300 nm detection wavelength) and non-resonance (300 nm excitation wavelength/317 nm detection wavelength). The detection limit for non-resonance operation was found to be about 12 ng/ml of tin, which was an order better than that observed for resonance, *i.e.*, 150 ng/ml of tin.

The LC/AAS Tandem Instrument Utilizing the Graphite Furnace Interface

The graphite furnace is a fairly obvious alternative interface to the flame for tandem LC/AS systems, and the off-line use of the furnace as an adjunct to a liquid chromatograph has been reported by a number of workers [18-20]. More recently, Nygren *et al.* [21] described an on-line LC/AAS tandem instrument that utilized a graphite furnace interface, in a relatively simple manner. A diagram of the critical part of the interface apparatus is shown in Figure 10.7 The thermo interface consisted of a graphite tube made from glassy carbon, 28 mm long, 1/8 in. O.D. and 1.5 mm I.D.. The cuvette was made from pyrolytically coated high-density graphite, 18 mm long, 4.5 mm I.D., and was side-heated with integrated electrical contacts. The injection hole in the tube was enlarged to 3.1 mm, to accommodate the interface. The column was connected to the interface by means of 25-30 cm of fused quartz capillary (0.32 mm I.D.). The upper part of the interface consisted of a 1/16 in. length of stainless steel tubing (15 cm long and 0.04 in I.D.), covered by a coiled, porcelain-insulated resistance wire and could be heated to 200°C.

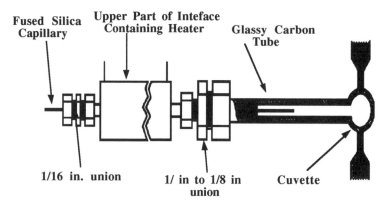

Reprinted with permission from O. Nygren, C. Nilsson and W. Frech, *Anal. Chem.*, **60(20)**(1988)2204, Copyright 1988 American Chemical Society

Figure 10.7 The Thermo Interface

The fused silica tubing terminated inside the wider part of the carbon tube, about 10 mm from the end. The fused silica tubing was secured in a 1/16 in. union, at the top of the stainless steel tubing, and the glassy carbon tube was secured to the 1/16 in. to 1/8 in. union, by means of graphite ferrules.

The graphite furnace was a simple modification of that previously described by Frech *et al.* [22]. The furnace body was made of brass, with enlarged water cooling conduits to maintain adequate cooling when heated over extended periods of time; *i.e.* during the development of a chromatogram. The temperature control was arranged to operate over a maximum period of 33 minutes. The furnace was mounted directly onto the burner holder of an SP 192 (Pye Unicam, Cambridge) atomic absorption spectrometer. The instrument was fitted with a deuterium lamp background correction system. For monitoring organotin compounds, a tin electrodeless discharge lamp was employed. Although a furnace temperature of 2000°C is usually recommended for tin, the maximum tin signal was observed at 1100°C with the interface described. The limit of detection was shown to be about 0.5-1 ng and the linear dynamic range of the response was between 1 and 100 ng.

Many wood-preserving paints and stains are based on a variety of organotin compounds, which are usually dispersed in a white spirit mixture, sometimes in conjunction with appropriate colored stains. Bis(tributyltin) oxide, tributyltin chloride or fluoride and some tributyltin naphthenates, are the most frequently used compounds and are usually present at the 2% level. The tributyltin compounds decompose on storage to dibutyltin and inorganic tin compounds. The tandem instrument described was also successfully used to assay the tributyltin content of different samples of wood preservative.

The graphite furnace interface appears to provide an overall sensitivity that is comparable to that furnished by many other LC/AA interfaces, and obviously could be applied to a number of analyses involving metal speciation. In addition, it would also appear that the integrity of the separation achieved by the liquid chromatograph, is not significantly impaired on passing through the somewhat complex interface.

Inductively Coupled Plasma LC/AS Interfaces

The high performance liquid chromatograph combined with the inductively coupled plasma atomic emission spectrometer furnishes an eminently flexible separating system supported by an element specific detector. However, besides element specificity, the inductively coupled plasma atomic emission spectrometer can also provide very low limits of detection, which, for most elements, will usually be in the sub ppm concentration levels. One of the main problems associated with the combination of the liquid chromatography with the inductively coupled plasma atomic emission spectrometer is the latter's intolerance to the commonly used solvents in liquid chromatography development. Consequently, most LC/ICP-AAS systems have been employed with ion exchange columns as this separation process largely involves aqueous mobile phases that are amenable to the ICP-AAS instrument. The use of acetonitrile or tetrahydrofuran in the mobile phase has usually evoked a change in interface design to accommodate the different solvents.

Dorn and Frame [23] utilized a SEC/ICP-AAS tandem instrument for determining traces of silicon compounds in water. Organosilicone compounds have been used for many years in foods, pharmaceuticals, health and beauty products, adhesives, sealants and many other commercial products. Silicones can appear in the environment in a number of forms, as polar silanols, as nonpolar poly(dimethylsiloxane) (PDMS) polymers or volatile cyclic compounds such as octamethylcyclotetrasiloxane. It follows, that a simple and effective method for determining silicon compounds in environmental samples (*e.g.* surface water) is required and Dorn and Frame assembled an apparatus specifically for this purpose the details of which are shown in Figure 10.8.

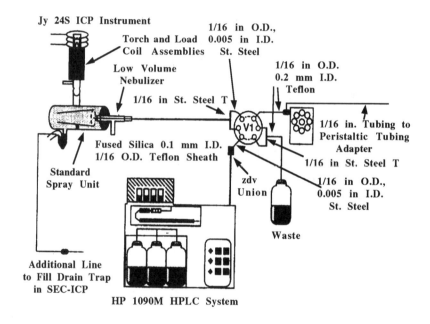

Figure 10.8 The LC/ICP-AAS Tandem Combination Used for The Determination of Silicones in Water (ref. 23)

As the silicon compounds had high molecular weights, the authors employed size exclusion chromatography as the separation technique. The chromatograph was the Hewlett-Packard Model 1090M fitted with an auto sampler. The spectrometer was a Jobin-Yvon sequential JY Model 24S which was equipped with dual monochromators. A standard Scott double-pass glass spray chamber was used with a low dead volume pneumatic nebulizer as an interface between the chromatograph and the ICP-AAS. The SEC separations were carried out on a Phenogel 5 Linear size exclusion column 30 cm long, 2.2 mm I.D. packed with 5 µm particles. Xylene was use as the mobile phase as the silicones were readily soluble and it was possible to maintain a stable plasma in the presence of this solvent. An example of the type of separation that was obtained is shown in Figure 10.9.

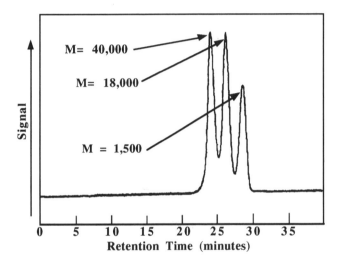

Figure 10.9 The Separation of Some Silicone Polymers by Size Exclusion Chromatography Monitored by ICP-AAS (ref.23)

It is clear that silicone polymers covering a wide range of molecular weights can be separated and quantitatively assayed by the technique. The analytical system was extremely stable, and the variation of the average peak area was less than 10% for 180 individual analyses continuously run over a period of 120 hr. A 10 µg sample (200 µl of a solution containing 50 ppm of the silicone polymer) run 13 times over a 2 week period gave average peak areas that varied by less than 3.7%. The sensitivity of the tandem instrument is illustrated by the elution curves for a pair of silicone samples shown in Figure 10.10. From the upper chromatogram it would appear that the detection limit was about 100 ng. This would be equivalent to a minimum concentration sensitivity of about 0.4 ppm of silicone.

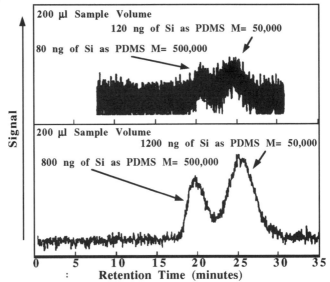

Figure 10.10 The Response of Different Sample of Silicone Polymers Separated by Size Exclusion Chromatography Monitored by ICP-AAS (ref.23)

The authors also demonstrated that by employing ion exchange columns with appropriate mobile phases the ionic silicon compounds could also be separated and detected at about the same level of sensitivity.

The LC/ICP-AAS tandem system has been used to monitor many elements of environmental or toxicological importance. Inorganic arsenic compounds, which are known carcinogens, are metabolized in the body (and in the process detoxified) by methylation to monomethylarsonic acid and dimethyl arsonic acid before renal excretion. It follows, that the determination of the metabolites in urine is a convenient way of monitoring inorganic arsenic exposure. Dietary intake of arsenic is primarily from seafood in he form of the arsenicals arsenobetaine arsenocholine. Intake of seafood results in a minor increase in hydride generating arsenic compounds and a much greater excretion of total arsenic. Another example of the use of the LC/ICP-AAS tandem arrangement for the determination of arsenic and selenium is afforded by the work of LaFrenier *et al.* [24], who used a direct injection nebulizer to infuse the column eluent from a liquid chromatograph into a ICP atomic spectrometer. They used a standard ICP torch, manufactured by Plasma-Therm. Inc., driven by a Model HFS-5000D: 27.12 MHz generator. The monochromator was the McPherson Model 2051, and the spray system was the direct injection nebulizer, manufactured by Ames Laboratory Inc. The nebulizer flow rate was 200 ml/ min., and the auxiliary argon flow 600 ml/min. An example of a separation obtained by the authors is shown in Figure 10.11.

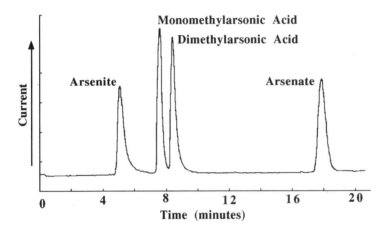

Reprinted with permission from K. E. LaFreniere, V. A. Fassel and D. E. Eckels, *Anal. Chem.*, **59(6)**(1987)879, Copyright 1987 American Chemical Society

Figure 10.11 The Separation of Some Arsenic Species Employing Ion Pairing Reagents

The separation was carried out employing a reversed phase column, with an ion pairing reagent present in the mobile phase. The column (Whatman Partisil 5 ODS 3) was 25 cm long and 4.2 mm I.D. This means the solutes were retained on the stationary phase by a mixture of dispersive and ionic interactions. The mobile phase consisted of 5 mM tetrabutylammonium phosphate in water. The flow rate was 0.75 ml/min, and about 15% of the eluent was passed to the nebulizer and into the plasma. The sample volume was 200 µl of a solution containing 10 µg/ml of each arsenic species. As a consequence, each peak represents the injection of 2 µg of the arsenic compound.

Employing the same chromatographic column, but with a mobile phase that consisted of 90% 5 mM tetrabutylammonium phosphate in water and 10% methanol, the authors easily separated a selenite from a selenate. The flow rate was again 0.75 ml/min, and about 15% of the eluent was passed to the nebulizer. Each peak represented a mass of 0.6 µg and the wavelength monitored to detect the selenium was 196.1 nm. The signal to noise ratio appeared from the chromatogram to be about 20 and so the lower limit of detection for the different selenium species appeared to be about 60 ng.

It is interesting to note that Mürer *et al.* [25] produced a similar separation of the arsenic compounds using an ion exchange procedure but monitored the eluents using a standard atomic absorption spectrometer (PE FIAS 200). the separation obtained is shown in Figure 10.12.

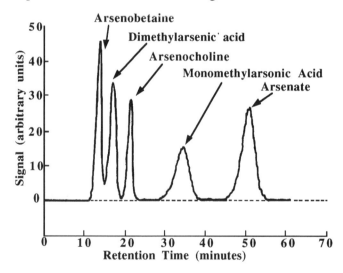

Figure 10.12 The Speciation of Arsenic in Urine by LC/AAS (ref. 25)

The SP-TSK 5PW column was 6 cm long, 8 mm I.D. and the sample was 200 μl of a solution containing 200 μg/ml of each arsenic compound. The signal to noise ratio for all he peaks appears to be at least 10 to 1 and thus the limit of detection would be about 8 μg mass, or a concentration of about 40 μg/ml. The higher sensitivity achieved by the ICP-AAS is clearly evident.

Another metal of considerable environmental interest is tin in its various forms. Tributyltin is added to paints used on ships to prevent 'fouling' but, unfortunately, the tin released into the sea water has been found to target various organisms and in particular molluscs and gastropods. Concentrations as low as 2-3 ng/l in water can produce shell deformation in oysters, and growth inhibition in gastropods. It follows that the estimation of tin in water is an important indication of water contamination. Tin has been assayed employing GC as the separating technique but requires the conversion of the organotin to volatile derivatives. However, in LC/ICP-AES the organic solvents that can be used in the mobile phase are severely limited due to resulting low signal generation and plasma instability. Rivaro *et al.* [26] solved this problem by inserting a hydride generator between the column and the ICP-AES.

The liquid chromatograph was the Varian 5000 fitted with a 200 μl sample valve. The ion exchange column was a Partisil SCX10 25 cm long, 4.6 mm I.D. packed with ion exchange particles 10 μm in diameter. The mobile phase was 0.1 M ammonium acetate, 80% methanol water, pH 7.4 and modified with 0.1% tropolone. The post column reactor used for hydride formation is shown in Figure 10.13. The column eluent passed through a T junction where it was joined by a 0.7 ml/min. flow of 0.3 M HCl provided by a peristaltic pump. The mixture then passed through a second T junction where it was joined by a 0.7 ml/min flow of 0.25M of sodium tetrahydroborate in 1 M NaOH provided by a second peristaltic pump. The mixture then passed into a vessel in which a stream of argon leached out the tinhydride derivatives and passed them to the ICP torch. The vessel also acted as a gas/liquid separator to remove the extracted liquid to waste. The ICP/AES instrument was a Jobin Yvon 24 operated at 1.1 kW.

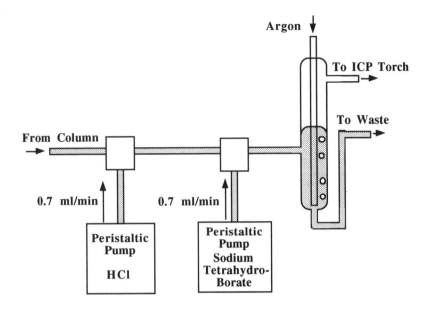

Figure 10.13 The Post Column Hydride Reactor

As with many biological samples, the sample preparation was a little complex. 3g of the shellfish muscle tissue were homogenized with 15 ml of a solution of a chelating reagent (0.05% tropolone solution in methanol). The extraction was carried out twice and the extraction aided by sonication. The tissue remains were separated by centrifugation at 3000 rev/min for 10 minutes. The supernatant liquid was diluted with 100 ml of deionized water and extracted with 30 ml of dichloroethane by shaking in a separating funnel for 5 minutes. The dichloromethane containing the organotins was evaporate to dryness in a rotary vacuum evaporator. The residue was then dissolved in methanol and 200 µl aliquots used for analysis. The method was tested using spiked samples of mussel tissue.

The results obtained are shown in Figure 10.14. It is seen that the components are well separated and the resolution quite adequate for the assay of tin compounds in marine animal tissue.

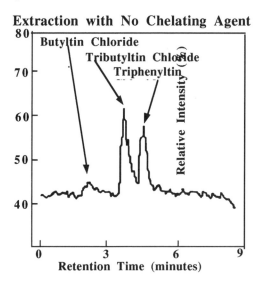

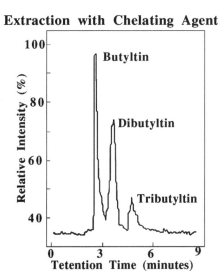

Figure 10.14 The Separation Butyltin Chloride, Tributyltin Chloride and Triphenyltin Chloride With and Without a Chelating Reagent (ref. 26).

The effect of the chelating ragent, however, is quite striking. The presence of the chelating reagent not only allows some compounds to be eluted and measured (without the agent they are held irreversibly on the column) but it also effects the sensitivity of the test very significantly. The authors also demonstrated that the retention times of the solutes were changed and the monobutyltin and dibutyltin were not eluted in the absence of the chelating agent. The detection limit of any specific tin species was reported to be 0.7 ng of tin.

An example of the assays of two samples of mussel digestive glands, one after exposure for 9 days to 5 ng/l of tributyltin, the other unexposed, is shown in Figure 10.15.

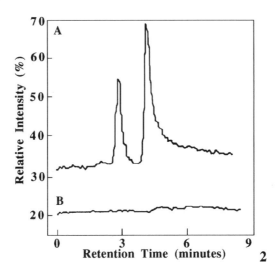

Figure 10.15 The Assay of Mussel Digestive Glands After Exposure to Tributyltin (ref. 26)

It is seen that the method is extremely useful and the amount of uptake of tin by the mussel from such a low concentration of tributyltin in the sea water is quite surprising. The LC/AES tandem instrument is clearly extremely useful for monitoring heavy metal contamination of the environment.

Ahmad *et al.* [27] utilized an alumina column in conjunction with a direct current plasma emission spectrometer (Spectraspan V. Beckman Instruments) to determine the speciation of chromium in aqueous samples. The column was 10 cm long, 4.0 mm I.D. and the outlet of the columns was connected directly to the peristaltic pump inlet system of the spectrometer. The arrangement was extremely simple and provided very useful sensitivity. The peaks for Cr^{III} and Cr^{VI} from four repetitive samples are shown in figure 10.16.

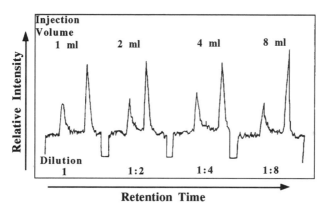

Figure 10.16 Separation of th Tri and Hexavalent Chromium Ions on an Alumina Column (ref. 27)

The same mass was injected in each case but the sample was diluted x2, x4 and x8 and appropriately increasing sample volumes injected onto the column to maintain a constant sample mass. The authors pointed out that the peaks were not dispersed by the increase in sample volume and approximately the same peak height was realized. This is hardly surprising, however, considering the relatively large diameter of the column.

A Liquid Chromatography Atomic Absorption Tandem System Involving Thermochemical Hydride Generation

As a result of the toxicity of arsenic compounds there has always been a considerable interest in methods that could quantitatively assay trace amounts of arsenic. As already discussed, the discovery of the presence of appreciable concentrations of arsenic in a variety of marine organisms has stimulated still further interest in the development of arsenic assays. The electrothermal [28] and cool diffusion flame [29, 30] quartz nebulizers are several orders more sensitive to arsenic than the conventional flame-AAS atomizers. However, these atomizers are restricted to the analysis of arsenic compounds that can form volatile hydride derivatives.

An LC/AAS tandem instrument, involving a novel thermochemical hydride generator for the identification of species of organoarsenic and inorganic arsenic compounds, was described by Blais *et al.* [31]. This on-line interface is based on the spray nebulization of a liquid chromatograph methanolic eluent, followed by the pyrolysis of the solute in a methanol/oxygen flame. The products are then subjected to a gas-phase thermochemical hydride generation, using excess hydrogen, and cool diffusion flame atomization of the resulting arsine into a quartz cell, which is mounted in the atomic absorption spectrometer optical beam. A diagram showing the basic principle of the thermochemical hydride generation interface is shown in Figure 10.17.

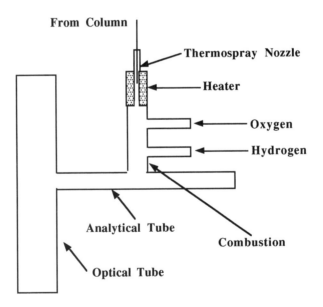

Figure 10.16 The Thermochemical Hydride Generation Interface

The all-quartz main body consisted of an optical tube (12 cm long, 11 mm O.D. and 9 mm I.D.), which was positioned in the optical path of the AA spectrometer. The thermospray nebulizer jet was situated in a side tube, joining the analytical flame tube, and around which was wound a heater coil. Between the heater and the analytical flame tube were two T junctions, 2.5 cm apart, the first carrying hydrogen gas and the down-stream port oxygen. The combustion chamber-thermospray assembly met the analytical flame tube at an angle of 45°. The arrangement had a somewhat complex routine to produce 'smooth' ignition.

It was shown that arsenic pentoxide was derivatized, and no signal was observed in the absence of the post-thermospray hydrogen or in the absence of the cool diffusion flame, which confirmed the nature of the thermochemically mediated arsine-generation mechanism. It was shown that the solvents used with both reversed phase and normal phase separations were compatible with the interface, and the system was relatively insensitive to small changes in the operating conditions. The system was used to successfully separate a mixture of arsenobetaine, arsenocholine and tetramethyarsonium.

The interface was shown not to denigrate the separation obtained from the column, and that a very useful sensitivity was obtained. The absolute limits for the arsenic compounds, arsenobetaine, arsenocholine and tetramethyarsonium, were claimed to be, 13.3, 14.5 and 7.6 ng respectively. Inherently, this arrangement has also the advantage of both a low purchase cost and low operating costs, and this makes it an attractive choice for routine analysis.

The Moving Wheel Liquid Chromatography Helium Microwave Induced Plasma Interface

This interface involves a 'wheel' transport system, used in conjunction with an atmospheric pressure helium plasma interface, specifically designed for halogen-selective detection. The device was described by Zhang et al.[32] and the layout of their apparatus is shown in Figure 10.18.

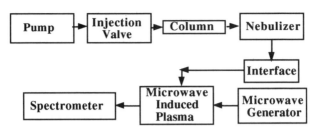

Figure 10.18 The Moving Wheel LC Helium Microwave Induced Plasma Interface

The layout of the apparatus was fairly straightforward and involved a standard LC pump, column, and nebulizer. The output from the nebulizer passed through the transport interface, which will be discussed separately, and then into the microwave–induced plasma torch which was driven by an appropriate microwave generator. Light from the torch passed through an optical system to the spectrometer, and thence to a photo–multiplier and its associate electronics. The halides and oxohalogenated materials were separated by ion exchange chromatography and the column eluent directed as a mist onto a continuously moving wheel interface. The aqueous solvent was evaporated in a stream of hot nitrogen leaving the solute as a solid residue on the surface of the wheel.

The moving wheel then carried the dry sample into the plasma where it was volatilized, atomized and excited. The plasma was a small-volume helium microwave-induced plasma, and was operated at 100 W, with a helium support gas flow of 3.1 l/min.. The interface consisted of a stainless steel wheel 77.7 mm O.D. and 62.9 mm I.D. and 1.5 mm thick. The wheel was friction–driven by a high-torque electric motor. There were a number of auxiliary parts attached to the transport system including a heating coil. The microwave resonant cavity was connected at right angles to the interface in such a way that the edge of the wheel penetrated into the resonant cavity structure. The plasma was produced by a low-power microwave generator, and the torch was constructed from a fused silica tube, 7 mm O.D. and 2.5 mm I.D.

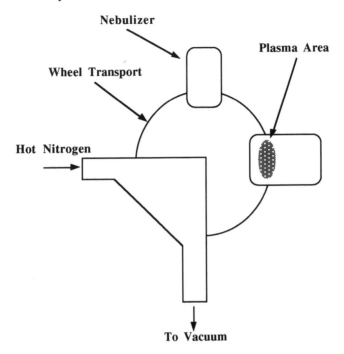

Figure 10.19 The Wheel Transport Interface

The authors demonstrated the efficacy of the device by separating a range of different halogen and oxy-halogen ions. The sensitivity realized varied widely from 4 ng/s for sodium iodide to 300 ng/s for sodium bromide. These limits of detection do not approach those of ion chromatography, but the system does offer selective detection, which would allow the halogen peaks of interest to be picked out of a complex group of coeluting ions containing different elements.

Synopsis

LC/AS techniques are almost exclusively employed in elemental analysis. The different forms of atomic spectroscopy offer high element sensitivity unambiguous element identification, and can accommodate samples contained in almost any type of matrix. The early attempts at developing LC/AS tandem instruments was to couple a standard liquid chromatograph directly to a flame atomic absorption spectrometer. The use of the flame

AA in this way was extended to laser excited atomic fluorescence, which was originally employed for measuring organic tin and manganese compounds eluted from an LC column. Sensitivities ranged down to 60 pg of the metallic element. The graphite furnace has also been used as an interface between the liquid chromatograph and the atomic spectrometer, but requires a specially fabricated thermo sample injector. The sensitivity of the system to tin was about 0.5–1.0 ng per peak. The atomic spectrometer has also been interfaced to a liquid chromatograph by a transport device involving a rotating wheel. The eluent from the column was sprayed on the perimeter surface of a rotating wheel and the solvent evaporated. The wheel rotated and placed the sample into a microwave induced plasma, where the element was ionized and measured in the usual manner. The sensitivity varied widely between different elements, but was of the order of about 10 ng of sample.

References

1. D. R. Jones, H. C. Tung and S. E. Manchen, *Anal. Chem.*, **48**(1976)7
2. D. R. Jones and S. E. Manchen, *Anal. Chem.*, **48**(1976)1897.
3. J. A. Koropchak and G. N. Coleman, *Anal. Chem.*, **52**(1980)1252.
4. E. D. Katz and R. P. W. Scott, *Analyst*, (**March**)(1985)253.
5. W. Holak, *J. Liq. Chromatogr.*, **8(3)**(1985)563.
6. K. T. Suzuki, H. Sunaga and T. Yajma, *J. Chromatogr.*, **303**(1984)131.
7. J. W. Robinson and E. D. Boothe, *Spectrosc. Lett.* **17(11)**(1984)689.
8. J. D. Messman and T. C. Rains, *Anal. Chem.*, **53**(1981)1632.
9. R. Ajlec and J. Stupar, *Analyst* **114**(1989)137.
10. J. C. Van Loon and R. R. Barefoot, *Analyst,* **117**(1992)565.
11. L. Ebdon, S. Hill and P. J. Jones, *Anal. At Spectrom* **2**(1987)205.
12. J. C. Van Loon, *Am. Lab.* **5**(1981)47.
13. I.S.Krull, *Liquid Chromatography in Environmental Analysis*, (Ed. J. F. Lawrence) Humana Press (1984)169.
14. Y. K. Chau, *Sci. Total Environ.*, **49**(1986)305.
15. L. Ebdon, S. Hill and R. W. Ward, *Analyst (London)*, **112**(1987)1.
16. K. S. Epler, T. C. O'Haver, G. C. turk and W. A. MacCrehan, *Anal. Chem.*, **60(19)**(1988)2062.
17. A. P. Walton, Guor-Tzo Wei, Z. Liang, R. G. Michel and J. B. Morris, *Anal. Chem.*, **63(3)**(1991)232.

18. L. Ebdon, S. Hill and R. W. Ward, *Analyst (London)*, **112**(1987)1
19. F. E. Brinkman, W. R. Blair, K. L. Jewett and W. P. Iverson, *J. Chromatogr. Sci.,* **15**(1977)493
20. K. L. Jewett and F. E. Brinkman, *J. Chromatogr. Sci.,* **19**(1977)583.
21. O. Nygren, C. Nilsson and W. Frech, *Anal. Chem.*, **60(20)**(1988)2204.
22. W. Frech, E. Lunberg and A. Cedergran, *Can. J. Spectrosc.* **30**(1985)123.
23. S. B. Dorn and E. M. S. Frame, *Analyst*, **119**(1994)1687.
24. K. E. LaFreniere, V. A. Fassel and D. E. Eckels, *Anal. Chem.*, **59(6)**(1987)879.
25. A. J. L. Mürer, A. Abildtrup, O. M. Poulsen and J. M. Christensen, *Analyst,* **117**(1992)677.
26. P. Rivaro, L. Zaratin, R. Frache and A. Mazzucotelli, Analyst, 120(1995)1937.
27. S. Ahmad, R. C. Murthy and S. V. Chandra, *Analyst,* **115**(1990)287.
28. B. Welz and M. Melcher, *Analyst*, **108**(1983)213.
29. D. D. siemer, P. Koteel and V. Jawiwala, *Anal. Chem.*, **48**(1976)836.
30. J. Dedina and I. Rubeska, *Spectrochim. Acta.*, **35B**(1980)119
31. J.S.Blaise,G. M. Montplaisir and W. D. Marshall, *Anal. Chem.*, **62(10)**(1990)1161.
32. L.Zhang,J. Carahan, R. E. Winnans and P. H. Neill, *Anal. Chem.*, **61(8)**(1989)895.

CHAPTER 11

LIQUID CHROMATOGRAPHY/NUCLEAR MAGNETIC RESONANCE SPECTROSCOPY (LC/NMR) TANDEM SYSTEMS

In principle, the association of the liquid chromatograph with the NMR spectrometer should be a very powerful analytical tool for the separation and identification of unknown substances. There are, however, some serious difficulties in the association of the two techniques, more so, perhaps, than with other spectroscopic techniques. The main problems that must be solved were outlined many years ago by Bayer *et al.* [1] who were one of the first research groups to combine the liquid chromatograph on-line with the NMR spectrometer. More recently the basic problems have been again reiterated by Nicholson *et al.* [2]. The general challenges facing the designer of an LC/NMR tandem instrument are as follows.

1/ As a rule, the sensitivity realized in NMR measurements has been significantly lower than that achieved in UV spectroscopy. However, with the increase in magnetic field strengths, and the introduction of analog/digital converters having higher dynamic ranges, there has been a marked increase in the sensitivity available for LC/NMR tandem systems. The recent introduction of over-sampling and digital filtering techniques in data acquisition regimes has allowed spectral windows to be contracted to include only regions of interest, without the problem of signals from outside the region folding in. Furthermore, it will be seen later that the use of NMR micro-cells, if constructed correctly and fabricated from the materials having appropriate magnetic properties, can also provide a significant increase in sensitivity.

2/ The intensity of the NMR signal with on-line monitoring depends on the flow rate, and as the flow rate increases, the signal decreases. However, according to Bayer *et al.* [1], the reduction in signal can be restricted to a reasonable level at flow rates between 0.5 and 2 ml/min.

3/ In order to realize high NMR resolution, the magnetic field throughout the sample must be very homogeneous and to achieve this, the sample tube is usually spun at fairly high speeds. So far, this has proved impossible in flow-through cells and consequently, in the past, considerable resolution has been lost in these types of cell.

4/ The dynamic range of the NMR measurements is impaired due to ^{1}H NMR signals from eluting solvents. Consequently, efficient solvent suppression techniques are necessary that will also accommodate the solvent composition changing during the period of chromatographic development. This problem, will also apply to ^{13}C nuclei and the solvent can again interfere with the spectrum of the solute. In the case of proton spectroscopy, solvents may be chosen that do not contain protons, such as carbon tetrachloride or deuterated solvents, but the former restricts chromatographic performance and the latter can become very expensive if standard LC columns are used.

One of the more common solvents used in reversed phase LC is acetonitrile, which gives rise to a single resonance in the ^{1}H NMR spectrum at about $\delta 2.0$, which can be easily suppressed. However, this suppression leaves the ^{13}C satellite peaks, from the 1.1% of molecules with the naturally abundant ^{13}C isotope at the methyl carbon. These satellite peaks are often larger than those from the sample and thus they must also be suppressed. Suppression has been achieved in two ways by setting the suppression radiation frequency over the central peak and the two satellite peaks in a cyclic procedure or, alternatively, if an inverse geometry probe is used which includes a ^{13}C coil, the ^{13}C decoupling is possible and this collapses the satellite peaks under the central peak, enabling conventional single peak suppression. The use of small-bore columns would significantly reduce the solvent consumption, and render the use of special and expensive solvents more economically viable, but such columns would demand very small cell volumes and minimum extra-column dispersion, which leads to the next problem.

5/ The use of long transfer lines and relatively large volume measuring cells can cause loss in chromatographic resolution and thus the transfer of column eluent to NMR cell can become difficult.

6/ Due to the physical arrangement of the components of the liquid chromatograph and the NMR spectrometer, the association of the two instruments is often physically difficult.

Nevertheless, over the years, with improved techniques such as higher-resolution NMR spectrometers, operating at 500 and 800 MHz and using superconducting magnets, together with ^{13}C spectroscopy, practical LC/NMR systems have been successfully developed, and are now commercially available.

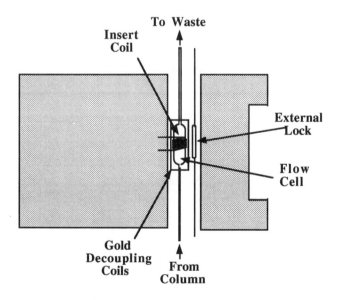

Figure 11.1 The Basic NMR Spectrometer Flow-through-sample Cell

For historical interest, a diagram of the original flow-through cell of Bayer [1], which incorporates the basic principles of the earlier LC/NMR systems that operated with electromagnets, is shown in Figure 11.1.

The cell-volume had to be kept as small as possible to minimize band dispersion, but, at the same time, the geometry of the flow cell had to provide optimum synchronization with the NMR sensing coil. Consequently, in the older electromagnetic instruments the wall of the cell had to be straight and parallel with the axis of the coil. Such flow-through cells had volumes in excess of 400 µl (too large for modern LC columns), and thus at a flow-rate of 1 ml/min the average residence time of the sample was about 25 sec.

Modern high-sensitivity, high-resolution NMR instruments have super-conducting magnets, and thus have entirely different sensing coils and cells. The column eluent cannot pass vertically through the field entering at the top of the instrument and exiting at the base and, furthermore, because of the geometry of the superconducting coils and cryostatic environment, the coils are no longer wound concentrically on the sample tube, but are often in the Helmholtz configuration (this configuration, however, is not always optimum as will be seen later). Consequently, the eluent from the column normally passed through a U-shaped conduit, one limb of which will be the actual sample cell.

Modern flow-through cells have volumes ranging from 20 to 50 µl but recently, Sweedler [3], has claimed the successful use of sample tubes with capacities that are measured in nanoliters, with an accompanying improved signal to noise ratio of about two orders of magnitude. The sensing coil is only 1 mm long, 0.5 mm O.D. and the capillary running through the coil has a sample capacity of 5 nl. The secret appears to be in surrounding the coil/capillary assembly in a commercial perfluorinated organic liquid that has the *same magnetic susceptibility* as the copper micro-coil. This device, which will be discussed in more detail later, is claimed to provide a more uniform magnetic field in the sample region leading to improved resolution and peak shape. This discovery could quickly lead to greatly improved LC/NMR systems.

The Modern LC/NMR Tandem Instrument

There are four methods of operating LC/NMR combination equipment and so the modern tandem instrument should be capable of functioning in all four modes. The four regimes are, 1. Continuous flow, 2. Stop-Flow monitoring, 3. Time-sliced stop-flow monitoring and 4. Peak collection and off-line monitoring.

1. Continuous Flow Monitoring

Continuous-flow measurement is the simplest method of monitoring an LC eluent. Unless enriched compounds are employed, this mode of operation is usually only suitable for ^{1}H or ^{19}F NMR. If gradient elution development is needed to separate the materials in a reasonable time, the resonance positions of the solvent peaks will continuously change. This usually means a preliminary run must be carried out to determine the optimum suppression program that will be necessary to obtain satisfactory spectra.

2. Stop-flow Monitoring

If either the retention times of the solutes are known or a UV detector is also used to monitor the eluent as the separation proceeds, the solute bands can be stopped when they are situated in the NMR measuring cell. Employing stop-flow monitoring, all the normal techniques of high-resolution NMR can be used, and as the diffusion of solutes in liquids is very slow, little chromatographic resolution is lost.

3. Time-sliced Stop-flow Monitoring

Time-sliced monitoring is similar to stop-flow monitoring, except that various portions of the peak are examined as they enter the NMR sensor cell. In this way the spectra from different parts of the peak can be examined which can be very helpful if the chromatographic resolution is not complete.

4. Peak Collection and Off-line Monitoring

This procedure is very similar to stop-flow monitoring, except that each peak is stored in its own respective sample loop as it is eluted, and later displaced into the NMR cell for NMR examination. This latter technique also allows all the normal techniques of high-resolution NMR to be used.

The Basic LC/NMR Tandem System

A diagram of a modern LC/NMR tandem system is shown in Figure 11.2. For the most part, modern LC/NMR systems are not really used as in-line devices, but more often function as automatic fraction collectors that pass the sample to the NMR spectrometer for examination by a normal static, or stop-flow, procedure. They are unique in that the cell is designed for a flow-through function and, as a consequence, must have a small sample volume.

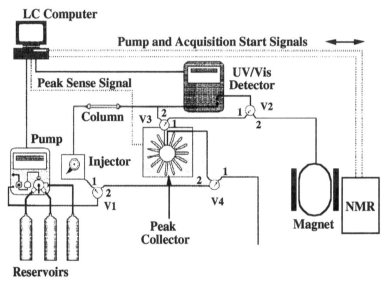

Courtesy of Varian Instruments Inc.

Figure 11.2 The Layout of Modern LC/NMR Tandem System

They also have special valving arrangements, some of which can be quite complex, to direct the column and solvent flows appropriately for fraction collection and sample monitoring. Nevertheless, as already stated, the spectra are generally run on a stop-flow principle; the sample being held in the cell while the spectra are obtained. This is often a direct result of the sample being generated by the chromatograph faster than the spectrometer can acquire the necessary data from it. The sample cell must be small enough so as not to destroy the chromatographic resolution, but still be large enough to contain sufficient sample to allow the spectrometer to acquire valid data in the residence time available. To date, these two criteria have not yet been efficiently achieved. It is seen in Figure 11.2, that the tandem system is basically a liquid chromatograph and a NMR spectrometer joined by a valving system. The valving system is designed to allow any given peak to be passed to the spectrometer or, if the spectrometer is acquiring data, to be stored in a sample loop until the spectrometer is free. A diagram of the flow control and sampling unit is shown in Figure 11.3.

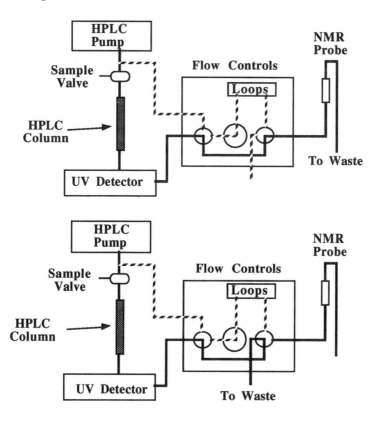

Figure 11.3 The Flow Control and Peak Sampling Unit for a LC/NMR System

The flow control and sampling unit is designed to provide three alternative methods of operation. First the eluent from the column can be made to flow directly from the UV detector to the NMR sample tube. Under these circumstances the spectra can be continuously monitored during the development of the separation. The success of this alternative will depend on a large number of factors: the cell volume (which will also determine the residence time of the solute and thus the measuring period), the sample size, the column flow rate, the resolution of the NMR spectrometer and the rate of data acquisition by the computer. In general, unless the new micro-cell facilities mentioned earlier are exploited, this procedure will rarely be successful, particularly if microbore columns are used and multi-component mixtures are being examined.

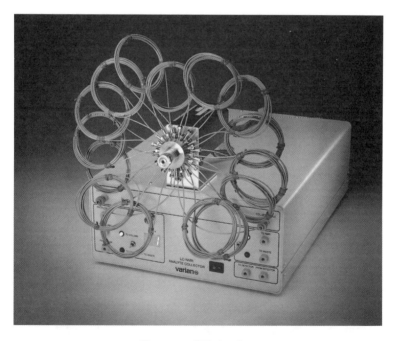

Courtesy of Varian Inc.

Figure 11. 4 The Flow Control Device for the LC/NMR System

The second alternative is to direct only those samples that are of particular interest to the NMR. Each sample can then be trapped in the cell and data acquired on a stop-flow basis, allowing an adequate number of pulses to be chosen that will provide the required resolution. Subsequently, the sample can be expelled from the cell, with solvent supplied directly from the chromatography pump. The third alternative is to direct the eluent from the column to a sample loop where it can be stored until the spectrometer is available to take data. This is basically a fraction collecting system. If necessary, a number of solutes can be stored in different loops and they can be examined when convenient. When the data has been acquired from one sample, the solute stored in the next loop can then be displaced into the NMR cell and examined. Samples that have been examined can either be displaced to waste or collected for further examination. A photograph of the Varian flow control device for their LC/NMR tandem instrument is shown in Figure 11.4.

An interesting example of the use of the Varian LC/NMR system in handling two unresolved peaks is shown in Figure 11.5.

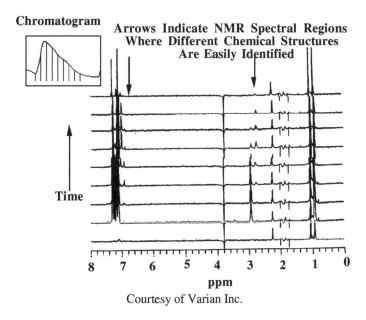

Courtesy of Varian Inc.

Figure 11.5 NMR Spectra Taken Across an Unresolved Peak Using the Varian Interface System the LC/NMR System

The spectra were obtained using Varian's new LC-NMR Microflow probe, which was developed to maintain good chromatographic resolution and also provide the necessary high NMR sensitivity. The cell had a sensor volume of 60 µl, which is far too large for microbore columns, but can be used successfully with wider bore columns. The spectra were obtained by stopping the column flow when the peaks were in the sensor cell, and running the spectra in the normal way. Using this technique, it is clear that a multiple peak can be scanned (a form of peak slicing). It is seen from the insert chromatogram obtained from the UV detector that there are two unresolved solutes present in the peak, and the NMR spectra taken across the peak clearly differ as shown in Figure 11.5.

Even though the components are not resolved, a shoulder on the main component peak yields an identifiable spectrum that is quite different from that taken from the front portion of the peak. The areas of the NMR spectra that demonstrate the change that results during the final elution of the first peak, and the start of the elution of the second peak, are depicted by arrows. This type of tandem system can provide unambiguous confirmation of solute identity that is ideal for forensic purposes.

Capillary LC-NMR Coupling: High Resolution ^{1}H NMR Spectroscopy in the Nanoliter Scale

Although coupling the liquid chromatograph with the NMR spectrometer produces one of the most powerful analytical systems available, it has done so in the past at a price. If the tandem combination is to be employed with the modern miniaturized LC systems, then the detection, or sensing volume, must be greatly reduced, and this reduction is usually accompanied by a dramatic loss in NMR resolution.

In general there are two types of sensing coils used in NMR probes, the 'saddle' coil and the 'solenoid' coil. These two types of coil take the form depicted in Figure 11.6A. There are also two basic types of flow cell that are used in LC/NMR and these are depicted in Figure 10.6B. The choice of coil is largely determined by the choice of flow cell. However, modern chromatographic systems are tending to smaller and smaller bore columns and thus the cell volume is required to be as small as possible. As a consequence the flow cell show on the right of Figure 11.6B used in conjunction with the solenoid type coil shown in Figure 11.6A is becoming the more popular combination for LC/NMR tandem instruments.

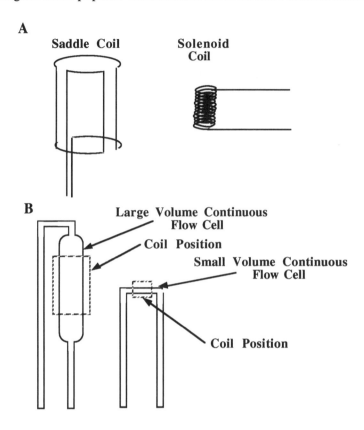

Figure 11.6 Probe Characteristics for LC/NMR Operation.

Behnke *et al.* [4] has investigated the different interfaces that have been proposed for LC/NMR tandem systems and reported on their advantages and disadvantages. In the past, on-line LC/NMR measurements have been made using 4.6 mm I.D. columns in conjunction with relatively large probe cells, which had a detection limit of about 5 nmol, when used with a 600 MHz spectrometer and a line width of 102 Hz. The major problem with such systems are, as already discussed, the elaborate solvent signal suppression that is necessary as part of the spectrum is masked by the solvent. Employing supercritical fluid chromatography allows the use of non–protonated solvents, and thus the whole ^{1}H spectrum can be obtained without the need to employ restricted solvent windows for measurement. The probes used in both these methods employed saddle-type radio frequency coils, as shown in Figure 11.6A. Over the years the volume of the NMR cells which was used with electromagnets have been reduced from 200 to 500 µl, to about 20 to 200 µl, which are the cell sizes commonly used with super conducting magnets.

A number of conditions point the way to increased NMR sensitivity and reduced sensor cell volume. On the assumption that the majority of the noise arises from the resistance of the sensing coil, the solenoid type of coil, as depicted in Figure 11.6A, should theoretically increase the signal to noise by a factor of three over that from the saddle type coil. Further, the use of microbore columns, with their very small volume flow rates, should allow any type of deuterated solvent to be used, without making the operation of the system unreasonably expensive. This would also eliminate the need for suppressing techniques and devices. The basic layout of the LC/NMR apparatus described by Behnke *et al.* [4] for use with small-bore columns is diagramatically depicted in Figure 11.7.

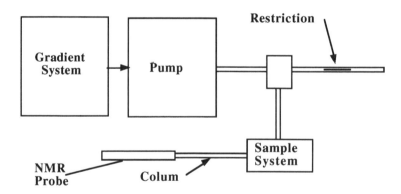

Figure 11.7 Apparatus for Capillary Column LC/NMR Tandem Systems

The liquid chromatograph was situated about 3 m from the NMR spectrometer and the column eluent was split by a stainless steel T piece between the column and a 15 cm of 50 mm I.D. fused silica capillary tube which had sufficient flow impedance to provided the required split ratio. The column flow then passed to an injection valve, and then to a length of fused silica tubing, 70 cm long, and 315 µm I.D., 12 cm of which was packed with reversed phase particles, 5 µm in diameter. A small portion of the polyimide coating was removed to act as the NMR detection window.

Optimization of LC and NMR Operating Conditions for Tandem LC/NMR Systems

The relationship between chromatographic peak width and NMR cell volume was investigated theoretically by Grifffiths, [5], who arrived at some conclusions that suggested that adjusting the chromatographic peak width to fit the NMR sample tube volume would be advantageous. This is a little difficult as the chromatographic peak width increases continuously with the retention time (or volume) of the solute and the resolution obtained from the chromatographic column increases as the peak width decreases. The more efficient the column, the more narrow the peak at any given retention time. For a column of given efficiency (determined, among a number of other parameters, by the particle size and method or preparation) the smaller the radius, the smaller the peak volume.

It was found that the interrelationship between peak width in time, peak width in volume, column dimensions and retention time can be very complex. The maximum sensitivity was realized when the chromatographic peak width was about one third of the cell width. It is clear that if only one solute, or a limited number of solutes that were eluted around the same retention volume or retention time, were the only peak or peaks of interest, then the adjustment of the chromatographic conditions to provide a peak one third of the cell volume in order to provide optimum sensitivity, might be practical and worthwhile. However, this would need to be arranged with a clear understanding that the chromatographic resolution must not be compromised, or the point of using liquid chromatography in the first place would be lost. If NMR spectra were required for a number of peaks that extended over a significant retention period, then adjusting the operating condition would be far more difficult and likely to be fruitless. Under such conditions, sample size, column diameter, and the shape and size of the NMR cell itself, may need to be changed to obtain greater sensitivity.

Employing 750 MHz Spectrometer in LC/NMR Tandem Systems

As the development of NMR spectrometers has continued, stronger and stronger magnetic fields have become attainable and, as a consequence, the sensitivity of the spectrometer has increased which, in turn, has rendered the LC/NMR tandem instrument more practical. Sidelmann *et al.* [6] coupled the liquid chromatograph to a 750 MHz NMR spectrometer to examine the positional isomers of 6,11-dihydro-11-oxodibenz[b,e]oxepin-2-acetic acid. NMR spectroscopy is perhaps the only simple technique for examining this type of problem. The details of the chemistry need not concern us here, but the sensitivity and spectra quality that was obtained are both interesting and exciting.

The chromatograph was fitted with a Bruker LC22C pump and an LC33 Variable wavelength UV detector operated at 200 nm. The column was a Spherisorb ODS-2, 25 cm long, 4.6 mm I.D., and packed with 5 μm particles. The mobile phase employed was acetonitrile : 0.2 M potassium phosphate (pH 7.4) : deuterium oxide (21:10:69) at a flow rate of 1 ml/min. and an excellent separation was obtained. The NMR spectrometer was the Bruker DMX 750 MHz model fitted with a ^{1}H flow probe. The results were obtained from 1.5 μl sample of a 0.15% solution of the solutes. The position of all the groups were unambiguously identified and it is probable that the tandem combination of liquid chromatography and high resolution NMR was only really effective way of examining this chemical system.

Sidelmann *et al.* [7] also employed a LC/NMR tandem system to characterize the positional isomers and anomers of 2-3- and 4-fluorobenzoic acid glucuronides in equilibrium mixtures. The apparatus was essentially that described previously, and the NMR chromatogram they obtained was by directly coupled continuous-flow 750MHz-LC-NMR of the O-(3-fluorobenzoyl)-D-glucopyranuronic acid mixture The NMR peak assignments were obtained in the same way as those previously. A preliminary separation by liquid chromatography allowed the spectra of each component to be identified, and then the spectra so obtained could be used to assign the NMR peaks present in the spectrum of the mixture.

Sweatman *et al.* [8] examined the general performance of a directly coupled liquid chromatograph, with a 600 MHz NMR spectrometer, and found that the ^{1}H sensitivity was about 85 ng for ethyl benzene, and the ^{19}F sensitivity about 40 ng for 4-trifluorothymine. These levels of sensitivity are more than adequate for the effective use of LC/NMR tandem instruments, but much still depends on the design of the interface and, in particular, the volume of the flow cell. The tendency has been to concentrate on micro–columns and packed capillary columns, which renders the interfacing procedure even more difficult. In the majority of samples presented for analysis there is usually adequate material available for the separation to be carried out satisfactorily on standard LC columns, although the extraction and concentration of some samples may be necessary. It follows that small-bore columns, and, in particular, packed capillary columns are not always necessary, and should only be employed when they are essential to the satisfactory handling of the sample. In this way less strain is placed on the interface design, and a far greater proportion of samples could be analyzed on LC/NMR tandem instruments. The main advantage of the small-bore columns is their economic use of solvents, and thus deuterated solvents can be used for the chromatographic separation without undue cost. Nevertheless, with a little imagination a phase system can often be chosen that involves solvents that are acceptable to the NMR spectrometer and, at the same time, provide the chromatographic selectivity that is required to achieve the desired separation.

LC/NMR Detection of Peptides Employing Micro-Cells

Wu *et al.* [9] developed a microbore system interface for the analysis of peptides and amino acids. Their radio frequency micro-coils were very simple in form and have been described previously by Wu *et al.*[10]. The sensor was constructed from a fused silica tube, around which a tiny radio frequency coil was wound. The tube served both as the sample chamber and the coil former. The actual detector coil was 1 mm in length, and the cells were fabricated from tubing that had internal diameters ranging from 75 to 530 µm, which provided detecting volumes that ranged from 5 to 200 nl. Employing such cells in the static mode, detection limits of less than 50 ng were achieved for amino acids for one-minute data acquisition times. Although the concentration sensitivity of the system was poor for the 75 mm I.D. capillary (*ca* 30 mM) such coils allow less than 100 ng (0.1-1.0 nmol) amounts of amino acids to be detected with the 60-second acquisition time. The NMR line widths were in the several 100 Hz range for thin-walled tubes (30 µm walls) but could be reduced to about 10 Hz using thicker walled tubes (140 µm walls). Since thermal noise in the windings of the coil, not sample noise, is the primary noise contributor, the coil resistance must be minimized. Coil resistance depends on winding geometry, inter-turn spacing and the specific resistance of the material from which the wire is constructed. The authors used 42 gauge copper wire in conjunction with spaced winding. Two wires were wound onto the tube in parallel, and after completion, one winding was removed, leaving the coil with evenly spaced windings equivalent to the diameter of the wire. It should be noted that the coil resistance could be reduced very significantly by using superconducting materials.

The coil was wound from varnished copper magnet wire with the aid of a pin vice, micro-manipulator and stereo-microscope. Typically the coil contained 14 to 17 turns of wire extending over 0.9 to 1.1 mm. This produces a cell having a volume of about 50 nl. Generally, as the sensor coil is reduced in size, the mass sensitivity improves, because of the increase in strength of the rf field per unit of current flow. It has been reported that as the coil diameter is reduced from 1000 µm to 50 µm, there is a twenty-fold reduction in the limit of detection. This would correspond to a 400-fold increase in measurement time for the larger coil to obtain the same signal to noise ratio as the smaller coil. The layout of the tandem apparatus is diagramatically depicted in Figure 11.16. The micro coil was contained within the NMR probe so that the micro-coil/capillary assembly could be positioned reproducibly in the bore of the

NMR spectrometer. The liquid chromatograph consisted of a syringe pump, a pressure gauge and an injector with a loop volume of 10 μ

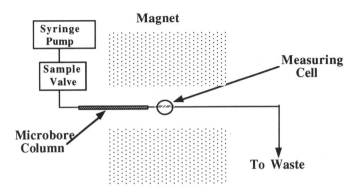

Figure 11.8 The LC/NMR Tandem System

The column was also located in the bore of the magnet, and was connected to the rf coil by means of a short length of Teflon tubing. The pump and mobile phase supply were separated from the column and magnet by a 3 m length of Teflon tubing. The chromatographic conditions were developed employing a separate UV detector. The microbore column was 15 cm long, 1 mm I.D., and packed with a 5 μm C18 reversed phase packing. The mobile phase was prepared by mixing appropriate amounts of 2% TFA in D_2O (pD = 2.4), and deuterated acetonitrile, 74:26 v/v. The mobile phase flow rates that were employed ranged from 10 to 50 μl/min. The samples were dissolved in 25% deuterated acetonitrile in deuterated water and 0.25 N D was added to render soluble. The measuring conditions were as follows.

a. Ley-Arg, 1.4 μg (28 nm) (512 scans).

b. An oxytocin fragment 6–9, Tyr–Pro–Leu—Gly–NH_2, 0.78 μg (15.6 nM), (512 scans).

c. Deca-peptide, Gly–His–Trp–Ser–Tyr– Gly–Leu–Arg–Pro–Gly, 1.2 μg (23 nM), (256 scans).

d. A hexa-deca-peptide, Val–Phe–Gly–Thr–Gly–Thr–Lys–Val–Thr–Val–Leu–Glu–Gin–Pro–Lys–Ala, o.8 mg (16nM), (256 scans).

The two-dimensional chromatograms (both surface and contour plots) shown in Figure 11.9 were obtained using the same cell. The flow rate used was 30 μl/min. Chromatogram (A) was obtained with 64 scans and a 0.06 sec pulse delay, (B), 128 scans and an 0.06 sec pulse delay and (C) 256 scans with a 0.03 sec pulse delay. Not surprisingly, the fewer the number of co-added scans, the better the chromatographic efficiency. It is seen from chromatograms (A) and (B) that increasing the number of scans from 64 to 128 (9-18 sec acquisition time) increases the sensitivity and only slightly reduces the chromatographic resolution.

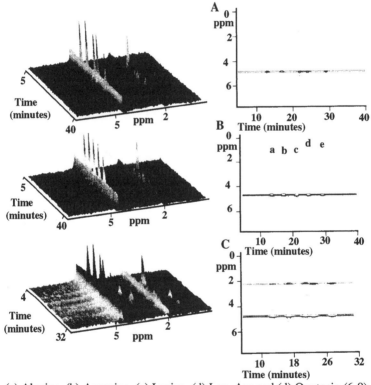

(a) Alanine, (b) Argenine, (c) Lysine, (d) Leu–Arg and (d) Oxytosin (6-9).

Reprinted with permission from N. Wu, A. G.Webb, T. L. Peck, and J. V. Sweedler, *Anal. Chem.* **67(18)**(1994)3101, Copyright 1994 American Chemical Society

Figure 11.9 Two-dimensional Chromatograms of Mixtures of Amino Acids and Peptides

Acquiring with additional scans, as shown in (C), decreases the sensitivity, because the additional scans are acquired after the solute has left the detection cell, but dramatically increases the apparent chromatographic efficiency.

Synopsis

NMR is the most difficult technique to operate on-line with a liquid chromatograph. The earlier NMR instruments had very limited sensitivity relative to other spectroscopic techniques. Even with the introduction of the 600 and 750 MHz spectrometers, sensitivity was still a problem with the in-line combination of the spectrometer with the liquid chromatograph. However, with the recent introduction of over–sampling, and digital filtering techniques, in data acquisition regimes spectral windows can be constructed to include only regions of interest, without the problem of signals from outside the region folding in. It is still necessary to accommodate the changing composition of the sample in the NMR cell due to the flow of mobile phase, and to design flow-through cells that do not interfere with homogeneity of the magnetic field immediately around the sample. The problem of interfering solvents still remains, although techniques have been developed to abrogate the interfering signals, which have certainly permitted more flexibility in the choice of the mobile phase. Modern LC/NMR systems are largely off-line devices, where the solutes are captured in a sample loop and passed to the NMR spectrometer for examination when convenient. The valving system can be quite complex, allowing on-line measurement if so desired, specific peak selection, or complete peak storage. Nanoliter flow cells have been developed to allow the use of microbore columns or packed capillaries that have provided complete spectra from 1 nmol of sample. These small cells also allow expensive deuterated solvents to be used, and thus eliminate solvent interference without excessive cost. For optimum sensitivity, the cell volume and peak volume must be matched, which is extremely difficult in practice. The use of the 600 and 750 MHz spectrometers with the micro-flow cell, has produced satisfactory spectra from as little as 50 ng of material. The quest for higher sensitivity continues and it is clear that the limit of detection has not yet been reached.

References

1. E. Bayer, K. Albert, M. Nieder, E. Grom and T. Keller, *J. Chromatogr.* **186**(1979)497.
2. J. C. Lindon, J. K. Nicholson and I. D. Wilson, couse on *Direct Coupling of Chromatogrphic Separations to NMR Spectroscopy*, Birkbeck College, London.
3. V. Sweedler, *Science*, **270**(1995)1967.
4. B. Behnke, G. Schlotterbeck, U. Tallarek, S. Strohschein, L. Tseng, T. Keller, K. Albert and E. Bayer, *Anal. Chem.*, **68(7)**(1996)1110.
5. L. Grifffiths, *Anal. Chem.*, **67(22)**(1995)4091.
6. U. G. Sidelmann, E. M. Lenz, M. Spraul, M. Hofman, J. Troke, P. N. Sanderson, J. C. London, I. D. Wilson and J. K. Nicholson, *Anal Chem.*, **68(1)**(1996)106.
7. U. G. Sidelmann, C. Gavaghan, H. A. J. Carless, M. Spraul, M. Hoffman, J. C. Lindon, I. D. Wilson and J. K. Nicholson, *Anal. Chem.*, **67(24)**(1995)441.
8. B. C. Sweatman, R. D. Farrant, P. N. Sanderson, I. Philippe, S. R. Salman, J. K. Nicholson and J. Lindon, J. Magh, *Rson, Anal.* (1995)9
9. N. Wu, A. G.Webb,T. L. Peck, and J. V. Sweedler, *Anal. Chem.* **67(18)**(1994)3101.
10. N. Wu, T. L. Peck, A. G.Webb, R. L. Magin and J. V. Sweedler, *Anal. Chem.* **66(22)**(1994)3849.

PART 4

OTHER TANDEM SYSTEMS

CHAPTER 12

THIN LAYER CHROMATOGRAPHY/ SPECTROSCOPY (TLC/S) TANDEM SYSTEMS

Tandem combinations, involving the *in-line* association of a spectroscopic instrument with a TLC plate, is not possible in the generally accepted meaning of the term. The separation must be completed before the plate can be examined. IR spectra, mass spectra and, if sufficient material is available, even NMR spectra can all be obtained from TLC fractions by scraping the material contained in the spot from the plate, and extracting the solute with an appropriate solvent, and using the solution for spectroscopic examination. The technique requires a little dexterity, but is not difficult, and by employing modern spectroscopic equipment, good spectra can usually be obtained. This off-line procedure is still that most commonly used in the contemporary examination of TLC fractions. There are, however, a number of tandem systems where the plate is used as a transport medium, which allows the spots to be examined *in situ,* by an appropriately modified spectrometer. One of the first of these methods to be developed was the technique called scanning densitometry. Scanning densitometry and other TLC tandem systems involve relatively complex equipment, which is perhaps incongruous in view of the intrinsic simplicity and low cost of the TLC analysis. Nevertheless, despite the apparent conflict between the unpretentious and inexpensive TLC plate and the complex and costly associated spectroscopic equipment necessary to scan the plate, TLC tandem systems have been actively developed over recent years. In view of the great similarity between the techniques of

TLC and LC, and that the main advantage of TLC is its simplicity and low cost, the need to develop both TLC and LC tandem instruments might appear unnecessary and perhaps wasteful.

Scanning Densitometry

Scanning densitometry can serve a number of purposes. It can be employed to precisely identify the position and size of the spot, for the accurate measurement of R_f values, and it can also provide transmission, reflectance or adsorption spectra to confirm spot identification. In addition, the density of the spot can also be employed to estimate the *quantity* of solute present in the spot. If, for some reason, the technique of TLC is chosen as preferential to LC for a particular analysis, then *in situ* scanning of the TLC plate, employing optical instrumentation, is now considered essential for both the accurate location of a spot and the precise quantitative estimation of its content. The surface of the plate can be examined using either reflected light, transmitted light, or fluorescent light. In addition, the incident light can be either adsorbed, diffusely scattered or transmitted.

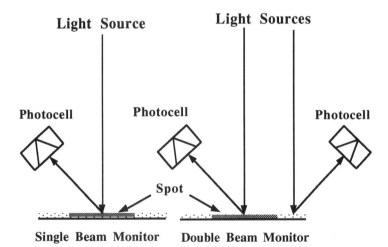

Figure 12. 1   Single and Double Beam Densitometers

It is usual to measure the light scattered, reflected, or generated by fluorescence from the spot, and compare it electronically with light from a part of the plate where no sample has passed (the channel between the spots). For this reason, single beam and double beam instruments are available, both of which are diagrammatically depicted in Figure 12.1. The double beam instrument is to be preferred, and in this configuration, one sensor monitors the sample lane (the strip of plate along which the separation has been developed), and the other monitors a blank region between the lanes. The difference signal is taken as that responding solely to the sample. In most instruments the incident light can be chosen to have wavelengths ranging between 200 and 700 nm. Halogen or tungsten lamps can be used to provide light at the higher wavelengths, whereas for light between 200 and 400 nm, a deuterium lamp is to be preferred.

In order to induce fluorescence, lamps with higher energy outputs are sometimes necessary, such as the mercury lamp, which generates most of its emission at 254 nm, and the xenon arc lamp, which generates light over a broader range of wavelengths, in a similar manner to the deuterium lamp, but at much higher intensities. Some samples may be measured with greater sensitivity by using light having a narrow band of wavelengths, in which case, a monochromator is introduced between the light source and the plate to select the appropriate wavelength. If a monochromator is employed, then either the tungsten or deuterium lamp are normally used as the light source.

The sensitivity of a scanning densitometer depends on a number of factors, including the basic instrument design and, in particular, the quality of the optics. The plate surface is viewed by the scanner through a slit and the major factor affecting the overall sensitivity is the slit height to spot diameter ratio. Although the slit dimensions are usually selectable, as the spots along the plate will be of different size it is not possible to adjust the slit to an optimum size for scanning the whole of the plate.

When measuring either the adsorbed light or the fluorescent light, the sensitivity is inversely related to the scan rate. It follows that the slower the scan, the greater the signal. However, carried to the extreme, this approach can extend the analysis time considerably. The relationship between the adsorbed light and the concentration of solute in the spot is not linear, and so the system must be calibrated, or the signal must be electronically modified by the appropriate use of a non-linear amplifier, to render the output linearly related to solute concentration. In any event, standard solutions must be run for calibration purposes, but with manual calibration, many more calibration samples are necessary. In contrast, when measuring fluorescence, the output is linearly related to solute concentration; in fact, the relationship can be linear over a concentration range of up to three orders of magnitude (*e.g.* 0.1 to 100 ng). Calibration is still necessary, but as linear curves are obtained, linear amplifiers can be used to electronically process the fluorescence signal.

As might be expected, the major sources of error that arise in scanning densitometry originate largely from sample manipulation. The sample must be carefully applied to the plate in a very reproducible manner and the diameter of the spot very carefully controlled. The chromatographic conditions must also be kept constant, and although this is relatively easy in GC and LC, due to the nature of chromatographic development, it is more difficult in TLC.

Courtesy of CAMAG Inc.

Figure 12.2 The CAMAG Thin Layer Chromatography Scanner

A photograph of a commercial TLC scanner manufactured by CAMAG is shown in Figure 12.2. Finally, in the measuring process, extreme care must be taken to ensure the spot is located in the exact center of the measuring beam or, again, errors will result

Most scanning densitometers are suitable for use with both TLC plates and also electrophoretic stains, and can deal with objects (plates or gel sheets) 200 x 200 mm in size. The spot can be monitored by either reflected light, emitted light, or transmitted light. The wavelengths available range from 190 nm to 800 nm and scanning speeds as fast as 100 mm/s are available. The spatial resolution can be selected from 25 to 200 mm. A diagram of the optical layout is shown in Figure 12.3.

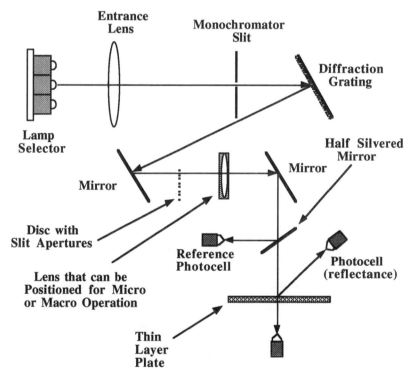

Courtesy of CAMAG Inc.

Figure 12.3 The Optical layout of the CAMAG TLC Scanner

There are three light sources from which to choose: a deuterium lamp providing light from 190 to 400 nm, a tungsten filament lamp providing light from 350 to 800 nm and a low-pressure mercury vapor lamp, which provides high-intensity line emissions at 254 nm and 578 nm. Light from the lamp passes through a lens that focuses the light through a slit, and onto a diffraction grating. The light from the diffraction grating is reflected by a plane mirror, through a selectable slit, and thence to another plane mirror and onto a half silvered mirror. Light is taken from the half silvered mirror to a reference photocell and the remaining light passes to the plate. The light reflected from the plate is monitored by a photocell aligned at 30° to the normal of the plate, and the transmitted light is monitored by a photocell placed directly under the plate. The stage is driven by stepping motors in both the (x) and (y) directions, and the positioning reproducibility of the monitor is ±50 μm in the (y) direction, and ±100 μm in the (x) direction, at a maximum scanning speed of 150 mm/sec. An example of the scan of a thin layer plate employing light at 220 nm, is given in Figure 12.4.

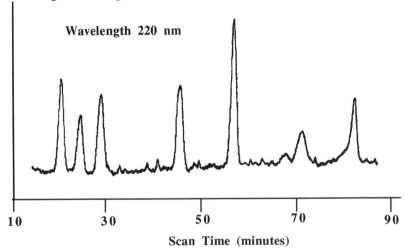

Courtesy of CAMAG Inc.

Figure 12.4 The Analog Curve Produced by Scanning a TLC Plate with the CAMAG Scanner

TLC Tandem Systems

The measuring points were taken as averages from consecutive sample pairs and the scan was taken over a period of 80 minutes. Nevertheless, a very respectable chromatogram is achieved, from which quantitative data can readily be obtained. In Figure 12.5, a chromatogram of the separation of several sulfonamides and antibiotics is shown that was obtained by scanning a TLC plate, employing a series of different wavelengths. The flexibility of the multi-wavelength scanning is clearly demonstrated. The wavelengths employed were previously identified as optimum for specific components, *i.e.* each component gave a maximum response to light of the chosen wavelength. The optimum wavelengths for all the compounds proved to be in the UV spectrum, *viz.*, 254 nm, 365 nm, 302 nm, 313 nm and 366 nm. This presentation can be compared to that produced by the diode array tandem system in the LC/UV instrument.

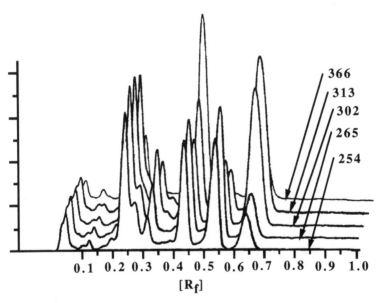

Courtesy of CAMAG Inc.

Figure 12.5 The Separation of Some Sulfonamides and Antibiotics from an Animal Feed Mix Monitored at Different Wavelengths

The three-dimensional presentation provides the maximum information, with chromatographic characteristics displayed on one axis and spectroscopic characteristics displayed on the other. Nevertheless, the scanning procedure is time consuming, and many of the advantages of TLC, relative to LC, are lost. It is true that operating costs are still lower, and solvent disposal problems will be significantly reduced, but the procedure will be slower, and the monitoring equipment, although costing less than the liquid chromatograph, will still be relatively expensive.

There are a number of ways in which scanning densitometry can be used for qualitative assessment. The incident light directed on the spot can have a single wavelength, derived from a specific light source such as a mercury lamp, or have a broad range of wavelengths such as that derived from a deuterium lamp. Either the reflected, transmitted, or fluorescent light can be arranged to pass through a monochromator, and the intensity recorded over a range of wavelengths. The spectra so obtained can be matched with the spectra of a reference compound, and the identity of the sample confirmed. Unfortunately, as UV or visible light are usually employed, the spectra provide insufficient information for the structural determination of a completely unknown solute. However, TLC spots can also been scanned by a number of IR spectroscopic techniques, including diffuse reflectance Fourier transform IR, and photo acoustic IR spectroscopy and some of these alternatives will now be discussed.

The IR Scanning of TLC Plates

FTIR measurements of a thin layer plate sample usually require a complementary measurement of the background optical properties of the plate, and the spectra of the material is taken as a difference value. Unfortunately, the infrared absorbance of the background matrix is often so strong that it obscures the spectrum of the material of interest. One of the first *in situ* IR measurements on a TLC plate was made by Percival and Griffiths [1], who employed silver halide plates coated with a thin layer of adsorbent. Silver chloride has many of the desirable features of a thin layer plate substrate. It is almost insoluble in water, and insoluble with the many other solvents used in the development of TLC plates, and is mechanically strong. Silver chloride has a high transmission in the infrared region above 450 cm^{-1}. Both circular discs, 1 in. in diameter, and silver plates 6 in. × 6 in. were examined, coated with silica gel and alumina powder, having a mean particle diameter ranging from 10–40 µm. The plates were prepared from a slurry containing 1 gm of adsorbent in 10 ml of a 50% v/v methanol/water mixture. The transmission spectra were measured using a Digilab FTS-14 FTIR spectrometer, and the samples were held in the focus of a Perkin Elmer, potassium bromide, refractive beam condenser, the average transmission of the beam condenser being 40%. The transmission of films of alumina and silica gel 100 µm thick were measured and the transmission curves are shown in Figure 12.6.

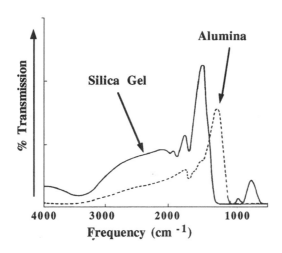

Reprinted with permission from C. J. Percival and P. R. Giffiths, *Anal. Chem.*, 47(1975)154, Copyright 1975 American Chemical Society

Figure 12.6 Transmission Curves for Alumina and Silica Gel on Silver Chloride. Plates

The measurements demonstrated that at least 10% transmission was obtainable through the coated discs between 3200 and 1250 cm^{-1}. It was found that scattering due to the alumina was much greater than that from the silica gel, as shown by the transmission at high frequency. However, its low energy cut-off is about 1050 cm^{-1} so that the useful range for compound identification is actually greater than that for silica gel.

The sensitivity of the system was studied by spotting different solutions of various compounds on to the disks, and it was found that most substances were easily distinguishable from their spectra, even when as little as one microgram of material was present on the plate. As an example of the sensitivity of the silver halide plate measurement technique, 1 µg of methylene blue, eluted 5 cm along the plate with 90:90:4 mixture of chloroform:methanol:acetic acid gave a spectrum in which all the major absorption bands could be clearly discerned.

There have been a number of other approaches to the problem of scanning the TLC plate to obtain IR spectra of the solutes, including the use of diffuse reflectance IR Fourier transform spectrometry and photo acoustic infrared spectrometry (PAS). Both these techniques will be briefly discussed.

Diffuse Reflectance IR Fourier Transform Spectrometry (DRIFTS)

The principle of diffuse reflectance IR Fourier transform spectrometry is depicted in Figure 12.7. When incident light strikes a surface, the light that penetrates is reflected in all directions and this is called diffuse reflectance. As the light that leaves the surface has passed through a thin layer of the reflecting material, its wavelength content will have been modified by the optical properties of the matrix. Consequently, the wavelength and intensity distribution of the reflected light will contain structural information on the substrate.

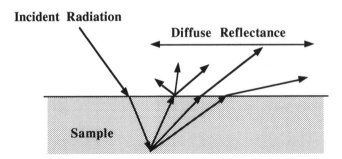

Figure 12.7 The Principle of Diffuse Reflectance IR Fourier Transform Spectrometry

TLC Tandem Systems

This has obvious application to scanning TLC plates and the possibilities were investigated by Zuber *et al.* [2]. In the system developed by Zuber *et al.* the TLC plates were inserted directly into the spectrometer sample chamber and the spectrum obtained by reflectance directly from the plate surface. A reference laser aids in spot alignment, and the plate contributions to the background are subtracted from the spectrum obtained in the usual manner. A number of different TLC plates were examined in this way, and it was found that spot identification was possible, providing reference spectra were available that had been obtained under the same operating conditions. The quality of the spectra and the useful size of the IR window available varied between different types of plate. The technique that was employed was an extension of some earlier work by Fuller and Griffiths [3], but smaller plates were used and great care was taken to ensure surface uniformity.

The preparation and handling of both the sample and background plates were be the key to the success of the method. It was found that although the selection of the solvent to achieve the separation was critical, providing the plate was completely dried before IR measurements were made, the solvent had no effect on the quality of the spectra that were produced. The spectrometer used was the Nicolet (Madison, WI) 6000 FTIR, which was equipped with a nitrogen cooled cadmium telluride detector, and a diffuse reflectance attachment was used to run the spectra. The interferometer was run at a mirror velocity of 0.586 cm/s and 2000 scans or less were found necessary to produce a good quality TLC/FTIR spectrum. In most experiments, small plates, (2.5 x 10 or 2.5 x 7.5 cm), were employed, so that they would fit easily into FTIR sample chamber. All plates were thoroughly dried before examination. The TLC spot diameter varied between 2 and 8 mm, and the diameter of the infrared beam, focused on each plate, was 1 mm. The sample and background spectra were run, and the spectrum of the sample obtained by difference. The analysis time was typically 15 minutes for the TLC separation, and about 30 minutes were needed to obtain the IR spectra. Examples of the results obtained from the tandem system are shown in Figure 12.8. It is seen that there is a distinct difference in the form of the spectra taken from the TLC plate compared with that from the KBr pellet (Figure

12.8). It follows that reference spectra that are to be used for solute identification should also be obtained from the TLC plate in the same manner. The mass of solute in each spot examined was about 10 µg but it was estimated that about 1 µg would be sufficient for a recognizable spectrum to be obtained.

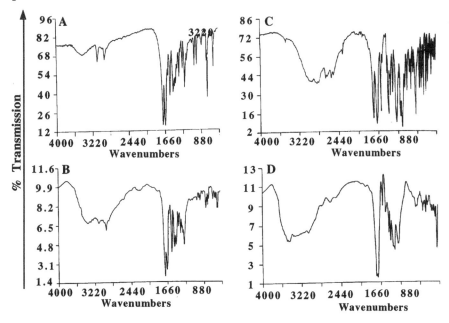

A and C, Spectra from KBr pellets of caffeine and aspirin respectively. B and D, Reflectance spectra from TLC plates of caffeine and aspirin respectively.

Reprinted with permission from G. E. Zuber, R. J. Warren, P. P. Begosh and E. L. O'Donnell, *Anal. Chem.*, **56**(1984)2935, Copyright 1984 American Chemical Society.

Figure 12.8 Transmission Spectra from KBr Pellets and Reflectance Spectra from TLC Plates of Caffeine and Aspirin

There is no doubt that direct measurements on the plate restricts the range of wavelengths that can be employed in the spectroscopic examination, whereas the removal of the solute from the plate allows the material to be examined over the normal range of wavelengths. Unfortunately, solute removal and recovery almost always involves losses, and sometimes the losses are accompanied by the decomposition or molecular rearrangement of labile materials. Chalmers *et al.* [4] chose to use FTIR diffuse reflectance spectrometry in an off-line manner, by extracting the material from the spot before measurement. The solute on the TLC plate was transferred to a KCl pellet, made directly from ball-milled potassium chloride. 0.7 g of the dried powder was pressed into a 13 mm disk die, at a pressure of 500 psi. The resulting pellet was about 4 mm high (± 0.5 mm). A metal backed TLC plate was employed for the separation, and the spot cut out and placed, metal backing downwards, in a tube 5 cm long and 17 mm I.D.. The KCl pellet was placed on the top of the disc, and about 2 ml of chloroform carefully pipetted down the inside of the tube. The chloroform is allowed to evaporate at room temperature, and after about an hour, the pellets are removed for IR examination. Good spectra were obtained, but the process was tedious, and considering the extraction and concentration processes that were involved, the overall methodology does not appear to have provided a very good sensitivity.

Danielson *et al* [5] investigated the use of zirconia as a suitable material for TLC plates, in the hope that it would also allow diffuse reflectance spectra to be obtained from solutes retained on its surface. They examined silica gel, alumina and zirconia to determine the relative merits of each medium. They used monoclinic zirconia microspheres 5 μm in diameter, having a surface area of about 25 m^2/g ,and a mean pore diameter of 220 Å. The silica and alumina microspheres were 7–12 μm in diameter. Silica and alumina slurries were prepared from 1 g of material and 3 ml of water, whereas the zirconia slurries were made from 1 g of material and 2 ml of water. The slurries were spread on plates 7.5 cm long and 2.2 cm wide and dried overnight. The coating was estimated to be about 200 μm thick.

The spectrometer employed was the Perkin Elmer Model 1800 FTIR, equipped with a Spectra-Tech microscope. In most cases the spot on the TLC plate was larger than the total sampling area of the microscope (*ca* 250 μm diameter). Diffuse reflectance spectra were taken by co-adding 256 scans at a resolution of 4 cm^{-1}. The spectral properties of the different materials examined are shown in Figure 12.9. It is seen that the spectrum of zirconia shows significantly higher IR reflectivity than either silica or alumina. In particular, strong IR absorption occurs below 2000 cm^{-1} in silica, and below 1600 cm^{-1} in alumina.

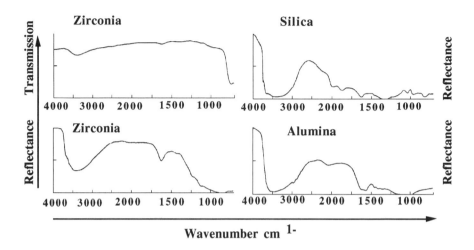

Reprinted with permission from N. D. Danielson, J. E. Katon, S. P. Bouffard and Z. Zhu, *Anal. Chem.*, **64**(1992)2183, Copyright 1992 American Chemical Society.

Figure 12.9 The Transmission and Reflectance Spectra of Zirconia, Silica Gel and Alumina Plates.

The 2000–1000 cm^{-1} region of zirconia has a relatively low IR absorption. It would appear that due to its advantageous transparency at the wavelengths of interest, zirconia on TLC plates does provide more useful IR spectra than silica or alumina. However, the advantages are not great and may not be sufficient to establish zirconia as a viable alternative to silica as a TLC stationary phase.

Scanning Thin Layer Plates by Photoacoustic Spectroscopy

In photo acoustic spectroscopy the energy absorbed from the infrared radiation is measured by the mechanical vibration produced, employing appropriate acoustic measuring devices. A diagram representing the photo acoustic spectroscopic sensing system is shown in Figure 12.10.

The incident radiation is allowed to fall on the sample contained in a suitable enclosure. When the modulated infrared radiation is absorbed by the sample, the substance heats and cools in response to the radiation received. Situated in the enclosure is an acoustic sensing device, which may be a simple microphone or a piezoelectric sensor. The sensor detects the acoustic pulses as they are generated by the different IR frequencies that are absorbed. The advantage of this type of IR measurement is that it can be used effectively with very black or highly absorbing samples.

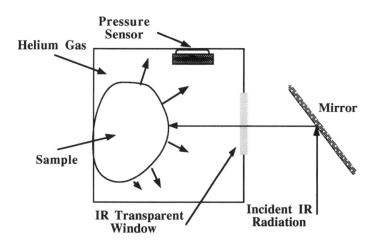

Figure 12.10 Diagram of the Photoacoustic Sensing System

Lloyd *et al* [6] employed this technique to scan thin layer plates. A simple microphonic detector was employed as the sensor, but the performance of a piezoelectric sensor was also tested. The thin layer sheets that were used were either aluminum or poly(ethylene terephthalate) backed, and both silica and alumina were used as the stationary phase, in the form of a thin layer about 250 µm thick. 1 cm diameter discs were excised from the plates and placed in the sample compartment of the microphonic cell. The cell was sealed in a glove bag after purging for 15 minutes with helium. The cell itself was fabricated from polished stainless steel with a sodium chloride window, and was supported on vibration-free mounts. It had a total volume of about 0.4 cm^3. The IR output from a Nicolet 7199 FTIR spectrometer was focused through the sodium chloride window onto the plate surface.

The acoustic waves were detected with a Brüel and Kjoer 4165 microphone, which was exposed to the helium in the cell, through a pipe 10 mm long and 1 mm I.D. By comparison with the spectra obtained from a KBr pellet sample, reasonably fine structure was disclosed. However, as might be expected, the signal to noise was not very good, and consequently the absolute sensitivity was also not as good as that obtained by other scanning procedures.

Alternative TLC/IR Interfaces

Other analysts have tackled the problem of linking TLC with IR spectroscopy in different ways. Shafer *et al.* [7] adopted a sample transfer approach to the problem, in which each separated component is removed simultaneously from the TLC plate to an IR transparent substrate, and then examined by diffuse reflectance IR spectrometry. The transfer is achieved with the minimum sample loss, decomposition or contamination, and spectra have been obtained from as little as 40 µg of material for identification. The separations were performed on EMR TLC plates, 8 cm x 2 cm, carrying silica gel 60 and cut from plates 50 cm x 50 cm. The plates were cleaned prior to use in a bath of acetone, and then reactivated by heating at 200°C for 2 hours. Each plate was spotted with 1 µl of sample, and placed about 1 cm from the side of the plate, the reason for which will be given later.

The separation was developed along the length of the plate. After the separation was complete, and the plate dried, it was turned sideways and attached to an aluminum strip, in which a row of cups had been drilled. The transfer system is diagramatically depicted in Figure 12.11. Each cup was 4 mm deep and 1.2 mm I.D., and packed with finely ground infrared-transmitting glass, composed of germanium, antimony and selenium. At the mouth of the holes, a bundle of glass fibers was placed which allowed a solution to be transferred by surface tension forces up to the powdered glass packing. The fibers were arranged to be in contact with the TLC plate, which was then developed at right angles to the original development with dichloromethane. This second development transferred the spot contents through the fibers and into the glass packing which took less than 2 minutes. After sample transfer, the aluminum block was placed in an oven at 40°C for 30 seconds to remove the dichloromethane, and the contents of each cup ware then examined by diffuse reflectance.

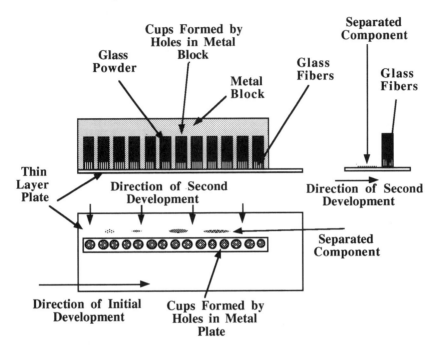

Reprinted with permission from K. H. Shafer, P. R. Griffiths and W. Shu–Qin, *Anal. Chem.*, 58(1986)2708, Copyright 1986 American Chemical Society.

Figure 12.11 Diagram of TLC Plate Attached to the Strip Plate for the Transfer of the Separated Spots

After the IR measurements have been made, acetone was passed through the cups to clean them. The diffuse reflectance measurements were carried out on an Analect FX-6200 FT-IR, equipped with a diffuse reflectance accessory. The strip was moved with a standard x/y drive mechanism. The location of each component was found by monitoring the individual spectra and/or the reconstructed chromatogram. The spectrum was taken at the most concentrated portion of the peak at 4 cm^{-1} resolution.

The percentage of infrared radiation reflected from the stationary phase of a TLC plate is significantly less than that reflected from an IR-transparent powder. The signal loss can be more than 70% over most of the spectrum and consequently, *in situ* measurements require as many as 2000 scans to be co-added, before the signal to noise ratio is increased to a level where solute identification is possible.

Another approach to the development of Tandem TLC/IR instrumentation was put forward by Bouffard *et al.* [8], which they termed micro-channel thin layer chromatography with infrared micro-spectroscopic monitoring. The authors prepared a TLC system comprising a brass plate in which three channels were cut, 5 cm long, 400 μm wide and 200 μm deep. The channels were packed with zirconia, particle diameters 5–10 μm, having a mean pore diameter of 220 Å, and a surface area of 25 mm^2/g. The channels were filled by pouring a zirconia–methanol slurry (0.2 g of zirconia in 3 ml of methanol) onto the face of the plate containing the channels. After the methanol had evaporated the excess zirconia was removed by scraping the surface with a razor blade.

Zirconia was employed to eliminate the strong background interference normally associated with silica gel. In practice, it was found that the minimum detectable mass was reduced to between 1 and 10 ng, which is a factor of 500 better than that reported for other methods. Diffuse reflectance were obtained by co-adding 256 scans (typically) at 4 cm^{-1} resolution. Very good spectra were obtained, and that obtained for 12.5 μg of sample exhibited sufficient detail for solute identification.

An example of the efficacy of the system is afforded by the separation of a mixture of aromatic compounds shown in Figure 12.12. *n*-Hexane was used as the mobile phase and the separation time was about 15 minutes. The spectra on the left are reference spectra, and those on the right are from, 3, 2, 2 and 2 μg of phenanthrene, 7, 8-benzoquinoline, phenanthradine and 1,10–phenanthroline respectively. It is seen that the sensitivity is high, and the resolution of the spectra more than adequate for solute identification. The increased complexity of TLC plate fabrication, coupled with the possible limitations in chromatographic selectivity imposed by the use of zirconia as an alternative to silica gel, might be a disadvantage for some types of analyses.

TLC Tandem Systems

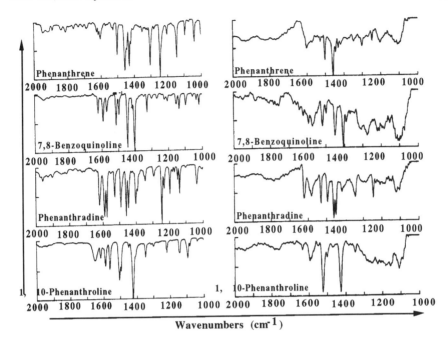

Reprinted with permission from S. P. Bouffard, J. E. Katon, A. J. Sommer and N. D. Danielson, *Anal. Chem.*, **66**(13)(1994)1937, Copyright 1994 American Chemical Society.

Figure 12.12 The Spectra Obtained for the Separation of a Series of Aromatic Compounds

Thin Layer Chromatography/Mass Spectroscopy

On–line TLC/MS systems must invariably involve either a transport procedure or an *in situ* desorption ionization interface. An interesting tandem system, recently reported by Gusev *et al.* [9], is the combination of the time of flight mass spectroscopy with thin layer chromatography using matrix assisted laser desorption. This technique takes advantage of the unique possibilities that time of flight mass spectroscopy can offer for this type of sample ionization, and was emphasized when the time of flight mass spectrometer was discussed in chapter 2.

The procedure is diagramatically depicted in Figure 12.13. Aluminum backed plates were employed, and after the separation was complete, the plates were dried. Small circles of foil were cut from the plate containing the solutes and each was then treated with a solvent (*e.g.* dichloromethane) to dissolve the solute.

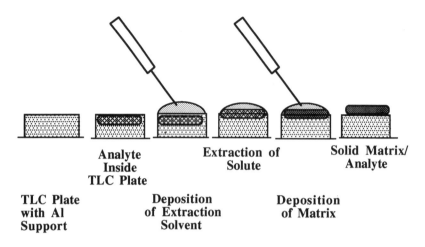

Figure 12.13 Extraction of Solute and Deposition of Matrix

Extraction was allowed to continue for a few minutes and then the matrix agent was added. The solvent was then allowed to completely evaporate leaving a solid mixture of the matrix agent and the solute.

A number of different substances were employed as matrix agents including 4-hydroxy-3-methoxycinnamic acid, 3,5-dimethoxy-4-hydroxycinnamic acid, 3-hydroxypicolinic acid and 2-(4-hydroxyphenylazo)benzoic acid. The samples were then transferred to a LAMMA 1000 time of flight mass spectrometer comprising a laser microprobe in conjunction with a nitrogen laser, and post-acceleration to 1.5 keV onto a discrete dynode type secondary ion multiplier.

The mass spectrometer system included an x/y manipulator and microscope to allow position adjustment of the sample and visual selection of the position of the laser. The system was found to provide spectra from as little as 2-4 ng of sample, but this minimum sample size varied somewhat with the nature of the solute and that of the matrix. An example of the spectrum obtained for 100 ng of Tyr-bradykinin extracted from a thin layer plate is shown in Figure 12.14.

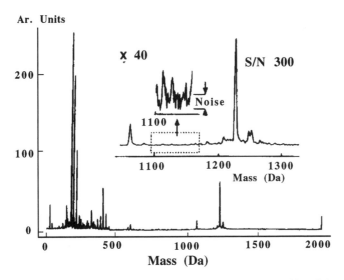

Reprinted with permission from A. I. Gusev, A. Proctor, Y. I. Rabinovich and D. M. Hervcules, *Anal. Chem.*, **67(11)**(1995)1805, Copyright 1995 American Chemical Society.

Figure 12.14 Mass Spectrum for 100 ng of Tyr-bradykinin Extracted from a Thin Layer Plate

It is seen from the insert that the signal to noise ratio is about 300 to one and thus, assuming a signal to noise of 3 is necessary to unambiguously identify the presence of a mass peak, then the minimum detectable mass is about 1 ng.

A further extension of the technique by Gusev *et al.* [10] involved the development of the TLC plate followed by mobile phase removal by drying in an appropriate manner. Second, the matrix ionization aid is formed separately as a thin film on a stainless steel plate. The reagents, dissolved in an appropriate solvent, are sprayed onto the steel plate and dried. The thin layer plate is then sprayed with an appropriate solvent and while wet, the stainless steel plate is pressed onto the thin layer plate (the coating being in contact with the thin layer plate).

The solvent on the TLC plate serves two purposes. First it extracts the solutes into solution so that they can come in contact with the matrix film. Second, the matrix film takes up the solvent and thus incorporates the solutes into the matrix layer. The steel plate is then removed, leaving the matrix layer containing the solutes on the surface of the thin layer plate. The thin layer plate is then placed in the laser desorption apparatus shown in Figure 12.15.

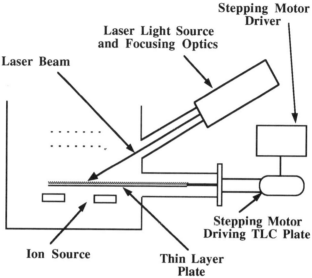

Figure 12.15 The TLC Plate Sampling Device for the Time of Flight Mass Spectrometer

The desorption device consists of a TLC plate holder, driven by a stepping motor, that can be traversed both in the (x) and (y) axis so that the plate can be precisely scanned in two dimensions. The plate is situated in the ion source of the time of flight mass spectrometer, and by pulsing the laser beam the spectra can be obtained and, if necessary, accumulated in the usual manner. This technique, inevitably causes each spot on the plate to spread to some extent, which might diminish the chromatographic resolution. It appeared, however, that the dispersion was not too serious and would not be important in the majority of TLC separations. Nevertheless, it must be pointed out that the TLC/MS tandem systems are tedious to use and sometimes difficult to operate. In addition, they are basically off-line procedures. Serious consideration should be given to the alternative use of LC/MS or LC/IR for an analysis before a TLC tandem system is chosen, and then perhaps it should be chosen only as a last resort, under very special circumstances.

Matrix peaks occurred at masses ranging from 250 to 400 Da and thus interfered with the spectrum. Consequently, a limit is set on the mass measurement of low molecular weight materials. The sensitivity quoted was 40 pg but the size of the spot was not given. Nevertheless, very useful mass spectra were obtained by directly scanning the thin layer plate and taking advantage of the unique properties of the time of flight mass spectrometer. The spot dispersion was examined by the authors using guineas green B and the three-dimensional image of the spot after processing is shown in Figure 12.16.

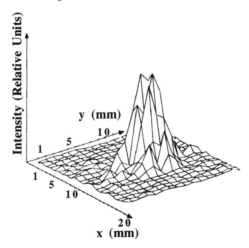

Reprinted with permission from A. I. Gusev, O. J. Vasseur, A. Proctor, A. G. Sharkey and D. M. Hercules, *Anal. Chem.*, **67**(1995)4565, Copyright 1995 American Chemical Society.

Figure 12.16 The Tree Dimensional Image of Guinea Green B (20 ng) Separated on a Plastic Backed TLC Silica Gel Plate

Synopsis

True in-line TLC tandem systems are not actually possible, as the TLC separation must be developed before the spots can be monitored. It follows that all TLC tandem instruments operate as either fraction collectors or off-line monitoring devices. The most common off-line monitoring system is that of scanning densitometry. The technique is used both to accurately locate the spot and also provide spectroscopic information to aid in identification. Double beam instruments are commonly used that compare the surface characteristics of the spot to those of the natural plate surface. Monochromatic and polychromatic light from lamps, such as deuterium tungsten, mercury, etc., can be used to measure the spectroscopic properties of the spot; absorbed light is not linearly related to solute concentration but fluorescent light is. Slow scanning speeds provide the greatest sensitivity, but extend the analysis time. Scanning densitometry is now very widely used in TLC plate measurement. Infrared spectrometric measurement, taken as the difference between those of the spot and those of the background, yields a very poor signal to noise ratio and is thus very insensitive. The sensitivity is improved by using a suitable IR transparent material as the plate, and employing a very thin layer of stationary phase. More recent techniques employ diffuse reflectance FTIR spectrometry, and using this technique spectra have been obtained from as little as 1 µg of material. The presence of the chromatographic stationary phase severely interferes with the quality of the spectra produced, and for this reason, a number of elaborate plate extraction procedures have been developed, all of which add to the complexity and tedium of the analysis. It was found that zirconia provided less spectrometric interference than silica, and consequently produced improved IR spectra. However, zirconia is a poor substitute for silica, from a chromatographic point of view. Photoacoustic spectroscopy has also been used to scan thin layer spots, using different materials for the supporting plate, but the quality of the spectra were, in general, not as good as those obtained by diffuse reflectance spectrometry. An elaborate solute extraction procedure, employing metal cups made by drilling holes in a metal plate, has provided good spectra, but again involves a rather complex experimental procedure, and one that would need considerable modification to be optimum for each specific sample. On-line TLC/MS must involve a transport system and extraction procedure, or an *in situ* monitoring technique. Extraction techniques have proved effective but are little more than complicated fraction collectors. Matrix assisted laser desorption mass spectroscopy has been carried out successfully on a metal disc transport system using a time of flight mass spectrometer very. However, for the majority of samples, there seems little advantage in choosing a TLC/MS tandem system over the simpler, and more sensitive, LC/MS tandem instrument.

References

1. C. J. Percival and P. R. Giffiths, *Anal. Chem.*, **47**(1975)154.
2. G. E. Zuber, R. J. Warren, P. P> Begosh and E. L. O'Donnell, *Anal. Chem.*, **56**(1984)2935.
3. M. P. Fuller and P. R. Griffiths, *Anal. Chem.*, **50**(1978)1906.
4. J. M. Chalmers, M. W. Mackenzie and J. L. Sharp, *Anal. Chem.*,**59**(1987)415.
5. N. D. Danielson, J. E. Katon, S. P. Bouffard and Z. Zhu, *Anal. Chem.*, **64**(1992)2183.
6. L. B. Lloyd, R. C. Yeates and E. M. Eyring, *Anal. Chem.*, **54(3)**(1982)549.
7. K. H. Shafer, P. R. Griffiths and W. Shu-Qin, Anal. CVhem., 58(1986)2708.
8. S. P. Bouffard, J. E. Katon, A. J. Sommer and N. D. Danielson, *Anal. Chem.*, **66(13)**(1994)1937.
9. A. I. Gusev, A. Proctor, Y. I. Rabinovich and D. M. Hervcules, *Anal. Chem.*, **67(11)**(1995)1805.
10. A. I. Gusev, O. J. Vasseur, A. Proctor, A. G. Sharkey and D. M. Hercules, *Anal. Chem.*, **67(24)**(1995)4565.

CHAPTER 13

TANDEM SYSTEMS INVOLVING CAPILLARY ELECTROPHORESIS AS THE SEPARATION TECHNIQUE

Capillary electrophoresis can provide extremely high plate efficiencies, and although the selectivity of the electrophoretic process is not as great as that attainable from liquid chromatography, the resolution provided by the system is still extremely high. It follows that the combination of an electrophoretic instrument with a spectrometer would be an extremely useful tandem system.

The first on-line capillary electrophoresis/mass spectrometer tandem instrument was reported by Olivares *et al.* [1] and Lee *et al.* [2] who coupled the capillary column of the electrophoretic system to a quadrupole mass spectrometer, using an atmospheric pressure ionization electrospray interface. A diagram that depicts the basic arrangement is shown in Figure 13.1. As a result of the small flow rates that are involved (usually about 1 µl/min) the design of the interface is critical. The electrophoresis is carried out employing a high-voltage power supply giving 0 to 60 kV. The high-voltage electrode, one end of the capillary tube, is contained in an insulated box fitted with remote manipulators. The fused silica capillaries are 10 cm long, 100 µm I.D.. The cathode end of the capillary (the low-voltage end), is terminated by a stainless steel sheath 300 mm I.D. and 400 mm O.D., the potential of which is controlled by a separate 0–5 kV power supply. The stainless steel end functions as both the electrophoresis cathode and the electrospray needle, electrical contact being maintained with the solution by means of the stainless steel collar.

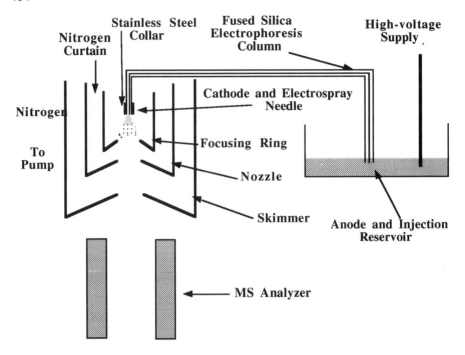

Figure 13.1 The First Capillary Electrophoresis Quadrupole Mass Spectrometer Tandem Instrument

The electrospray ionization is carried out in a stainless steel tube, 2.54 cm long and 2.3 cm I.D. The end of the cylinder contains a hole 0.475 cm in diameter, that acts as a focusing ring and is biased at 190 V dc. The copper ion-sampling nozzle is held at ground potential, and has a 0.5 mm sampling orifice. The whole spray system is heated to 60°C by a number of appropriately placed cartridge heaters. A curtain of nitrogen (at a flow rate of 2.5 l/min) is maintained between the focusing ring and the nozzle that provides a counter flow as an aid in solvent evaporation. The vacuum system has three differentially pumped stages, which maintains the necessary pressures in the different parts of the mass spectrometer, and ensures that a pressure of 2×10^{-6} torr is held in the mass analyzer by means of a 500 l/sec turbomolecular pump.

In most cases the solvent used for the electrospray was composed of 50/50 methanol water containing 10^{-4} M KCl. The initial test mixture contained five ammonium salts, tetramethyl-ammonium bromide, tetraethyl-ammonium perchlorate, tetrapropyl-ammonium hydroxide and trimethyl-phenylammonium iodide. The dominant peaks were found to be due to the quaternary ammonium cations of tetramethylammonium bromide (m/z = 74), tetraethylammonium perchlorate (m/z = 130), trimethylphenyl-ammonium iodide (m/z = 136), tetrapropyl-ammonium hydroxide (m/z = 186), and tetrabutylammonium hydroxide (m/z = 242), together with a background of Na-MeOH$^+$ (m/z = 55). The system functioned well, providing clearly resolved, unambiguous spectra, with a high signal to noise ratio. An indication of the absolute sensitivity that was obtainable from the system, is given by the electropherogram shown in Figure 13.2.

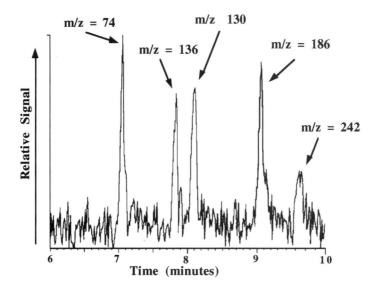

Reprinted with permission from J. A. Olivares, N. T. Nguyen, C. R. Yonker and R. D. Smith, *Anal. Chem.*, **59**(1987)1230, Copyright 1987 American Chemical Society.

Figure 13.2 Reconstructed Total Ion Current Electropherogram of Five Quaternary Ammonium Salts Demonstrating the Limit of Sensitivity

The sample placed on the column was about 0.7–0.8 fmol. It is seen that the signal-to-noise was about 3 to 4, so taking the solute tetramethylammonium bromide with a molecular weight of 159, the mass contained in the peak was,

$$139 \times 0.75 \times 10^{-15} = 0.14 \times 10^{-12}$$

Now assuming the signal to noise ratio for the peak was about 3, and the signal to noise ratio required to unambiguously identify solute is 2,

$$\text{Then the minimum detectable mass} = (2/3)\, 0.14 \times 10^{-12} \sim 0.1 \text{ pg}$$

Sensitivity Enhancement Techniques

Wahl *et al.*, [3] claimed to have increased the sensitivity of the system described by Smith *et al.* [1] by reducing the diameter of the electrophoresis tube to 5 µm. Unfortunately, the advantages or disadvantages that result from the use of the smaller diameter tubes, are confused by the method employed to report sensitivity. As already discussed, the use of molecular units to report sensitivity can be deceptive. The mass sensitivity of a measuring device, which is the basic specification of all analytical measurements, will be different for solutes that have different molecular weights but the *same* molar sensitivity. The results of this muddle are well illustrated by these two publications. The sensitivity was quoted for a number of different proteins. Taking myoglobin with an m/z value of about 770 which, in the electropherograms published, carried 22 charges, the actual molecular weight would be about ,

$$770 \times 22 = 16940$$

Consequently, the sensitivity published as 600 amol of the protein would be equivalent to

$$16940 \times 600 \times 10^{-18} \text{ g, or } 10.1 \times 10^{-12} = 10.1 \text{ pg}$$

It is shown that the sensitivity, in terms of absolute mass, is nearly an order less than that described for the smaller molecular weight compounds discussed in a previous report, albeit the molar sensitivity is much greater. The problems associated with defining sensitivity have been discussed extensively elsewhere [4,5], and as instrumental sensitivity continues to become more important in environmental analysis, pharmaceutical analysis and in the many facets of biotechnology, the need to understand and clearly define the measurement becomes extremely urgent. In general, the units that are most important to the majority of scientists and, in particular, analysts, are the units of mass and concentration (mass/unit volume). If the mass and concentration sensitivity are known, the values can be converted to any other units that may be of interest. If the sensitivities of all sensing and measuring devices are defined in terms of mass and concentration, then they can all be compared on a rational basis.

Concentration Enhancement Techniques

One of the problems that arise when associating an electrophoretic separation technique with the mass spectrometer, or for that matter, any other spectroscopic instrument, is the low masses and concentrations of solute that are involved. Consequently, a number of concentrating techniques has been developed to circumvent this difficulty. One of the early methods, developed by Cai and Rassi [6], was to use a capillary coated with an octadecyl bonded phase, to concentrate the sample by dispersive interactions (hydrophobic) onto the surface of the tube. The device was, in fact, the micro-form of a solid state extraction cartridge. Two fused silica tubes were used, joined by a Teflon sleeve. The first contained the dispersive coating of octadecyl chains, and the second was used for the electrophoretic separation. In one example, the concentration tube was 20 cm long and 50 µm I.D., and the electrophoresis tube was 60 cm long (30 cm to the point of detection) and 50 µm I.D. The inner surface of the pre-concentration tube was roughened by etching with a 5% solution of ammonium hydrogen bifluoride, at 250-300°C, under sealed conditions. The roughened tube was then filled with a 20% w/v solution of dimethyloctadecyl chlorsilane and heated at 110°C for an hour. The silanizing process was repeated twice. The sample was accumulated on the walls of the tube from an aqueous electrolyte solution, and was subsequently displaced into the electrophoresis tube, with an electrolyte solution containing acetonitrile. This process is similar to the recovery of sample from a solid state extraction cartridge. The recovery was claimed to improve the lower detection limits by a factor of 10 to 35.

Another method of solute concentration was introduced by Thompson *et al.*, [7], who employed an isotachophoretic procedure for pre-concentrating the sample before separation by capillary electrophoresis. The different types of ionic separation techniques are described at the end of Chapter 1, and it is recalled that in isotachophoresis, there is a leading and terminating electrolyte. On application of the driving potential, the solutes line up between the two electrolytes according to their individual ionic mobilities. The system is self-focusing, and if some ions disperse away from their equilibrium position, they will suffer differential forces that cause them to move back to their respective ionic mobility position. Consequently, if the sample is dilute, each solute will be compressed between the leading and terminating electrolytes and will be concentrated.

The arrangement of the apparatus used by Thompson *et al.* is depicted in Figure 13.3.

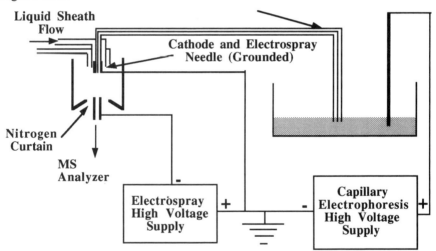

Figure 13. 3 The Capillary Electrophoresis Electrospray Interface

The electrophoretic system comprised a power supply, a fused silica separating capillary, and an anode reservoir. The capillary could be used, separately or sequentially, both for isotachophoretic concentration and for electrophoretic separation. The separated components were driven under electro-osmotic flow into an electrospray nebulizer which was operated from a separate high voltage supply. The electrospray nebulizer carried a liquid nebulizing sheath-flow as well as a gas sheath-flow.

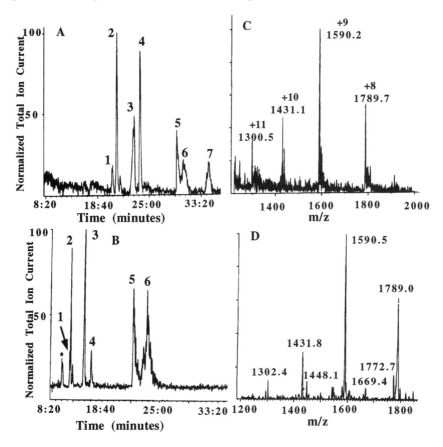

1, lysozyme, 2, ribonuclease, 3, cytochrome c, 4, myoglobin, 5, β lactoglobulin A, 6, β lactoglobulin B, 7, carbonic anhydrase.

Reprinted with permission from T. J. Thompson, F. Foret, P. Vouros and B. L. Karger, *Anal. Chem.*, **65**(1993)900, Copyright 1993 American Chemical Society.

Figure 13.4 Proteins Separation Using Capillary Electrophoresis with and without Isotachophoretic Concentration

The cylindrical electrospray electrode carried a nitrogen curtain gas flow and the ions passed through a channel into the analyzer of the mass spectrometer. The concentrating system was tested by first separating a sample of peptides by simple capillary electrophoresis. The results obtained are shown in Figure 13.4(A). The sample concentration was about 10^{-5} M for each protein. If the sample was further diluted by a factor of 10 (10^{-6} M), it was found that most of the proteins were not detectable. An example of the mass spectra obtained are shown in Figure 13.4(C). The preliminary isotachophoretic concentration and subsequent capillary electrophoretic separation was accomplished as follows. First the column was filled with the background electrolyte followed by injection of 750 nl of the sample dissolved in ammonium acetate buffer. The end of the column was then returned to the background electrolyte reservoir. The ammonium ions (which have high mobility) moved ahead of the sample when the field was applied. At this stage, the sample ions stacked behind the ammonium zone in a narrow band and began moving at a constant velocity, as in isotachophoresis. The ammonium ions continued to move through the slower background electrolyte, and consequently, the concentration of ammonium ions began to decrease below the point necessary to sustain isotachophoretic migration. At this point the zones continued to separate by capillary electrophoresis. The separation obtained is shown in Figure 13.4 (B). Each protein is present at a concentration of 5×10^{-7} M. Consequently, by employing isotachophoretic concentration, the practical concentration range for this particular separation was reduced by one to two orders of magnitude. It is clear that the technique is useful, and could be used with the majority of electrophoretic equipment presently available.

Capillary Electrophoresis in Tandem with the Ion Trap Mass Spectrometer

Capillary electrophoresis has also been combined with the ion trap mass spectrometer for the separation and identification of some isoquinoline alkaloids by Henion *et al.* [8]. A arrangement of their interface is depicted in Figure 13.5 . The low dead volume T that delivers the sheath flow, was made from PEEK, and fitted with finger-tight unions. PEEK tubing was also used to act as a seal round the capillaries, and as a guide between the sheath flow tube and the CE capillary. The CE capillary was housed in a stainless steel sheath tube (254 μm I.D. and 508 μm O.D.), through which the make-up buffer or solvent flow was introduced. A larger stainless steel tube (584 μm I.D. and 902 μm O.D.), placed concentrically round the inner capillaries, served as the nebulizing gas conduit, through which nitrogen was passed at a pressure of 50 psi.

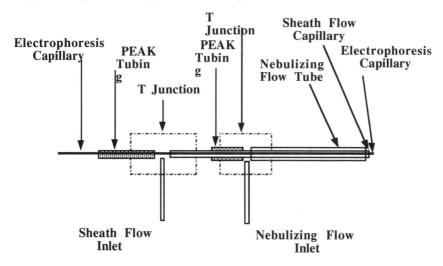

Figure 13. 5 The Ion Spray CE/MS Interface

The sheath flow liquid was 2 or 5 mM ammonium formate in methanol, and delivered at a flow rate of 10 μl/min. The ion trap mass spectrometer was the bench-top Varian Saturn II equipped with an atmospheric ionization source. An example of a total ion current capillary electropherogram is shown in Figure 13.6. It is seen that the mass of sample represented by each peak was, on average, about 45 fmol (*i.e.* 45 x 10^{-15} mol). Now assuming an average molecular weight for the different alkaloids to be similar to that of berberine (*i.e.* 326), the actual mass of the solute present will be,

$$45 \times 326 \times 10^{-15} = 1.47 \times 10^{-11} = 14.7 \text{ pg} = 0.015 \text{ ng}$$

For analysts familiar with the mass spectrometer, the mass of 14.7 pg will permit a rational comparison with other instruments, whereas for those employing UV detection, the value of 0.015 ng will have more meaning.

An example of the use of the CE/MS tandem system is afforded by the example given in Figure 13.6. The two chromatograms obtained by single ion monitoring were from a solid phase extract of urine from a female patient who had been treated with o.5 mg/day of Haloperidol.

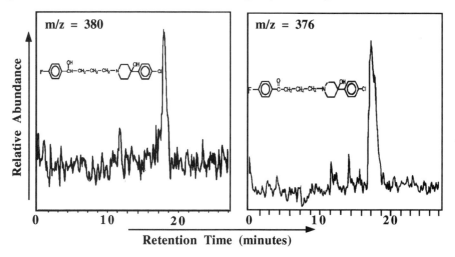

Figure 13.6 The Chromatogram Obtained by Single Ion Monitoring from a CE/MS Analysis of Haloperidol and its Metabolite Contained in an Urine Extract.

The capillary zone electrophoresis conditions were 50 mM NH_4OAc containing 10% methanol and 1% of acetic acid separated at 15 kV in an uncoated silica capillary (65 cm long and 50 µm I.D.). The sheath liquid for the electrospray interface was a mixture of isopropanol, water and acetic acid (60:40:1) and the electrospray voltage was -3.4 kV. The eluents were scanned from 50 to 400 D at 2 sec/decade and a resolution of 1200. The ions at 376 and 380 corresponded to unmetabolized Haloperidol and reduced metabolite respectively.

Capillary Electrophoresis Operated in Tandem with the Magnetic Sector Mass Spectrometer

Capillary electrophoresis has also been operated in tandem with a magnetic sector mass spectrometer by Perkins and Tomer [9], who employed both bare fused silica tubes, and those with surfaces carrying 3-

(aminopropyl)-trimethoxysilane derivatives, for the electrophoretic separation. The tandem system was used to separate some simple peptide mixtures and the small proteins that naturally occur in snake venom. A potential problem encountered in the electrophoretic separation of the basic peptides and proteins results from the net positive charge that they acquire, even in solvents at high pH values. This leads to their retention on the negatively charged column walls, and consequent peak broadening. If the walls of the tube are derivatized with 3-(aminopropyl)trimethoxysilane, then the amino-propyldimethoxy groups on the column surface will cause the surface charge to become positive.

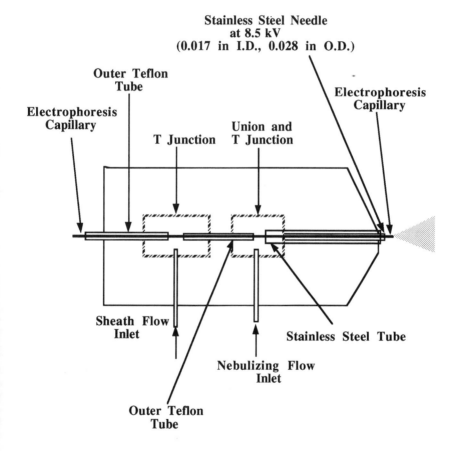

Figure 13.7 Modified Probe for Capillary Electrophoresis Interface with Magnetic Sector Mass Spectrometer

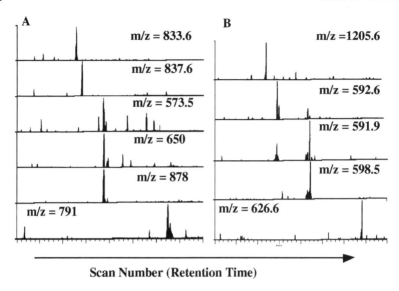

A. The separation of α–Melanocyte (1 pmol, M = 1665.1), Neurotensin,(1.8 pmol, 1673.2), Met-enkephalin (2 pmol, 877.2), β-Casomorphin (2.5 mol, 790.0), Proctolin (2.5 pmol, 648.8) and Met-Enkephalinamide, 3.0 pmol, 572.8) in a bare silica tube.
B. Five releasing hormones separated on a derivatized silica tube.

Reprinted with permission from J. R. Perkins and K. B. Tomer, *Anal. Chem.*, **66**(1994)2835 Copyright 1994 American Chemical Society.

Figure 13.8 Extracted Electropherogram of Some Bioactive Peptides (A) and Some Hormones (B)

This will also cause the direction of the osmotic flow to be reversed, so that it will flow from a high negative potential to ground. The use of buffers below pH 10.6 will result in the wall having a positive charge and will thus repel the positively charged analytes. The spectrometer was the Kratos Concept ISQ, fitted with an electrospray attachment. The probe was modified to accommodate a zero dead volume T, which permitted the coaxial introduction of the make–up flow (50% methanol and 50% water containing 3% of acetic acid). A diagram of the modified probe is shown in Figure 13.7. The electrospray needle comprised a piece of stainless steel tubing, 0.028 in. O.D. and 0.017 in. I.D. (22 gauge). The use of the larger needle allowed columns with greater outer diameters to be employed, which were much more resistant to 'electro-drilling' and subsequent column breakage. The electrophoresis column was 1.1 m long, 75 µm I.D. and 375 µm O.D., and buffers, of 0.01 M ammonium acetate (pH 8.0) and 0.01 M acetic acid (pH 3.5), were used with the native silica, and the derivatized silica tubes, respectively.

The reconstructed electropherograms shown in Figure 13.8, clearly demonstrate the advantages that are obtainable from the use of derivatized tubes. The sharper elution curves depicted in Figure 13.8 B, show that the dispersion that occurs in the bare silica tube (A), due to ionic retention by the surface charges (indicated by the dispersion at the base of the larger peaks), has been virtually eliminated by the use of the derivatized tube. Indirectly, the use of the derivatized tube has also increased the resolution attainable from the capillary column.

Capillary Electrophoresis Operated in Tandem with the Time of Flight Mass Spectrometer

Fang et al., [10] employed a capillary electrophoresis instrument, in tandem with a time of flight mass spectrometer, to examine a series of peptide and protein mixtures. The system was claimed to have a separation efficiency of 50,000 theoretical plates, with a concentration detection limit of about 1 to 2×10^{-6} molar. Thus, in units of molecular weight this would be equivalent to a sensitivity of 40–80 fentamol. As already discussed, defining sensitivity in this way can be difficult to appreciate for analysts, and this is another good example of the confusion that can arise. It was shown in their publication that the egg-laying hormone, used by these authors, had a measured mass of about 1000 m/z, when carrying a quadruple charge. Consequently, this value of m/z would mean that the actual molecular weight of the egg-laying hormone will be about 4,000.

Thus, 40–80 fentamol of egg hormone will be equivalent to,

$$4000 \times (40 \text{ to } 80) \times 10^{-15} \text{ g}$$

Therefore the mass sensitivity of the system will be *ca* 0.2 to 0.3 ng. In a similar way the concentration limit can be defined as

$$4000 \times 10^{-6} \text{ g/l} = 4000 \times 10^{-9} \text{ g/ml} = 4 \times 10^{-6} \text{ g/ml}$$

From the point of view of the analyst, it is seen that the actual *concentration* sensitivity in terms of modern instrumentation is relatively poor. However, due to the very small volumes involved, the actual *mass* sensitivity is commensurate with that of other modern detection methods.

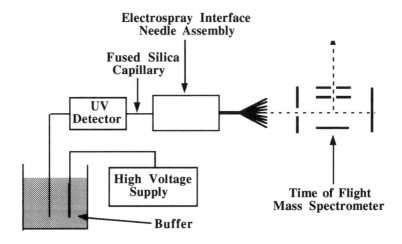

Figure 13.9 The Basic Tandem Instrument Incorporating the Time of Flight Mass Spectrometer

The general arrangement of the apparatus is shown in Figure 13.9. It consisted mainly of three sections, the usual high-voltage electrophoresis power supply, the buffer reservoir, fused quartz capillary conduit and electrospray injection system, and finally the mass spectrometer. The time of flight mass spectrometer was that manufactured by R. M. Jordan and Co. The details of the interface are shown in Figure 13.10.

The fused quartz capillaries were 110 to 120 cm long, 200 μm O.D., and 100 μm I.D., and the running buffer was composed of a 50 mM solution of acetic acid in 50/50 methanol water mixture, having a pH of approximately 4.0. The electrospray unit was a modified form of that manufactured by Analytica Inc., the fused quartz capillary replacing the stainless steel needle of the original device. A thin gold wire, 25 μm in diameter, was inserted about 2 mm into the exit orifice of the quartz capillary (the use of a microscope was found necessary for this operation), and the other end connected to the stainless steel outer tube with silver paint.

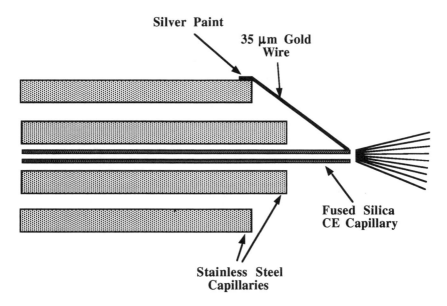

Figure 13.10 The Capillary Electrophoresis/Time of Flight Mass Spectrometer Interface

This arrangement applied the appropriate potential to the exit liquid to cause electrospray. Care had to be taken to ensure that good contact was made between the gold wire and the edge of the capillary, to avoid the excessive production of methanol ion clusters. The electro-osmotic stream amounted to a flow rate of about 0.15 to 0.26 µl/min, which, with a potential of 3 to 4 kV, provided a stable spray. No extra solvent from a sheath-flow was found to be required. A stream of heated nitrogen gas flowed in the counter direction to the electrospray to facilitate drop evaporation, and consequent ion formation.

The lay-out of the time of flight mass spectrometer is shown in Figure 13.11. In the Jordan instrument, an orthogonal geometry was employed, and the 'ion-packet' was generated for measurement by pulsing the continuous beam at 90° to the ion path from the electrospray.

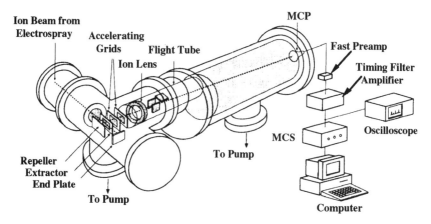

Courtesy of the R. M. Jordon Co.

Figure 13.11 The Configuration of the Time of Flight Mass Spectrometer Chamber

This was accomplished by applying 400V pulses at 10 kHz (10 ns rise time) to a repeller plate situated at the side of the ion path from the electrospray. The potential on the second skimmer plate of the electrospray had to be kept relatively high (100–120 V) to obtain a usable ion current (5 pA). This resulted in the ions, in the ion packet and ion storage regions, having relatively large translational energies (i.e. 100–130 eV) and a large energy spread. As a consequence, such ions were difficult to turn through 90°. to carry out mass analysis. Due to the high energy, applying a positive potential to the end plate could only slow a portion of the total beam. The positive potential applied on the end plate was carefully chosen to ensure that, when combined with the 400 V transient pulsing potential, a 90° ion-turning mirror was formed, to deflect the majority of the ions into the time of flight tube. In addition a long focal length ion lens was employed, immediately after the acceleration grids, to refocus the ion packet.

A pair of x and y plates were used to fine tune the ion trajectories to the ion sensor. The data acquisition system comprised the usual fast preamplifier, filter amplifier, computer and oscilloscope presentation. The

Capillary Electrophoresis

results obtained from the apparatus in the separation of the six bag cell peptides, from the sequencing of the egg laying hormone, are shown in Figure 13.12.

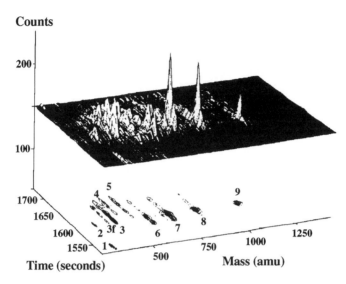

Peptide Name	Amino Acid Sequence
Egg-laying Hormone (ELH)	ISINQDLKAITDMLLTEQIRERQ–RYLADLRQRLLEK
α bag cell (1–7) (α1)	APRLRFY
α bag cell (1–8) (α2)	APRLRFYS
α bag cell (1–9) (α3)	APRLRFYSL
β bag cell	RLRPH
γ bag cell	RLRPD

Reprinted with permission from L. Fang, R. Zhang, E. R. Williams and R. N. Zare, *Anal. Chem.*, **66(21)**(1994)3696, Copyright 1994 American Chemical Society

Figure 13.12 A Two-dimensional Contour Presentation of the Separation and Mass Spectra of a Sample of Egg-laying Hormone

The authors modestly claimed that the present study was rather primitive, and that both mass resolution and sensitivity could be improved.

Nevertheless, it is seen that very useful results were obtained and the system obviously has a wide field of uses to which it could be applied.

The Capillary Electrophoresis/Quadrupole Mass Spectrometer Tandem System Employing an Atmospheric Pressure Ionization Source

Although present methods of ionization are very successful for certain types of sample, they are not universally applicable, and to improve the scope of the capillary electrophoresis/mass spectrometer tandem system, Takada *et al.* [11] introduced the atmospheric ionization interface. In addition, other types of interface placed severe restrictions on the amount, and type, of buffer that could be used and inhibited the use of electrokinetic micellular chromatography. The basic interface used by Takada *et al.* consisted of an electrospray nebulizer, a heated vaporizer and an atmospheric chemical ionization source fitted with a needle electrode. The layout of a commercial API source for operation with capillary electrophoresis is shown in Figure 13.13. The mass spectrometer employed was the Hitachi M-1000. The interface originally supplied with the spectrometer employed a heated pneumatic nebulizer, but this was replaced by a sheath flow-assisted electrospray nebulizer. One end of the fused silica capillary tube (50 μm I.D. and 150 μm O.D. and 40 cm long) was inserted into the electrospray probe (0.25 mm I.D. and 0.40 mm O.D. stainless steel capillary), which was maintained at 2.8 kV by an external power supply. The sheath flow was delivered by a syringe pump at a rate of 5 μl /min. The droplets produced passed into the heated vaporizer, which consisted of a stainless steel block with a hole 5 mm I.D. and 6 cm long, and was heated uniformly to 300°C by appropriately placed cartridge heaters.

The needle electrode was situated between the vaporizer and the sample aperture of the mass spectrometer and was maintained at 3 kV. The vaporized sample was ionized by corona discharge which resulted in a number of different types of ion–molecule reactions. The ions that were formed were introduced into the mass spectrometer using the usual differential pumping arrangement. The first and second apertures were heated to 120°C, and the distance between the first aperture and the electrospray probe was about 8 cm. The background spectra from the atmospheric ionization interface were very simple, in which the protonated methanol peak (m/z = 33) was dominant. However, spectra from the electrospray interface also exhibited a strong background of sodium ions.

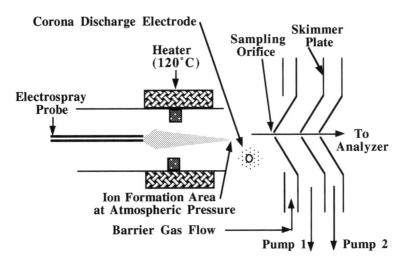

Figure 13.13 The Capillary Electrophoresis/API–MS Tandem Apparatus.

Capillary Electrochromatography Employing the Electrospray Interface

Electrochromatography is not a form of electrophoresis, and so strictly, the subject should not be included in a chapter on electrophoresis tandem systems. Electrochromatography is a term given to a liquid chromatographic separation that employs a capillary column, but utilizes electro-osmotic flow to drive the mobile phase through the column. Physically, the system is very similar to capillary electrophoresis, and employs a very comparable type of apparatus. Consequently, the electrochromatograph/mass spectrometer tandem system will be included in this chapter.

The use of electro-osmotic flow, to replace the high-pressure pump in a liquid chromatograph, not only simplifies the mobile phase supply system, but is also provides plug flow through the column, as opposed to the parabolic velocity profile normally associated with fluid-flow through tubes. The lack of a parabolic velocity profile can have a dramatic effect on the column efficiency, as shown in Chapter 1. One of the early reports on the successful use of electrosmotic flow in LC, was that of Schmeer *et al.* [12], and the layout of their apparatus is depicted in Figure 13.14.

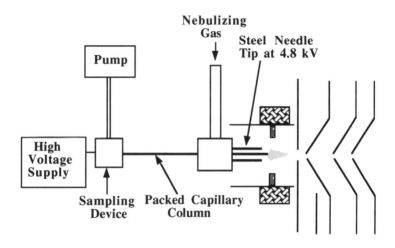

Figure 13.14 The Capillary Electrochromatography Apparatus

Electro-osmotic flow is obtained by applying a strong electric field across the length of the column. The velocity of the flow is independent of the particle size of the packing, and so columns 0.5 mm in diameter, packed with particles 1.2 μm in diameter, can readily be used if so desired. Employing columns of this diameter, packed with such small particles, efficiencies of 200,000 theoretical plates can easily be obtained. In practice, to minimize resistance heating, columns of 50 to 100 μm I.D. are usually employed, with flow rates up to 2 μl/min. This magnitude of the flow rate is ideal for electrospray nebulizing and consequently, no sheath flow is necessary. However, in order to assure stable flow conditions a supplementary flow of mobile phase was provided by a mechanical pump as shown in Figure 13.13. The fused capillary column 20 cm long, 100 μm I.D., and 360 μm O.D., was fitted with a terminal fused silica gel frit (pore diameter 5 μm) by sintering. The tubes were then slurry packed with 1.5 μm particles of reversed phase, and the column terminated with another silica gel frit. The supplementary pressure, applied by the pump to the column during operation, was about 2250 psi.

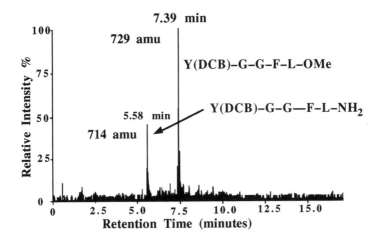

Reprinted with permission from K. Schmeer, B. Behhnke and E. Bayer, *Anal. Chem.*, **67(20)**(1995)3656, Copyright 1995 American Chemical Society

Figure 13.14 Selected Ion (m/z, 714, 7129) Chromatogram of a Peptide Separation

To avoid bubble formation from resistive heating, the concentration of TFA in the mobile phase was kept to 0.07 ml/l, and the mobile phase consisted of a mixture of 40% acetonitrile and 60% TFA solution in water. The applied potential gradient, used to produce electro-osmotic flow, was the difference between the overall potential 25 kV and that of the electrospray potential 4.8 kV, which was equivalent to about 20 kV. The accompanying electro-osmotic current was 0.8 μA. The concentration of the solutes in the sample placed on the column was about 20 μg/ml. The high selectivity and separating power of the technique is demonstrated by the separation of the enkephalin methyl ester and the enkephalin amide shown in Figure 13.14. It must pointed out, however, that the selectivity of the separation was enhanced by the inevitable accompanying electrophoretic migration. The mass spectra indicated a high abundance of sodium and potassium adducts. These adducts significantly reduce the intensity of the simple protonated species. In fact, in order to detect the small amount of peptide injected (3 pmol) it was found necessary to decrease the resolution of the mass spectrometer to increase its sensitivity.

In theory, at least, the combination of the separating power of chromatography with that of electrophoresis used in conjunction with the mass spectrometer, should provide a very powerful tandem analytical technique indeed. In practice, however, it would appear that there is some way to go before the real potential of this particular tandem system is fully realized.

Off-line Matrix Assisted Laser Desorption Ionization Time of Flight Mass Spectrometric Monitoring of Capillary Electrophoresis Separations

Although several in-line electrophoresis/mass spectrometer tandem systems have been developed and described, the in-line approach, employing a suitable interface, has some distinct disadvantages as well as advantages. First, in most in-line systems, only a relatively small proportion of the sample is ionized, and reaches the mass spectrometer, thus the sensitivity of the overall system is often relatively poor. Second, the buffer solutions that can be used with in-line systems are limited, both in type and concentration, including micellar additives. As a consequence, the optimum performance of many forms of capillary electrophoresis cannot be realized when using in-line interfaces. In order to eliminate some of these disadvantages, Walker *et al.* [13] introduced an off–line fraction collecting system, to allow the total sample to be ionized by matrix assisted laser desorption ionization (MALDI), in conjunction with a time of flight mass spectrometer (TOF-MS). The collection device was miniaturized, and ultimately provided mass spectra from as little as 25 fmol of protein. In assessing this sensitivity in standard analytical terms, however, it should be noted that the molecular weights of proteins generally are very high. Theoretically, the combination of (MALDI) with (TOF-MS) can be made to provide a very high sensitivity, and can

Capillary Electrophoresis

obviously tolerate a wide range of diverse analytical conditions. However, many of the MALDI/TOF-MS in-line interfaces that have been developed have met with limited success for the reasons previously mentioned. If off-line measurements are made, there is little or no restriction on the choice of mass spectrometer, probe or ion source.

Source contamination with inorganic residues does not occur to anything like the same extent, and any type of buffer can be used (including inorganic buffers) if so desired. In addition, the use of off-line procedures allows the two instruments to be physically separated, and are therefore more convenient to operate, and each instrument can be optimized without affecting the performance of the other.

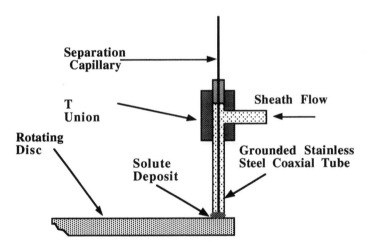

Figure 13.15 The Capillary Interface and Fraction Collector

The details of the transport fraction collecting system has been described in a previous report, [14]. It consists of a rotating disk on the heated surface of which is deposited the eluent from the electrophoretic capillary. After deposition, the disk is placed in the TOF-MS and rotated such that each spot is sequentially exposed to the laser beam. The device allows the electrical connections to be established with the separating capillary by one of two procedures. Either a coaxial sheath flow is employed through which the electrical connection is made or the outer surface of the capillary is coated with a gold filled epoxy resin. Both arrangements allow closely spaced fractions to be collected with minimal cross-contamination and dilution. Sample recoveries are better than 80% and virtually independent of the nature of the sample. A diagram of the fraction collector is shown in Figure 13.15. The coaxial capillary interface (to the fraction collector) was made from a 1/16 in. Swagelok T, and a 2 cm length of 700 μm O.D. and 530 μm I.D. fused silica tubing. Fluid was continuously pumped through the T by a mechanically actuated syringe, at a flow-rate of about 100 μl/ min. If the coaxial sheath flow was dispensed with the capillary interface was could be rendered conductive by coating it with gold-filled epoxy resin, and curing for 15 minutes at 120°C.

The electrophoretic separations were carried out on capillary tubes, 50 cm long, 52 to 75 μm I.D., having a detection window for the UV detector situated 20 cm from the end. The activity of the capillary walls was reduced by derivatization with (3-aminopropyl)triethoxysilane. Samples were introduced both by electro-kinetic and/or hydrodynamic procedures. During fraction collection, the disk was moved by a stepping motor as each peak was eluted. The separation was monitored by a variable wavelength UV detector at either 200 or 254 nm, and besides monitoring the separation, the output was used to actuate the disk, and move it on, to allow the next peak to be deposited onto a clean portion of the disk. The time taken for the solute to pass from the detector cell to the end of the capillary was taken into account in the fraction collector program. The eluates were collected sequentially on the perimeter of the disk, and after each peak was eluted, the same portion of disk was replaced under the capillary tube exit to collect between-peak eluent, or any zones that were not specifically required.

The MALDI TOF-MS instrument was that manufactured by Laser/Tec Research and was equipped with a neodymium–YAG laser. Under typical conditions, a spot of 50 μm diameter was sampled with 30 μJ pulses. All spectra were measured in the linear time-of-flight mode (1.3 m flight path). The ions were accelerated by a potential of about 30 kV, and the resulting ion current monitored at 2 ns intervals. Spectra were generated from about 12 to 75 single consecutive acquisitions with 2–10 point smoothing. Separation efficiencies were reported to be 130,000 theoretical plates.

Off-line fraction collecting procedures produce excellent results and, due to the method of fraction collection, the tandem arrangement is very similar to a disk transport interface. Until there has been significant improvements made to the present in-line interface devices, to make them more amenable to a wide range of CE operating conditions, the off-line procedure developed by Walker *et al.*, might well be the preferred starting point for those analysts interested in utilizing CE/MS tandem techniques.

Interface for Capillary Electrophoresis and Inductively Coupled Plasma Mass Spectrometry

Lu *et al.* [15], also employed a coaxial conductive liquid sheath as a return ground in capillary electrophoresis, in a tandem apparatus incorporating an inductively coupled plasma mass spectrometer. The overall tandem instrument was evaluated in the separation of metallothionein and ferritin. A diagram of the interface is shown in Figure 13.16. The basic system was built round the ICP/MS (Perkin Elmer Sciex ELAN 5000a), but the normal double-pass spray chamber, and cross-flow nebulizer were replaced by a conical chamber, with an inner impact bead and a commercial glass nebulizer. A Teflon adapter was machined to connect the torch box to the spray chamber outlet. Except for nebulizer flow-rate adjustment, the operating parameters of the ICP-MS remained unchanged. In normal CE/MS operation, the capillary tip is held at 3–5 kV to operate the electrospray ionization–evaporation process. However, the capillary in the nebulizer must be held at *ground* potential, because any electrical disturbance of the plasma will cause it to extinguish. The contact with ground potential was achieved using an electrically conducting sheath liquid. A concentric glass nebulizer (manufactured by Meinhard) was used and the CE capillary was inserted into the nebulizer central glass tube.

Since the glass tube tapered towards the tip of the nebulizer, the CE capillary did not reach the tip, and terminated about 2 cm from the end.

Another fused silica capillary, 530 μm I.D. and 800 μm O.D., was also inserted coaxially into the nebulizer central glass tube, around the CE capillary. This outer capillary only extended to the joint with the nebulizer gas tube. The maximum distance between the ends of the CE capillary and the outer make-up capillary was 18 mm. The other end of the outer capillary was sealed in a stainless steel tube (1/16 O.D.) that was connected to a stainless steel T. The coaxial sheath-flow (10 mM sodium chloride) was supplied through the outer capillary.

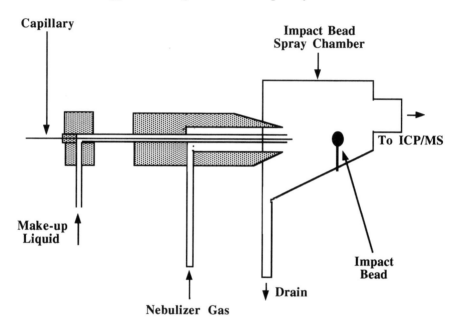

Spray chamber volume 35 ml, impact bead diameter, 4 mm.

Figure 13.16 The CE/ICP-MS Interface

Metallothionein (MT) is a small, metal-binding protein, the mass of which varies from under 6000 to 7000 Da, and contains 61 amino acid residues, 30 of which are cysteine. In most animal species MT can take two forms (MT1 and MT2) that are different in charge due to amino acid composition. Up to seven divalent transition ions, usually, zinc, cadmium and copper, can bind to metallothionein through metal-thiolate bonds. Ferritin is one of the major iron storage proteins with a mass between

460 and 480 Da. These materials were use to evaluate the tandem system and the results obtained are shown in Figure 13.17.

It is seen that a good separation was obtained and the different metals are easily speciated on the basis of their m/z value. Little if any resolution is lost in passage through the interface, and with single ion monitoring, the individual metal elements can be selected for exclusive display.

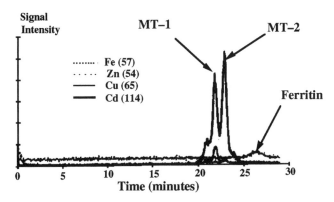

Reprinted with permission from Q. Lu, S. M. Bird and R. M. Barnes, *Anal. Chem.*, **67(17)**(1995)2949, Copyright 1995 American Chemical Society.

Figure 13.17. The Separation of Some Metallothionein Isoforms and Ferritin

Capillary Electrophoresis Coupled with Inductively Coupled Plasma Ionization and Atomic Emission Spectroscopy

Olesik *et al.* [16] developed an interface to connect a capillary electrophoresis instrument to an inductively coupled atomic emission spectrometer. They developed a device that produced a fine aerosol from the end of an electrophoresis capillary, that could be introduced directly into the ICP. In practice, they employed element selective detection, as a trade-off for electrophoretic resolution, to increase the speed of analysis. The detection limit obtained was about 0.06 ppm.

The capillary was made of fused silica, about 40–50 cm long, 97 μm I.D., and 130 μm O.D. About 10 kV was employed to develop the separation.

The inlet of the capillary was held at a positive potential, while the outlet was grounded. Typical electrophoretic currents were 5–10 µA, and the volume of the capillary was about 4.3 µl. The basic nebulizer was the Meinhard TR–30–A3, the nebulizing argon flow being accurately governed by a precision flow controller. 4 to 5 cm of the external surface of the capillary was coated with silver paint which served two purposes. First it provided an electrical connection to ground, and second it acted as a cushion so that it fitted snugly inside the center tube of the pneumatic nebulizer. The end of the electrophoresis tube was placed within 0.5 mm of the tip of the two tubes forming the concentric pneumatic nebulizer. It was noted that under certain conditions contamination with silver ions could occur. The spray chamber is usually employed to prevent large droplets entering the plasma, and to restrict the aerosol transport to about 20 µl/min. The conical spray chamber with an impact bead is typically used to prevent excessive aerosol loading. However, as the flow from the capillary was so small, a spray chamber similar in shape but devoid of an impact bead was employed.

The interface was coupled to a standard ARL ICP torch and an ARL 34000 direct-reading optical emission spectrometer. The outer argon gas flow was 15 l/min, and the intermediate flow was about 1 l/min. A power of 1.1 kW was used to form the plasma. The system was used to monitor chromium and the total chromium and ion concentrations were about 100 µg/ml, and the electrolyte employed was 0.04 M sodium acetate at pH 8.2. The different elements were exclusively displayed by monitoring the light emission at 259.9 nm (iron) and 267.7 nm (chromium) respectively. In addition, the electrophoresis clearly separated the different species containing each of the two elements.

The separation of the two valency states of tin was also demonstrated in employing basically the same interface but with the ICP torch coupled to a mass spectrometer. It was found possible to monitor 1 ppm of tin in each valency state which was separated in 0.04 M sodium chloride, at pH 7.0. The electrophoretic potential employed was 10 kV. The separation of the two valency states of tin was complete in under a minute with more than baseline resolution, indicating that the system could be developed to produce exceedingly fast analyses if so desired. In addition the CE/ICP-AES tandem system was employed successfully to separate the different copper species. The samples contained Cu^{2+} ions, $Cu(EDTA)^{2-}$ and a mixture of Cu^{2+} and $Cu()EDTA)^{2-}$ ions. The separation was carried out in 0.06 M calcium chloride solution, pH 5.0, under a potential of 4 kV.

The advantage of the element selectivity of the ICP–AES, and the separating efficiency of the capillary electrophoresis, is seen to be a very powerful analytical tandem combination. In addition, for routine repetitive analyses, the speed at which the separations can be carried out also makes the combination the ideal technique for monitoring element speciation.

NMR Spectroscopy on the Nanoliter Scale in Tandem with Capillary Electrophoresis

Finally, one of the most exciting, recent developments in tandem techniques must be described, and that is the design and construction of nanoliter cells for NMR spectroscopy by Wu *et al.* [16]. Such cells can allow the NMR spectrometer to be used in tandem with capillary electrophoresis, and provide the unique CE/NMR combined instrument. The cells were formed by wrapping a radio frequency coil directly round the fused capillary tube that was to be used for the electrophoretic separation. The cell so formed was about 1 mm in length, and for capillaries ranging in I.D. from 75 to 530 µm, the volume would be between 2 and 200 nl. Using such cells, spectra for amino acids could be obtained from as little as 50 ng of material after one minute of data acquisition. The NMR lines were several hundred Hertz wide, but could be reduced by choosing fused silica tubes of the appropriate wall thickness.

The magnitude of the NMR signal detected by the sensor coil is directly proportional to the volume of the cell. However, to obtain maximum sensitivity the device must be designed to provide optimum coupling between the coil and sample. The advantage of the use of smaller coils is to obtain greater mass sensitivity, and the ability to work with volumes as small as those produced in capillary electrophoresis separations. Although sample masses of less than 100 ng (0.1–1 vmol) can provide good spectra, the concentration sensitivity of the device is relatively poor. The experimental arrangement of the micro cell and electrophoresis apparatus is shown in Figure 13.18.

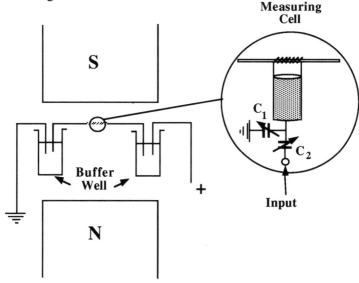

Reprinted with permission from N. Wu, T. L. Peck, A. G.Webb, R. L. Magin and V. Sweedler, *Anal. Chem.*, **66(22)**(1994)3849, Copyright 1994 American Chemical Society.

Figure 13.18 The Experimental Arrangement of the NMR Microprobe Electrophoresis Combination

The micro-coil is wound directly onto the polyimide-coated fused silica tubing, 75 mm I.D. and 355 µm O.D. The capillary serves as both the sample chamber and the former for supporting the micro-coil. The coil was wound with the aid of a pin vice, micro-manipulator and stereo microscope. The copper wire was 42 gauge (63.1 µm diameter) insulated with varnish, and typically 14–17 turns of wire were placed on the tube. To minimize susceptibility-induced static magnetic field distortions the coils were formed with single tightly wound turns, and the bifilar winding technique was not employed. Initially, one end of the coil was attached to the silica tube wall with a drop of epoxy cement, and held under constant tension. The wire was then slowly wound on the tube by gentle rotation and finally secured with more epoxy cement. The epoxy cement does cause line broadening, but its effect was minimized by only using a very small quantity at the ends of the coil. The impedance matching capacitors also have some susceptibility, and to reduce any broadening effect that might result, they are placed at opposite ends of the 3 cm length of coaxial cable.

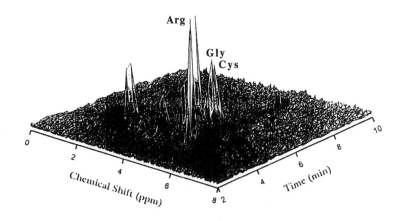

Reprinted with permission from N. Wu, T. L. Peck, A. G.Webb, R. L. Magin and V. Sweedler, *Anal. Chem.*, **66(22)**(1994)3849 Copyright 1994 American Chemical Society.

Figure 13.38 A Two-dimensional Presentation of the Separation of Some Amino Acids Employing the Capillary Electrophoresis/NMR Spectrometer Tandem System

The micro-coil assembly was mounted in a modified NMR probe and oriented so that the coil was perpendicular to the static magnetic field over the detection region. In order to avoid interference and damage from the high voltages necessary in capillary electrophoresis, the potentials used were kept below the breakdown voltage of the silica tube in the sensor volume section. The NMR spectrometer that was employed was the GN-300(7.05T)/89 mm, and the capillary electrophoresis system was situated in an acrylic enclosure at the end of the NMR probe. The reduction in sample volume is not without sacrifice the reduction in sensor volume of nearly three orders of magnitude is accompanied by a twenty-fold loss in sensitivity.

An elegant demonstration of the use of the CE/NMR tandem technique is shown in Figure 13.18. The two-dimensional presentation has the electrophoretic separation time represented on one axis and the chemical shift on the other. The sample separated is a mixture of arginine, glycine and cysteine. A complete spectrum was acquired every 16 seconds, with each being eight scans of a 512 point, 4000-Hz spectral width.

Synopsis

Capillary electrophoresis is capable of generating extremely high plate efficiencies but, at the same time, also produces solute bands of extremely small volume. Consequently, the association of the technique with a spectrometer demands special interfaces and techniques. Because of its readily available high sensitivity, the mass spectrometer was one of the first tandem instruments to be associated with the capillary electrophoretic system. An electrospray interface was used in conjunction with a quadrupole mass spectrometer. The combination was reported to have a minimum detectable mass of about 0.1 pg. The sensitivity of the system was subsequently improved by reducing the diameter of the electrophoresis capillary. A number of concentration enhancement techniques were tried, including micro-cartridge extractors and isotachophoretic concentration. Capillary electrophoresis has also been satisfactorily associated with the ion trap mass spectrometer, the magnetic segment mass spectrometer and the time of flight mass spectrometer, using both the electrospray interface and the atmospheric ionization interface. Several modified jet systems have been examined for the CE/MS interfaces, to optimize on sensitivity without denigrating the electrophoretic resolution. The basic CE/MS tandem system has also been used with microbore columns, using electro-osmotic flow as an alternative to hydraulic flow. The in-line association of capillary electrophoresis with a spectrometer has a number of disadvantages, as only a small proportion of the solutes are ionized and reach the spectrometer, and the in-line system places restrictions on the operating conditions that can be chosen for the capillary electrophoresis. Consequently, some of the best CE tandem results have been obtained off-line, using a hybrid fraction collector-transport interface. Capillary electrophoresis has also been successfully coupled to both the mass spectrometer and the atomic emission spectrometer, using an inductively coupled plasma interface. In addition, by employing very small sensor cells and coils, CE/NMR tandem systems have also been demonstrated to be effective, sensitive, and provide results in which the resolution of neither the capillary electrophoresis separation, nor the NMR spectra has been seriously impaired.

References

1. J. A. Olivares, N. T. Nguyen, C. R. Yonker and R. D. Smith, *Anal. Chem.*, **59**(1987)1230.
2. E. D. Lee, W. Muck, J. D. Henion and T. R. Covey, *J. Chromatogr.*, **458**(1988)313.
3. J. H. Wahl, D. R. Goodlett, H. R. Udseth and R. D. Smith, *Anal. Chem.*, **64**(1992)3194.
4. R. P. W. Scott, *Liquid Chromatography Detectors*, Elsevier, Amsterdam-Oxford-New York (1986)22.
5. R. P. W. Scott, *Chromatography Detectors*, Marcel Dekker, New York-Base-Hong Kong, (1996)36.
6. J. Cai and Z. El Rassi, *J. Liq. Chromatogr.*, (**6&7**)(1992)1179.
7. T. J. Thompson, F. Foret, P. Vouros and B. L. Karger, *Anal. Chem.*, **65**(1993)900.
8. J. D. Henion, A. V. Mordehal and J. Cal, *Anal. Chem.*, **66(13)**1994)2103.
9. J. R. Perkins and K. B. Tomer, *Anal. Chem.*, **66**(1994)2835.
10. L. Fang, R. Zhang, E. R. Williams and R. N. Zare, *Anal. Chem.*, **66(21)**(1994)3696.
11. Y. Takada, M. Sakairi and H. Koizumi, *Anal. Chem.*, **67(8)**(1995)1474.
12. K. Schmeer, B. Behhnke and E. Bayer, *Anal. Chem.*, **67(20)**(1995)3656.
13. K. L. Walker, R. W. Chiu, C. A. Monnig ans C. L. Wilkins, *Anal. Chem.*, **67(22)**(1995)4197.
14. R. W. Chiu, K. L. Walker, J. J. Haggen, C. A. Mornig, and C. L. Wilkins, *Anal. Chem.*, **67**(1995)4190.

15. Q. Lu, S. M. Bird and R. M. Barnes, *Anal. Chem.*, **67(17)**(1995)2949.
16. J. W. Olesik, J. A. Kinser and S. V. Olesik, *Anal. Chem.*, **67(1)**(1995)1.
17. N. Wu, T. L. Peck, A. G. Webb, R. L. Magin and V. Sweedler, *Anal. Chem.*, **66(22)**(1994)3849.

Index

absorbance ratioing, peak purity 259
acylation, derivatizing technique 209
additives, post column for LC/MS 375
aflatoxins, by LC/FS 275
alkaloids, by LC/UV 263
alkylbenzimidzole, analysis by GC/IR 159
alumina, IR transmission 465, 470
amino acids, by CE/NMR 513
anabolic steroids by GC/MS 179
analgesics LC/UV analysis 265
analytical instruments
    evolution of 3
    history of 3
anthracene, LC/FS analysis 271
anti-inflammatory drugs, LC/UV 266
API interface 384
archaeological potsherds, lipids by GC/MS 184
argon matrix, for IR measurement 149
aromatic hydrocarbons
    by GC/IR 145
    by LC/UV 260
    by TLC/IR 475
arsenic by LC/AS 421
aspirin, by TLC/IR 468
atmospheric ionization interface 384
atomic
    absorption spectrometer, flame 75
    absorption spectrometry 75
    emission spectrometer 73
        helium plasma 74
        inductively coupled 244
    spectroscopy 73
Babington nebulizer 410
basil oil, GC/IR analysis 143
beef fat, dioxins by GC/MS 178
belt transport interface 344
benzodiazepines, LC/UV analysis 265
benzyl benzoate, IR spectrum of 310
Bieman concentrator 167
blood plasma
    4,4'-methylene bisaniline by GC/MS 192
    anabolic steroids by GC/MS 181
    medroxyprogesterone by GC/MS 181
blood, iron by LC.FAAS 408

bonded phases 22
bovine albumin, spectra of 338
bradykinin, TLC/MS assay 477
brush phases 23
bulk phases 25
caffeine, by TLC/IR 468
cancer cells, waxes and lipids by GC/MS 183
capillary
    columns 14
        fused silica 14
    coupling, LC/NMR 444
    electrophoresis 35
        apparatus 40
carboxylic acids, by LC/UV 256
cardiovascular drugs, by LC/UV 263
carousel transport interface, LC/IR 291
CE/ tandem techniques 483
CE/AES
    ICP interface 509
CE/MS
    analysis of egg laying hormone 499
    API interface 500
    electrospray interface 488
    ferritin, iron speciation 509
    haloperidol in urine 492
    ICP interface 507
    ion spray interface 491
    of proteins 489
    peptides, bioactive 494
    sensitivity enhancement 486
    structure of quaternary ammonium salts 485
    transport interface 505
    using the ion trap MS 490
    using the quadrupole MS 484
    with magnetic sector MS 492
    with time of flight MS 495
CE/MS concentration techniques 487
CE/NMR 511
    amino acid separation 513
    microprobe interface 512
cell
    flow, fluorescence 270
    gas, IR 136
    micro IR 315
chemical ionization 94, 171
    effect of reagent gas 172

mechanism of 171
reagent gas supply 173
chiral
   detector 81
   instruments 80
   optical systems 76
cholesterol purity by GC/MS 186
chromatography 7
   classification of 8
   liquid 18
   thin layer 25
chromium
   by LC/AS 427
   by LC/MS 391
   in urine by GC/MS 196
   speciated by LC/MS 391
circular dichroism spectrometer 79
circular transport interface for LC/FAAS 411
classification 8
cloxacillin, LC/MS analysis of 371
coal gasification 246
coiled tubes, dispersion in 125
columns
   capillary 14
   efficiency 29
   function of 27
   GC 12
      dispersion in 106, 112
      packed, dispersion in 108
   glass, 60 ft 137
   LC 19
      dispersion in 114, 116
   packed 12
   SCOT 17
combustion interface 219
concentration techniques, for CE/MS 487
concentric flow nebulizer 300
conduits, interface 106
cone voltage, effect of 373
connecting tubes
   dispersion in 114
   GC, dispersion in 106
   low dispersion 123
continuous flow monitoring, C/NMR 439
convoluted peaks, resolution of 281
cryostat interface 147
   design of 148
cut-off, UV of solvents 264
cypermethrin, LC/MS analysis 373
cytochrome c
   by micro column electrospray 383
   LC/MS analysis 365

delay time due to dead volumes 321
derivatizing techniques 208
   acylation 209
   esterification 208
   post column 211
detector
   EC 12
   FID 12
   gas chromatographic 11
   katherometer 12
   liquid chromatographic 19
   NPD 12
dextrose IR spectrum of 312
diffuse reflectance FTIR spectroscopy 70
diffuse reflectance IR, in TLC/IR 466
dimethyl naphthalenes, GC/IR analysis 150
diode array
   fluorescence spectrometer 282
   resolution 257
   spectra and peak purity 262
   UV spectrometer 53
dioxins
   analysis by GC/IR 151
   in beef fat, GC/MS 178
direct inlet LC/MS interface 349
dispersion
   GC 106
      in open tubes 106
      in packed columns 108
   in coiled tubes 125
   in LC columns 114, 116
   longitudinal 32
   low, connecting tubes 123
   multipath 31
   open tubes 106
   peak 30
   resistance to mass transfer 32
distribution coefficient 28
diuretic drugs, LC/UV analysis 266
drug
   contaminants by GC/MS 216
   precursor, analysis by LC/IR 311
drugs of abuse, GC/IR analysis 142
egg laying hormone, by CE/MS 499
electro-endosmosis 39
electro-osmotic flow 39
   profile 39
electrochromatography, electrospray interface for 501
electromagnetic spectrum 47
electron
   capture detector 12

Index

impact ionization 93, 168
impact source 94
electrophoresis
  capillary 35
  zone 36
electrospray interface 359
environmental analysis
  by GC/MS 196
  organotin by GC/AS 235
esterification, derivatizing technique 208
evolution of analytical instruments 3
exhaust, analysis by GC/IR 146
extraction interface for LC/IR 317
extraction voltage, effect on fragmentation pattern 370
FAB 331
fast atom bombardment 331
fatty acids, in rat liver by LC/FS 277
ferritin by CE/MS 509
FID 12
field desorption ionization 338
field strengths and frequency in NMR 83
flavone analysis, by GC/MS 213
flow cell
  fluorescence 270
  LC/NMR 437
fluorescence
  spectrometer 58, 269
  spectrometer, flow cell 270
  spectroscopy 55
focusing, isoelectric 38
fragmentation pattern, effect of extraction voltage 370
frequency and field strength in NMR 84
furazolidone
  LC/MS analysis 354
gas cell IR 136
gas chromatograph
  basic 10
  detectors 11
gas chromatography 9
  columns 12
gasoline
  aromatics in 145
  GC/IR analysis 145
  lead by LC/FAAS 409
  stack plots of 145
GC/AS 227
  coal gasification 246
  gold amalgam interface 239
  ICP torch 245
  inductively coupled plasma 242
  mercury in natural gas 240
  MIP layout 231
  multielement analysis 232
  organotin analysis 234
  sediment analysis 237, 238
  selenium in coal 247
  SFC extraction 248
  siloxane analysis 249
GC/IR 133
  analysis of alkylbenzimidzole 159
  analysis of basil oil 143
  analysis of gasoline 145
  argon matrix 149
  aromatics in gasoline 145
  cryostat interface 147
  dimethyl naphthalene analysis 150
  dioxin analysis 151
  exhaust analysis 146
  modern systems 138
  optical system 139
  sensitivity to methacrylates 141
  stack plots 145
  with multidimensional analysis 152
  with TGA and MS 153
GC/IR/MS 134
  pneumatic system 134
  triple system for PCB analysis 203
GC/MS 165
  4,4'-methylene bisaniline 191
  anabolic steroids 179, 181
  biological particulates 220
  cholesterol purity 186
  chromium in urine 195
  combustion interface 219
  dioxins in beef fat 178
  drug contaminants 216
  early system 178
  environmental analysis 196
  flavone analysis 213
  grapefruit juice analysis 214
  head-space analysis 215
  herbicides in air 207
  herbicides in water 201
  ion sources 168
    chemical ionization 171
    electron impact 168
    ICP 174
  lemon juice analysis 213
  lipids 183
  lipids in archaeological potsherds 184
  medroxyprogesterone 181
  membrane interface 216
  microwave-induced plasma 228
  orange juice analysis 213
  organic chlorine in ground water 197
  PCBs

520   Index

in water 203
on dust 206
pesticides in water 199
silane purity 188
tellurium in urine 195
testosterone metabolites 180
trans-resveratol 189
urine analysis 190
vapor concentrators 165
    Bieman 167
    Ryhage 166
vapor sampling apparatus 198
waxes 183
GC/UV
    advantages 258
germanium disk, interface for LC/IR 308
ginsenoside, LC/MS analysis 367
glass column, 60ft 137
Golay equation, GC 33
grapefruit juice analysis by GC/MS 214
graphite furnace. in LC/AAS interface 415
guinea green B, TLC/MS assay 479
halogens
    by GC/MS 241
Haloperidol in Urine by CE/MS 492
head-space analysis
    for drug contaminants 216
    for GC/MS 215
helium plasma atomic emission spectrometer 74
herbicides
    contaminants in air 207
    in water by GC/MS 201
HETP 31
    curve 34
    equation, GC 33
    GC minimum 109
    LC minimum 115
high
    efficiency LC 21
    speed LC 20
history of analytical instrumentation 3
human growth hormone, LC/MS analysis 386
hydride reactor for LC/AS 424
hydrocarbons, aromatic by LC/UV 260
ICP
    interface for CE/AES 509
    interface for CE/MS 507
    interface for LC/AS 417
    interface for LC/MS 389
    ion source 174
        modified ion lens 176
    source for LC/MS 333
    torch 245
identification techniques, introduction 45
inductively coupled plasma
    interface 389
    source 333
infrared
    spectra
        interpretation of 62
    spectrometer
        Fourier transform 64
        grating 63
    spectroscopy 59
injector
    GC 13
    septum 13
    split flow 15
    valvless for GC 155
interface 382
    API for CE/MS 500
    atmospheric ionization 384
    belt transport 344
    carousel transport system 291
    circular transport for LC/FAAS 411
    combustion, for isotope ratio monitoring 219
    comparative performance 400
    concentric flow nebulizer for LC/IR 300
    conduits 106
    cryostat for GC/IR 147
    cryostat, design of 148
    direct inlet 349
    disc transport for LC/IR 295
    dual ionization belt 348
    electrospray 359
    electrospray for capillary electrochromatography 501
    electrospray for CE/MS 488
    extraction for LC/IR 317
    for LC/FAAS 407
    germanium disk for LC/IR 308
    gold amalgam for GC/AS 239
    ICP
        for LC/MS 389
    ICP for CE/AES 509
    ICP for CE/MS 507
    ICP for LC/AS 417
    ion spray for CE/Ms 491
    light pipe 138
    light pipe design 140
    linear transport 303
    membrane for GC/MS 216
    membrane for LC/MS 398

Index   521

micro-electrospray 365
microprobe for CE/NMR 512
moving wheel for LC/AS 429
nebulizer for disk transport 296
particle beam for LC/IR 307
particle beam for LC/MS 394
phase separator for LC/IR 319
spray deposition on belt 347
thermochemical hydride generator 428
thermospray 350
transport development of 292
transport for CE/MS 505
transport for LC/IR 290
transport for TLC/IR 473
triaxial probe 372
ultrasonic nebulizer 380
wire transport for LC/MS 341
interrupted elution development 135
ion sources 168
   chemical ionization 171
   dual ionization belt 348
   electron impact 168
   electrospray 359
   FAB 331
   field desorption 338
   ICP 174, 333
   laser desorption 335
   matrix assisted laser 337
   plasma desorption 332
   SIMS 330
   thermospray 350
ion spray, interface for CE/MS 491
ion trap
   mass spectrometer 98
   use with CE/MS 490
ionization
   chemical 94
   electron impact 93, 94
   methods 93
IR gas cell 136
iron in blood, by LC/FAAS 408
isoelectric focusing 38
isotachophoresis 37
katherometer detector 12
laser desorption source 335
laser induced fluorescence for LC/FS 286
LC/AS 405
   arsenic assay 421
   chromium assay 427
   circular transport interface 411
   hydride reactor 424
   ICP interface 417
   iron in blood 408

   iron in wine 410
   laser enhanced ionization for LC/FAAS 412
   LC/AAS with graphite furnace 415
   LC/FAAF
      interface for 407
   LC/FAAF systems 406
   lead in gasoline 409
   manganese detection 414
   moving wheel interface 429
   organotin detection 412
   post column reactor 424
   silicon assay 418
   tin in mussels 426
   tin in water 413
   wine, iron content 410
LC/FAAS
   interface 407
   iron in blood 408
   iron in wine 410
   laser enhanced ionization 412
   lead in gasoline 409
   systems 406
   transport interface 411
LC/FS 267
   aflatoxin analysis 275
   analysis by metal complexation 278
   anthracene analysis 271
   diode array system 282
   fatty acids in rat liver 277
   Hewlett-Packard system 268
   laser induced fluorescence 286
   milk analysis 272
   multiwave fluorescence 279
   polycyclic aromatic hydrocarbons 283
   post column reactor 276
   priority pollutants 273
LC/FTIR
   interface 301
LC/IR 289
   analysis of orange juice 313
   analysis of phenyl ureas 322
   analysis of polynuclear aromatics 305
   carousel transport system 291
   commercial system 308
   concentric flow nebulizer 300
   different types of nebulizers 299
   disk for transport interface 296
   disk transport interface 295
   drug precursor analysis 311
   extraction interface 317
   germanium disk interface 308
   linear transport interface 303
   micro cell 315

particle beam interface 307
phase separator 319
polymer additive analysis 294
spectrum of pyrene 306
transport interface 290
    development of 292
LC/MS
    analysis of bovine albumin 338
    API interface 384
    belt transport interface 344
    chromium by ICP-LC/MS 391
    cloxacillin, analysis of 371
    cypermethrin analysis 373
    cytochrom c
        analysis 365
    cytochrome c analysis 383
    different techniques 340
    direct inlet interface 349
    dual ionization belt interface 348
    electrospray interface 359
    fast atom bombardment 331
    field desorption 338
    furazolidone analysis 354
    ginsenoside analysis 367
    heavy metals in water 393
    human growth hormone analysis 386
    ICP interface 389
    ICP source 333
    insecticides in sweet potatoes 378
    laser desorption source 335
    levamisole analysis 357
    LSD in urine 388
    matrix assisted laser desorption 337
    membrane interface 398
    metal chloride additives 376
    micro-column electrospray interface 382
    micro-electrospray 365
    micro flow nebulizer 397
    momentum separator 395
    multiple charged ions 363
    oligosaccharide analysis 376
    particle beam interface 394
    penicillin G, analysis of 371
    peptide analysis 375
    peptides analysis 366
    plasma desorption 332
    post column additives 375
    propellants analysis 356
    SIMS 330
    spray deposition on belt 347
    systems 329
    thermospray interface 350
    transport interface 341

triaxial probe 372
tryptic digest analysis 362
ultrasonic nebulizer for electrospray 380
LC/NMR 435
    750 kH spectrometers 448
    basic system 440
    capillary coupling 444
    conditions for success 435
    continuous flow monitoring 439
    flow through cell 437
    modern system 439
    off-line monitoring 439
    optimum conditions 447
    peak sampling unit 441
    peptide analysis 449
    probe characteristics 445
    stop flow monitoring 439
    time-sliced monitoring 439
    two dimensional presentation 452
    unresolved peak identification 443
LC/RAMAN
    analysis of purine bases 325
    instrument design 324
LC/UV 255
    absorbance ratioing 259
    alkaloid analysis 263
    analgesic analysis 265
    analysis of carboxylic acids 256
    anti-inflammatory drugs 266
    aromatic hydrocarbon analysis 260
    benzodiazepines analysis 265
    cardiovascular drugs 263
    diode array 257
    diuretic analysis 266
    peak purity from spectra 261
    sulfur drugs analysis 266
LC/UV/FS 284
    instrument layout 285
lead, in gasoline by LC/FAAS 409
lemon juice
    flavone analysis 213
levamisole, LC/MS analysis 357
light pipe interface 138
light scattering, different forms of 67
linalool, spectrum of 144
linear transport interface for LC/IR 303
lipids by GC/MS 183
liquid chromatography 18
    detectors 19
longitudinal dispersion 32
LSD in urine, by LC/MS 388
lysozyme analysis LC/MS 362
magnetic sector MS with CE/MS 492

Index

MALDI 337
manganese, by LC/AS 414
mass spectrometer
  ion trap 98
  sector 91
  suppliers 222
  time of flight 100
matrix, argon for IR measurement 149
medroxyprogesterone by GC/MS 181
membrane interface
  for GC/MS 216
  for LC/MS 398
mercury
  in natural gas by GC/AS 240
  in sediments 238
metal chlorides as additives 376
metal complexes, analysis by complexation 278
micro-column, electrospray interface 382
micro-electrospray 365
micro-wave-induced plasma
  torch 228
microprobe, interface for CE/NMR 512
microwave-induced plasma 228
milk analysis, by LC/FS 272
minimum HETP
  GC 109
  LC 115
MIP 228
momentum separator 395
moving wheel transport interface for LC/AS 429
MS/MS identification of peptides 375
multi-element analysis
  GC/AS 231, 232
  in coal gasification 246
multidimensional analysis by GC/IR 152
multipath dispersion 31
multiple charged ions 363
multiwave fluorescence for LC/FS 279
mussels, tin content by LC/AS 426
natural gas, mercury by GC/AS 240
nebulizer
  Babington 410
  different types for LC/IR 299
  for disk transport interface 296
  micro flow 397
Newtonian flow profile 39
NMR spectroscopy 82
NPD 12
off-line monitoring, LC/NMR 439
off-line trapping 133
oligomeric phases 24
oligosaccharides, LC/MS analysis of 376

open tubes, dispersion in 106
optical rotation, measurement of 82
optimum velocity
  GC 109
  LC 115
orange juice
  analysis by LC/IR 313
  flavone analysis 213
organic chlorine by GC/MS 197
organotin
  analysis by GC/AS 234
  assay by LC/FAAS 412
  in mussels by LC/AS 426
  in sediments 237
packed columns
  GC 12
  dispersion in 108
parabolic velocity profile 107
particle beam
  interface 394
  interface for LC/IR 307
particulates, biological by GC/MS 220

PCBs
  by GC/IR/MS 203
  on street dust 206
peak
  dispersion 30
  purity by absorbance ratioing 259
  purity from UV spectra 261
  standard deviation 29
  unresolved, LC/NMR 443
peak sampling unit, LC/NMR 441
penicillin G, LC/MS analysis 371
peptides 366
  bioactive by CE/MS 494
  identification by MS/MS 375
  LC/NMR analysis 449
  separation by capillary electrochromatography 503
pesticides
  in water by GC/MS 199
phase separator for LC/IR 319
phenyl ureas, analysis by LC/IR 322
photo acoustic spectroscopy, TLC/IR 470
plasma
  desorption 332
  inductively coupled 242
    torch 242
  microwave-induced 228
    torch 228
pneumatic system, GC/IR/MS 134
polarization modulation 80

pollutants, priority by LC/FS 273

polycyclic aromatic hydrocarbons
  by LC/IR 305
  LC/FS 283
polythene, by pyrolysis and GC/IR 156
post column
  additives 375
  derivatization 211
  reactor 276
  reactor for LC/AS 424
potsherd, lipids, by GC/MS 184
priority pollutants, by LC/FS 273
probe
  characteristics for NMR 445
  NMR 90
propellants, LC/MS analysis 356
proteins, by CE/MS 489
purine bases, analysis by LC/RAMAN 325
pyrene, IR spectrum of 306
pyrolysis, polyethylene 156
quadrupole
  mass spectrometer 95
quaternary ammonium slats by CE/MS 485
Raman spectroscopy 67
rat liver, fatty acids by LC/FS 277
reactor, post column 276
references
  chapter 1 43
  chapter 10 432
  chapter 11 454
  chapter 12 481
  chapter 13 515
  chapter 2 104
  chapter 3 130
  chapter 4 163
  chapter 5 224
  chapter 6 250
  chapter 7 288
  chapter 8 327
  chapter 9 402
resistance to mass transfer 32
resolution
  diode array 257
  of convoluted peaks 281
retention 28
  volume 28
reversed phases 22
Ryhage concentrator 166
sample injector, valveless 155
scanning densitometry 458

scanning densitometry
  CAMAG scanner 461
scattering, light, different forms of 67
SCOT columns 17
secondary ion MS source 330
sector MS 91
sediment analysis
  mercury 238
  organotin by GC/AS 237
selenium in coal gasification 247
sensitivity
  fluorescence to anthracene 271
  of disc transport interface 297
sensitivity enhancement, CE/MS 486
separation techniques for tandem systems 7
separator, momentum 395
septum injector 13
serpentine tubes 127
  LC/FAAS interface 407
SFC
  sample preparation for GC/AS 248
silane purity, by GC/MS 188
silica gel
  IR transmission 465, 470
  pore distribution 22
silicon, by LC/AS 418
siloxane analysis by GC/AS 249
SIMS 330
solvents
  UV cut-off 264
spectrometer
  atomic emission 73
  circular dichroism 79
  fluorescence 58, 269
  fluorescence, diode array 282
  fluorescence, flow cell 270
  inductively coupled plasma atomic emission 244
  infrared
    Fourier transform 64
    simple grating 63
  mass
    ion trap 98
    quadrupole 94
    sector instrument 91
    time of flight 100
    triple quadrupole 97
  NMR 88
    probe 90
    super conducting magnet 88
  Raman 72
  UV
    diode array 53

Index 525

dispersive 52
spectroscopy
  atomic absorption 75
  diffuse reflectance IR 70
  fluorescence 55
  infrared 59
  mass 90
  NMR 82
  UV 48
  visible 48
spectrum
  bovine albumin (MS) 338
  drug precursor 311
  effect of different procedures 312
  electromagnetic 47
  Fourier transform 66
  high resolution NMR 87
  infrared
    interpretation of 62
  IR of benzyl benzoate 310
  IR of dextrose 312
  IR of pyrene 306
  IR of testosterone cypionate 310
  IR, drugs of abuse 142
  IR, linalool 144
  low resolution NMR 85
  lysozyme, MS 362
  multiple charged ions 363
  of orange juice components 313
  RAMAN of purine bases 325
split flow injector 15
spray deposition, on belt interface 347
stacked plots
  of alkylbenzimidzole 159
  of gasoline 145
standard deviation, peak 29
stationary phases
  bonded 22
  brush type 23
  bulk 25
  oligomeric 24
  resin 22
stop flow monitoring, LC/NMR 439
street dust, PCBs by GC/MS 206
sulfur drugs, LC/UV analysis 266
sulfonamides, separation by TLC 463
suppliers, mass spectrometers 222
sweet potatoes, determination of insecticides 378
synopsis
  chapter 1 41
  chapter 10 431
  chapter 11 453
  chapter 12 479

chapter 13 514
chapter 2 101
chapter 3 129
chapter 4 162
chapter 5 223
chapter 6 249
chapter 7 288
chapter 8 326
chapter 9 400
tellurium
  in urine by GC/MS 195
testosterone cypionate, IR spectra of 310
testosterone in urine by GC/MS 180
TGA/GC/IR/MS 153
thermochemical hydride generation 427
thin layer chromatography 25
time-sliced monitoring, LC/NMR 439
TLC
  analog curve from densitometer 462
  sulfonamide separation 463
  tandem systems 457
TLC/IR
  aromatic hydrocarbon separation 475
  caffeine and aspirin analysis 468
  diffuse reflectance IR 466
  photo acoustic spectroscopy 470
  plate scanning 464
  transmission of alumina and silica gel 465
  transport interface 473
TLC/MS 475
  assay of guinea green B 479
  Tyr-bradykinin assay 477
torch
  ICP 245
  inductively coupled plasma 242
  microwave-induced 228
  microwave-induced, venting system 230
trans-resveratol, by GC/MS 189
transmission, IR through silica and alumina 465
transport interfaces
  belt 344
  circular for LC/FAAS 411
  disc for CE/MS 505
  dual ionization belt 348
  for TLC/IR 473
  LC/IR 290
  moving wheel for LC/AS 429
  spray deposition 347
  wire 341
triaxial probe, LC/MS 372
triple quadrupole mass spectrometer 97

tryptic digest analysis 362
tubes
    coiled, dispersion in 125
    low dispersion 123
    open, dispersion in 106
    serpentine 127
ultrasonic nebulizer, interface for LC/MS 380
urine analysis
    4,4'-methylene bisaniline 191 by LC/MS 388
    chromium by GC/MS 195
    GC/MS 190
    haloperidol by CE/MS 492
    steroids in equine plasma 179
    tellurium by GC/MS 195
    testosterone metabolites 180
UV cut-off, solvents 264
UV spectrometer
    diode array 53
    dispersive 52
UV spectroscopy 48

valvless sample injector 155
Van Deemter equation 31
vapor concentrator 165
    Bieman concentrator 167
    Ryhage Concentrator 166
vapor sampler 154
vapor sampling, for GC/MS 198
velocity
    GC optimum 109
    LC optimum 115
    profile, parabolic 107
visible spectroscopy 48
water
    heavy metals by LC/MS 393
    silicones by LC/AS 418
    tin content by LC/AS 413
waxes by GC/MS 183
wine, iron by LC/FAAS 410
wire transport interface 341
zirconia, IR transmission 470
zone electrophoresis 36